CERAMIC-TO-METAL JOINTS & SEALS

EDITED BY

GREG EASTER

Wexford Press
2008

Table of Contents

Brazing

Ceramic-to-Metal Joints:

An Introductory Survey

by

Greg Easter

I. Applications of Ceramic-to-Metal Joints

Ceramic materials have countless applications in technology and industry today, with still more applications being developed constantly. They are relatively low in cost, and able to withstand nuclear radiation, extreme heat, and chemical corrosives. They can be honed to extremely sharp edges that will remain sharp much longer than alloy steel. The term *ceramic* refers to a broad range of materials such as aluminum oxide, titanium carbide, titanium nitride, sapphire, zirconium dioxide, silicon nitride, silicon beryllium oxynitride, etc. Aside from their physical robustness, many ceramics have other useful properties such as barium titanate (piezoelectric properties), ferrite (ferromagnetic properties), uranium dioxide (a nuclear fuel), yttrium barium copper oxide (a superconductor), zinc oxide (a semiconductor), and so on. For most applications, the ceramic needs some attachment to the rest of the device or machine, or an electrical contact, which requires a ceramic-to-metal joint. Even in the exotic instance of an all-ceramic device, the parts are not manufactured as a single block and must therefore still be assembled together, most often by brazing. Indeed, the methods employed for brazing metals to ceramics are also applicable to brazing ceramics to each other.

Aside from brazing, older methods of attachment, whether metal-to-ceramic or ceramic-to-ceramic, included mechanical means (such as screws and clamps) and/or adhesives. Today cyanoacrylate adhesives are frequently employed for joining ceramics in the fields of medical and biological technology, where temperature resistance is not an important factor, since living organisms exist well below the thermal breakdown point of the adhesive.

Other applications cover every aspect of modern technology, such as electronics, manufacturing, metalworking tools, turbine and other engine parts, nuclear reactors, and more. Nearly every application requires some kind of ceramic-to-metal joint or seal. The cost of that connecting point ranges from less than a penny to more than $50 per square inch, depending on the method and materials chosen.

Despite decades of technological advances and the advent of many alternative methods of forming secure ceramic-to-metal joints and seals, brazing remains the most frequently employed method in production.

This first section will provide the reader with a basic understanding of the brazing process, and introduce the vocabulary of common abbreviations and terminology most often encountered.

II. Soldering vs. Brazing

Soldering and brazing are closely related processes. In either method a filler material, either solder or a brazing alloy, is heated to the point of it melting such that it flows evenly between both of the materials being joined. Upon cooling, this filler material remains joined to both surfaces, thus securing that the parts remain in close contact with each other. One distinction between soldering and brazing is the melting point of the filler material (solder or brazing alloy). Anything below about 450°C is considered *soldering*, and anything above about 450°C is termed *brazing*.

Soldering is nearly always conducted at normal atmospheric pressure with the only precaution against the surrounding air contaminating the joint being a small amount of flux material, such as a paste. This is because at the relatively low temperatures needed for soldering, air is not particularly reactive with metal. At least not for the short duration required for making a soldered joint.

Flux is also employed in brazing ceramics to metals in order to reduce the oxidation on the brazing alloy and metal substrate (ceramics are generally immune to oxidation). Trade name fluxes applicable to ceramic-to-metal brazing include Handy Flux and Sure Flo Flux. There seems to be no consensus about which flux works best, even for seemingly identical operations. Different workers report both good and bad results with the same brand of flux, according to journal articles. Therefore, if the results are substandard then one should next experiment with a different flux.

The active component in flux is boron, which has the ability to bind to oxygen at high temperatures (between about 950° and 1700°C). You might wonder what the purpose of flux would be if the brazing is being performed in a vacuum, for example. What oxygen is it protecting the surface from then? First of all, a vacuum is never complete. There is always some oxygen present, particularly on the surface of the materials being brazed. Flux will remove this gas from the surface as it is heated. Second, flux greatly aids the flow of braze alloys—and flow is crucial in obtaining a secure and smooth joint.

Because brazing is conducted at much higher temperatures, and also because brazed joints are often placed in service under extreme stress where microscopic imperfections lead to shortened life, measures must be taken to prevent the oxidation of either surface being joined, or the brazing alloy itself. For nearly all applications this means that the braze must be performed in vacuum (10^{-4} Torr or greater) or in inert gas such as argon or nitrogen.

Ceramics cannot be brazed directly with ordinary brazing alloys because of their inherent difference from metals at the molecular level, being held together by covalent bonding. The surface of the ceramic must be modified to make it capable of bonding to a metal. Most ceramics are notoriously unreactive. Indeed, that is one of their most useful properties to industry, as mentioned in the previous chapter. This surface-modification of the ceramic can take place in advance of brazing, or thanks to the advent of newer specialized brazing alloys, simultaneously as the brazing material contacts the surface of the ceramic. Each method has its own advantages and disadvantages, and the requirements of the final product will largely determine the method. If the joint is part of an inexpensive tool to be mass-produced for consumer applications where it will not be exposed to very high temperatures or much physical stress, then less expensive methods will certainly be employed. On the other hand, if the joint is expected to remain reliable under extreme conditions, such as certain critical applications in nuclear reactor sites, then performance will obviously take precedence to the cost.

Another important difference between soldering and brazing is that soldering is nearly always performed by application of heat to the joint being soldered, or in the case of PC board manufacture, dipping the circuit board into a bath of molten solder. While brazing of metal-to-ceramic seals is sometimes performed with an open flame (torch brazing) it is far more frequently carried out in a kiln, or the atmosphere-controlled equivalent of a kiln. Parts are placed together with the flux and brazing metal, then baked at an experimentally determined time and temperature.

The next section will explore the most important options. One other method of establishing a ceramic-to-metal or ceramic-to-ceramic seal that is a type of brazing, but not in the traditional sense, is the explosive metal film method. In this approach, the parts are sandwiched together with a thin sheet of metal between them. A powerful high voltage discharge is applied to the metal, vaporizing it, and forming both the coating and the braze metal in a single step. The will be discussed in more detail later in this book. The next section deals only with methods of true brazing.

III. Methods of Preparing Ceramics for Brazing

All of the various methods of treating the surface of the ceramic for brazing can be divided into two basic groups: coating the surface with metal beforehand, or use of a special brazing alloy that treats the surface as it is brazed. Cleaning the surfaces of the materials to be joined is the first step in any joining process. The details of how the surfaces are cleaned is beyond the scope of this introduction, as there are numerous factors including the characteristics of the materials to be joined, the method that will be used to join them, and to a lesser extent, the final application of the joined parts. The following are all of the common methods for ceramic-to-metal joints and seals used today:

1. MOLY-MANGANESE, as it is commonly referred to, is a fairly complex procedure that begins with the application of a mixture of molybdenum, manganese and powdered glass. This mixture is made into a kind of paint and frequently applied in a process akin to silk screen printing, which has the advantage of enabling careful control over where the braze will adhere. After this mixture is applied to the ceramic, it is heated to 1500°C in a controlled moist atmosphere of hydrogen gas, then nickel plated, then heated in hydrogen a second time to 950°C before a metal alloy foil is applied. Despite the number of steps involved, this method is old now and automated methods of production are well established.

2. DIRECT ELECTROPLATING of a metal (usually nickel) onto the ceramic before brazing. A metal alloy foil containing titanium is often applied to the plated nickel before brazing takes place. This method is much easier, but only possible if the ceramic to be joined is electrically conductive (a requirement for the electroplating of any material.)

3. VAPOR SPUTTERING in which a layer of titanium just a few microns thick is applied to the ceramic under vacuum. Titanium promotes spreading of the braze, which is accomplished with any brazing alloy desired.

4. METAL HYDRIDES (usually titanium or zirconium) are reduced and brazed at the same time under vacuum, or an inert gas (usually argon or nitrogen). The braze alloys used in this process generally contain silver, which adds still more to the cost. The advantages are that the process is complete in a single step, and it can be accomplished at a lower temperature than most other brazing. However, the other side of that coin is that these brazing alloys melt at a relatively low temperature, so this type of joint cannot be used under high temperature conditions, as of the time of this writing.

5. ACTIVE BRAZE ALLOYS usually contain about 2% titanium, which enables them to form a "reaction layer" between the ceramic substrate and the rest of the brazing filler alloy. While it is this layer that enables the joint to be formed in a single step, it is also the weakness in the same joint if it is too thick. The braze alloy in this scheme is generally a silver alloy, and is still more expensive in production than some other methods. This is an area of technological development that is promising, however. Eventually active braze alloys may be developed with a higher melting point and at a lower cost.

There are other novel and exotic methods of coating a ceramic with a metal before brazing, but these are merely variations on the same theme. Undoubtedly more methods will be developed in the years ahead.

Another recent development in braze alloys is the idea of incorporating ceramic material such as alumina into the alloy (typically 5% by volume). This increases the strength of the joint and improves the reliability.

There is also promise of a new solder (melting point of 420°C) that can join ceramics and metals without the need for a protective atmosphere. The drawback is primarily the intolerance of the joint to high temperatures, since it would simply melt off above 420°C. Since one of the most important reasons for using ceramics in the first place is their ability to operate at high temperatures, this type of low-temperature joint will be of limited use in most fields.

IV. Joint Design

The importance of joint design cannot be overemphasized.   Aside from the rule that the greater the surface area between the parts being joined, the stronger the joint will be, there is another less obvious consideration. Because the CTE (coefficient of thermal expansion) of the ceramic is almost always quite different from the CTE of the metal being joined to it, and also different from the CTE of the brazing metal alloy, these parts will contract different amounts as the joint cools back down to room temperature. The slight difference in room temperature size of these components places internal stress on the joint.

Joints should be design whenever possible such that the stress is in the form of compression on the ceramic. Computer modeling using FEA (Finite Element Analysis) is an important tool for complex parts being joined, especially if they will be placed under mechanical stress.

This is an advanced topic that will be covered in more detail later in this book. The following references will also be instructive:

[1] O.C. Paiva, M.A. Barbarosa, J. Mater. Sci. **35** (2000) 1165

[2] K. Suganuma, Mater. Res. Soc. Symp. Proc. **314** (1993) 51

[3] J.A. Howe, Intern. Mat. Rev. **38**, 5 (1993) 51

V. Concise Glossary of Terms and Acronyms

alumina, is the ceramic Aluminum Oxide.

braze alloy (or brazing metal, or filler metal), is the metal that is melted to form the joint. In the analogy to soldering, this is the solder.

cermet, is a metal-ceramic alloy formed by sintering and compacting. The name comes from cer- (ceramic) and –met (metal). Examples are TiN and TiC, which are titanium nitride and titanium carbide, respectively. Cermets are a modern replacement for tungsten carbide in machining tools.

CTE, is an acronym for the Coefficient of Thermal Expansion. Since the size of materials change slightly with temperature, the difference in the CTE between two materials is directly proportional to the amount of inherent stress there will be in the brazed joint once it has cooled. This affects both joint design choices as well as the selection of materials used.

DLC, is an acronym for Diamond Like Coating, also sometimes translated as Diamond Like Carbon. This is a thin film of carbon that has been sputtered or applied through CVD (chemical vapor deposition), ion beam, RF plasma, or another related method. For certain applications, this material is more practical than a metal-to-ceramic joint to a cermet. One curious note is that the hardness of DLC cannot be stated with scientific precision, because it does not have a fixed crystalline structure (thus there is no single edge to measure the hardness of.)

eutectic (or eutectic mixture), is the proportion of elemental metals that melt at the lowest temperature.

SiBON, is an acronym for the ceramic, Silicon Berylium Oxynitride. Also referred to as Sibeon.

sintered (or sintering), is when a metal powder is heated to the point that particles stick together to form a "crust" but still below the temperature at which the powdered metal would melt into a puddle.

TBC, is an acronym for Thermal Barrier Coating. Ceramics are joined to metals for this function, particularly in modern high-performance engine designs.

TiCN, is an acronym for Titanium Carbonitride, a cermet. In many industrial applications this is replacing tungsten carbide, owing to its improved performance. Related materials include TiC/AL203 and TiN (Titanium Nitride).

wetting (or wettability), is the ability of a braze alloy filler to spread out on the surface of the ceramic when it is being applied. The purpose of coating the ceramic with a metal such as nickel or titanium before brazing is to increase the wettability.

zirconia, is the ceramic, Zirconium Dioxide.

V. Online Resources

The following article is an excellent source of general information:

Recent advances in metal-ceramic brazing

http://www.scielo.br/scielo.php?pid=S0366-69132003000400002&script=sci_arttext

One frequently encounters procedures in the literature cited only by their patent number. One useful resource for obtaining more information online is the URL:

http://www.freepatentsonline.com/

By entering the patent number, most of the time an abstract will be retrieved instantly, along with other useful information about the process and what it does. Many entries contain detailed stepwise procedures. The following are some patents that may be of interest to the reader concerning ceramic-to-metal joints:

6322871	**Method of treating ceramics for use as tips in saws**
4471026	**Ternary Alloys in Brazing Ceramics**
20070089563	**Braze Alloy containing Particulate Material**

Note that a similar resource for Canadian patents also exists online at:

http://patents.ic.gc.ca/cipo/cpd/en/search/number.html

Navigate to the search engine from there using the menu items on the left side of the page, if you do not have a Canadian patent number to enter directly.

VI. Disclaimer and Safety

Please note that there are many hazards involved in the procedures discussed here, and some of the materials employed are dangerous and/or highly toxic. This book is not intended as an instructional guide, and must not be interpreted as such. This is an overview of the basic concepts. <u>Etching and brazing of metals and/or ceramics should only be attempted by experienced professionals following the guidelines established by OSHA and the EPA</u>.

JOINING CERAMICS

AND GRAPHITE

TO OTHER MATERIALS

by

H.E. Pattee
R.M. Evans
R.E. Monroe

Contents

CHAPTER 1

Introduction

Ceramic-to-metal seals and joints are assuming increasing importance in today's technology, and it is safe to predict that their uses will increase in the future as the advantages of ceramics as construction materials are realized. For example, when the first space station is established in outer space, ceramic-to-metal joints will play a significant role in its construction and operation. Nuclear- and solar-power devices will probably be used to provide energy, and ceramic-to-metal joints will be needed in the equipment that converts nuclear and solar energy to electrical power. Similarly, ceramic coating may be used to provide the emissivity required to control and maintain the proper temperature level within the space station structure. The performance of the space vehicles now depends on the ability of ceramic-to-metal joints to provide strength, erosion resistance, and high-temperature protection where needed.

BACKGROUND

Since its beginnings in the late 1930's, the technology of joining ceramics to metals and other materials has progressed steadily. Like other materials-joining processes, the joining of ceramics to other materials was initially more an art than a science; the proprietary nature of developments in this field aggravated the situation. During and after World War II, however, efforts to develop procedures that could be used to produce reliable joints between ceramics and metals became serious. Much has been accomplished during the intervening years toward establishing a joining technology based on sound fundamental principles and an understanding of the reactions that occur during joining. However, the basic nature of the bond-producing mechanism still eludes investigators, although they have advanced many theories to explain it.

Ceramic-to-metal joining, as we know the process today, had its industrial origins during World War II in Germany, where, under wartime conditions, it was recognized that vacuum tubes with improved performance and reliability could be produced by substituting ceramics for glass. Pulfrich (Telefunken) and Vatter (Siemens and Halske, A. G.) established procedures to metallize ceramics, thus producing surfaces that, in subsequent joining operations, could be wet by brazing alloys. Numerous domestic and foreign patents on metallizing and joining techniques were issued to Pulfrich, Vatter, or the firms that employed them in the middle and late 1930's; some of the most significant of these patents are listed in references 1 to 9.* While the results obtained by Pulfrich and Vatter are somewhat crude in the light of current developments, their basic concepts are still valid, and their metallizing procedures, or modifications thereof, are used extensively to prepare ceramic surfaces for joining. The advantages associated with the use of ceramics for the envelopes, internal structural supports, and output windows of vacuum tubes are as follows:

1. Ceramic tubes can be outgassed at higher temperatures than glass tubes. For ceramic tubes, the permissible outgassing temperature depends on the melting temperature of the filler metal used in the ceramic-to-metal seal and the degree to which the thermal expansion coeffi-

---

* References appear at the end of this report.

1

cients of the joint members are matched. Because of the increased vapor pressure of the residual impurities, outgassing at high temperatures is more effective and promotes longer tube life and increased emission.

2. Because of the high-temperature seal, ceramic tubes withstand higher temperatures than glass tubes of similar dimensions.

3. Ceramic tubes are mechanically stronger and less sensitive to thermal shock than glass tubes. Thus, performance is less affected by rigorous service conditions.

4. Ceramic components can be ground to the precise tolerances required for vacuum-tube construction.

5. Ceramic materials have very low electrical losses at high frequencies. They are well suited for use as output windows in microwave tubes, because they can pass maximum amounts of power while separating evacuated and unevacuated areas of the tubes.

However, the use of ceramics in vacuum-tube construction is not without disadvantages, some of which are cited below:

1. The joint configurations for ceramic-to-metal seals are somewhat limited because ceramics have high compressive, but low tensile strengths. Compression seals are used wherever possible; butt-seals are satisfactory if tensile stresses are minimized.

2. Careful selection of the seal's ceramic and metal members is necessary to obtain a good match of the thermal expansion coefficients and the expected temperature range.

3. Most ceramics are opaque, so that the alignment of electrode assemblies cannot be inspected after assembly or operation; however, some high-purity alumina ceramics are nearly transparent.

4. Flaws initially present or produced during machining are difficult to detect.

5. Metallizing and joining operations are costly, because multi-step procedures must take place at high temperatures in controlled-atmosphere or vacuum furnaces.

Nevertheless, the advantages of ceramics for vacuum-tube applications far outweigh the problems. Most problems can be eliminated or minimized by proper materials selection and careful joint design; in many instances, ceramics are the only materials that fit a particular application.

Ceramic-to-metal joints were first used extensively in the construction of vacuum tubes for critical applications. The electronics industry is still the largest user of such seals, but other industries use ceramics increasingly as structural components, coatings, and strengthening agents in metals. When used as structural components, ceramic-to-metal joints may be required to complete an assembly or provide sealing. Some applications that indicate the versatility of ceramics in other than the vacuum-tube industry are outlined in the following paragraphs.

Because of their inertness in many corrosive environments, ceramics are used as seals in fuel cells and other devices that convert chemical, nuclear, or thermionic energy to electricity. In a current program to develop a nuclear power source to produce electrical energy for space applications, a large thin-walled ceramic cylinder is being used in the fabrication of a "bore" seal. The ceramic-to-metal seal is required to isolate the electrical sections (windings, armature, and poles) of a turbine-driven generator designed to operate in corrosive liquid-metal vapors (ref. 10). Bristow, Grossman, and Kaznoff have also developed special ceramic-to-metal seals for a cesium vapor-filled thermionic converter (ref. 11).

In addition to high-temperature strength and resistance to corrosive media, certain ceramics possess properties that make them attractive for use in high-temperature nuclear reactors of advanced design. Since some of the metals used for metallizing are unsuitable for reactor environments, Fox and Slaughter investigated the characteristics of several experimental filler metals to be used in joining the oxides of aluminum, beryllium, and uranium to themselves and to metals without previous metallizing (ref. 12). Simulated fuel-element assemblies were fabricated from alumina (Al_2O_3) and beryllia (BeO) using a Ti-49Cu-2Be brazing filler metal.

Extensive investigations of ceramics as structural components in gas turbines and rocket engines were conducted during the 1950's. In attempting to use oxide ceramics and cermets for turbine blades and buckets, the direct substitution of ceramics for metals was largely unsuccessful, because ceramics have poor impact and tensile properties. For such applications, ceramics have proved most useful in providing oxidation and wear-resistant metal coatings.

Extensive research has been conducted on the use of ceramics for the dispersion-hardening of metals; thoriated nickel (TD) and sintered aluminum powder (SAP) are the best known alloys of this type. For example, SAP is an aluminum material that maintains its strength at higher temperatures than other aluminum alloys. It contains a dispersed phase of up to 15-weight percent aluminum oxide. SAP can be extruded, rolled, forged, and drawn by conventional metalworking techniques. While care in surface preparation must be observed, SAP can be soldered, brazed, and welded to itself and other metals. SAP has been used in various forms by the automotive, aircraft, and nuclear industries. Currently, there is great interest in strengthening metals with single-crystal alumina fibers or whiskers (ref. 13).

Because of their transparency to microwaves, ceramics are used extensively for radomes to enclose radar antenna equipment on supersonic aircraft. Plastic radomes were suitable until sonic speeds were achieved when air friction and impact from dust and raindrops in the atmosphere created problems. Materials for radome fabrication are generally limited to high-alumina and some refractory glasses. Small radomes can be cast or formed; large radomes are more difficult to produce because of close tolerances on shape, wall thickness, and dielectric constant. In a recently completed program, engineers at the Whittaker Corporation used mosaic techniques to fabricate a large radome over a removable form; the radome was 37 inches in diameter and 8 feet high (ref. 14). Precisely cut 97.6-percent alumina tiles were assembled around the form and cemented in place. After assembly, the form was removed and the radome was placed in a firing tool. Differential tooling was used to apply pressure during firing.

Ceramics are also used as friction materials for brakes, clutches, and other energy-absorbing devices (ref. 15); coatings for nuclear fuel particles (refs. 16 and 17); constituents in high-temperature adhesives (refs. 18 to 20); and ablation materials and coatings for metals.

While some degree of joining is required in ceramic-to-metal seals, ceramic-coated metals, ceramic-filled metal structures, ceramic-fiber reinforced metals, cermets, and composites, the mechanism of joining varies from chemical and/or electronic bonding to mechanical interlocking of ceramic particles with a metal substrate.

PURPOSE AND SCOPE

In discussing the joining of bulk ceramics to metals and other materials, this report will emphasize ceramic-to-metal joining as applied in the electronics industry, because more research has been initiated by this industry than any other; as a result, extensive information is available. However, the techniques developed for joining ceramics to the metals of interest to the electronics industry can usually be applied to other materials as well.

The report will summarize data on joining ceramics to other materials available from Government and industrial research. Much of the research was prompted by a need to develop joining or sealing techniques for specific applications which will be related throughout the report in hope that the information can be applied to applications that are similar or involve unusual materials or service conditions.

In addition to an extensive review of the important considerations of ceramic-to-metal joining, the technology of joining graphite to other materials will be discussed in some detail. Ceramics and graphite have much in common. Both materials have excellent high-temperature properties and are chemically inert in many corrosive media; as a result, ceramics and graphite have been considered for missile components where materials that maintain their strength at high temperatures in a corrosive or erosive environment are required. Brazing filler metals do not wet either material readily, so special joining techniques are required.

The sources of information for this report included reports of Government and industry filed by the Defense Ceramics Information Center (DCIC) and the Defense Metals Information Center (DMIC), technical literature on joining ceramics and graphite, various engineering indexes, libraries of the Columbus Laboratories of Battelle Memorial Institute, and personal files. The literature searches included a Defense Documentation Center (DDC) search, a Redstone Scientific Information Center (RSIC) search, a NASA Scientific and Technical Information Facility search, and a DD–1498 Data Bank search. These searches were reviewed and about 100 reports on Government-sponsored research activities were requested and studied. Selected industrial contacts were also made to acquire information on other current investigations.

Vatter (ref. 25) and Jenkins (ref. 26) have published excellent reviews of the historical development of ceramic-to-metal seals. In 1953, a symposium on ceramics and ceramic-to-metal seals was held; the papers presented there appeared in various issues of *Ceramic Age* in 1954 and offer a good picture of the technology of that date. More recent reviews were published by Van Houten (ref. 27) in 1959 and by Clark, Ritz, and Girard (ref. 28) in 1965. Kohl has discussed developments in ceramic-to-metal joining on several occasions; his latest publication was dated in 1967 (ref. 29).

CHAPTER 2

Materials for Ceramic-to-Metal Joints

CERAMICS

It is easier to discuss the characteristics and properties of ceramics than to define them. Kohl states that ceramics is a "term used to describe a variety of solids of different compositions that have attained a crystalline state by the firing of inorganic nonmetallic materials" (ref. 29). The raw materials that form ceramics have a crystalline structure before firing and retain this property after firing. Glasses are not ceramics according to this definition, because a noncrystalline solid is formed after the glass constituents are fired. By recently developed processes some glasses can be converted to ceramics when certain nucleating agents are added to the glass constituents.

The technology of formulating ceramics has developed tremendously in the last decade in response to the requirements of industry. The selection of ceramics was once limited mainly to multiphase silicate ceramics; now, one may select carefully from a wide variety of oxides, borides, nitrides, carbides, and silicides on the basis of specific service requirements.

The raw materials from which ceramics are made have changed significantly in recent years. The first ceramics used in vacuum tubes were the steatites, multiphase silicate ceramics of the type $MgO \cdot SiO_2$; the quality of these ceramics varied according to the raw materials. Because the reactions during the firing process were not well understood, the proportions of the ceramic constituents were not exact, and precise control of the production process was lacking. Pulfrich noted that metallizing and brazing tests with representative samples from a production run of ceramics were necessary to ensure sound ceramic-to-metal seals; mechanical and electrical tests were conducted also as a quality control measure (ref. 20).

Many of the problems associated with the quality of early ceramics were related to the proprietary nature of developments in this field. The sources of raw materials were kept secret, and the ceramic compositions and processing details were not openly discussed. The Government-sponsored research undertaken after World War II to investigate and improve the techniques of ceramic-to-metals sealing did much to dispel this secrecy. More recent research has been concentrated on the effects of impurities and firing cycle on the basic properties of ceramics; the steady improvement in ceramic quality has demonstrated the value of disseminating the results of research.

A number of relatively inexpensive multiphase silicate ceramics were developed over the years. In addition to the low-loss steatites, these included forsterite ($2MgO \cdot SiO_2$), mullite ($3Al_2O_3 \cdot 2SiO_2$), and zircon ($ZrO_2 \cdot SiO_2$). About 35 years ago, the single-phase oxide ceramics were developed, the most important of which are alumina (Al_2O_3) and beryllia (BeO). High-alumina materials were originally used as sparkplug bodies because of their resistance to thermal shock. After World War II, the superior properties of alumina for vacuum-tube applications were recognized, and studies were begun to develop techniques for using alumina effectively. Unlike the multiphase silicate ceramics that are formed from plastic clays, the metal-oxide ceramics are produced in useful shapes by sintering a loose mass of oxide grains. During sintering, molecular forces acting be-

5

tween adjacent particles result in the formation of a bonded-crystal network. Strong bonds between the initially loose grains occur without gross melting or additional binders.

Alumina is attractive for vacuum-tube applications because of its high mechanical strength and resistance to thermal shock, its ability to be used at high temperatures, its imperviousness to gases, and its excellent electrical properties. However, the properties of alumina ceramics are affected significantly by their composition, microstructure, and sintering temperature and time. Table 1 lists the mechanical, electrical, and thermal properties of several high-alumina ceramics with Al_2O_3 contents ranging from 85 to 100 percent as investigated by Rigterink (ref. 31). LaForge obtained similar results in an earlier investigation and also noted a large variation in the dielectric loss factor among ceramics of the same general type and even of similar composition (ref. 32). Research has also been conducted to determine the effect of various parameters on the ability to produce high-quality ceramic-to-metal seals with high-alumina ceramics. LaForge conducted microscopic studies of alumina ceramics used in ceramic-to-metal seals. He found that the particle size and the sintering temperature had a marked effect on the alumina microstructure and the occurrence of porosity and grain growth. Cole and Hynes examined the properties of an alumina ceramic containing 94.5

percent Al_2O_3 as a function of firing temperature (ref. 33). They noted that stronger ceramic-to-metal seals and lower electrical losses were obtained at higher firing temperatures; however, these favorable properties were obtained at the expense of increased porosity and a lower rupture modulus. Floyd studied the effect of ceramic-flux compositions and crystal size on the strength of ceramic-to-metal joints (ref. 34). These investigations indicated the need for the precise control of each step in the production of alumina ceramic bodies, from analyzing the raw materials to inspecting the finished product. The conclusions apply equally to other ceramic materials.

Several other oxide ceramics have been developed; the most important are discussed below.

1. *Beryllia (BeO)*.—In many respects the properties of beryllia are superior to those of alumina; however, beryllia is not nearly so widely used as alumina, probably because of its possible toxicity. The melting temperature of beryllia is about 900° F higher than that of alumina, and its chemical stability surpasses that of most oxides. Beryllia is stable in most gases and in a vacuum at temperatures up to about 3100° F. The heat conductivity of beryllia far exceeds that of all oxide and silicate ceramics; its electrical resistivity is very high as well. Because of its high heat conductivity, beryllia is used in electron tubes where large quantities of heat must be dissipated.

TABLE 1.—*Some Properties of High-Alumina Ceramic Materials*

[From ref. 31]

Material	Percent Al_2O_3	Mechanical		Electrical		Thermal	
		Compressive strength, lb/sq in., 25° C	Tensile strength, lb/sq in., 25° C	Dielectric constant, 1 mc, 25° C	Dissipation factor, 1 mc, 25° C	Coefficient of linear expansion, 25° to 700°C	Softening temperature (°C)
Sapphire (single crystal)	100	30×10^4	6.5×10^4	10.3	0.00004	8.5×10^{-6}	2040
Alumina A	99+	42×10^4	3.4×10^4	10.0	.0001	8.0×10^{-6}	>1600
Alumina B	97	28×10^4	2.7×10^4	9.5	.0001	9.0×10^{-6}	>1600
Alumina C	96	30×10^4	2.6×10^4	9.0	.0003	9.0×10^{-6}	>1600
Alumina D	94	19×10^4	1.5×10^4	9.2	.0004	7.3×10^{-6}	>1600
Alumina E	85	20×10^4	1.8×10^4	8.2	.0009	7.9×10^{-6}	1400

2. *Magnesia (MgO).*—The stability of magnesia in gaseous environments is similar to that of alumina. Magnesia is not as stable as alumina when in contact with most metals at high temperature. It is reduced by carbonaceous atmospheres at elevated temperatures and vaporizes in a vacuum at about 3400° F. Magnesia is not used extensively in vacuum tubes because of its high thermal expansion, poor resistance to thermal shock, and low mechanical strength.

3. *Thoria (ThO₂).*—Thoria is the most stable oxide chemically; it is reducible by the most aggressive alkali metals only under special circumstances. It has the highest melting temperatures of all the oxide ceramics and a correspondingly low vapor pressure. Its use is limited principally by its high cost.

4. *Zirconia (ZrO₂).*—Zirconia is similar to thoria in its chemical stability; however, it is unstable in halogen, sulfurous, and carbonaceous environments. The electrical resistivity and thermal conductivity of zirconia are very low. The poor resistance of zirconia to thermal shock and spalling can be overcome in part by stabilizing zirconia with additions of CaO or MgO.

Additional data on the characteristics and properties of these and other refractory oxides, such as urania (UO_2) and titania (TiO_2), can be obtained from standard texts on ceramics and from articles by Kingery (ref. 35), McClelland (ref. 36), Schneider (ref. 37), and Ryshkewitch (ref. 38).

In addition to the conventional metal oxides, an impressive number of refractory ceramics have been developed for high-temperature applications. Among these materials are borides, carbides, nitrides, sulfides, and some silicides and aluminides. The properties of these ceramics are not nearly so well documented as those of the metal oxides. As can be seen in figure 1, most of these materials melt at exceptionally high temperatures (ref. 39). Many have excellent oxidation resistance. The properties of some refractory ceramics are shown in table 2 (ref. 40). These materials are used in jet and rocket aircraft, missiles, rockets, and satellites as structural components or coatings to provide oxidation resistance to other structural materials; their uses will increase as their characteristics become better known. Property data on many of these materials were presented at the Asilomar Symposia on high-temperature technology in 1959 and 1963 (refs. 41 and 42). Data

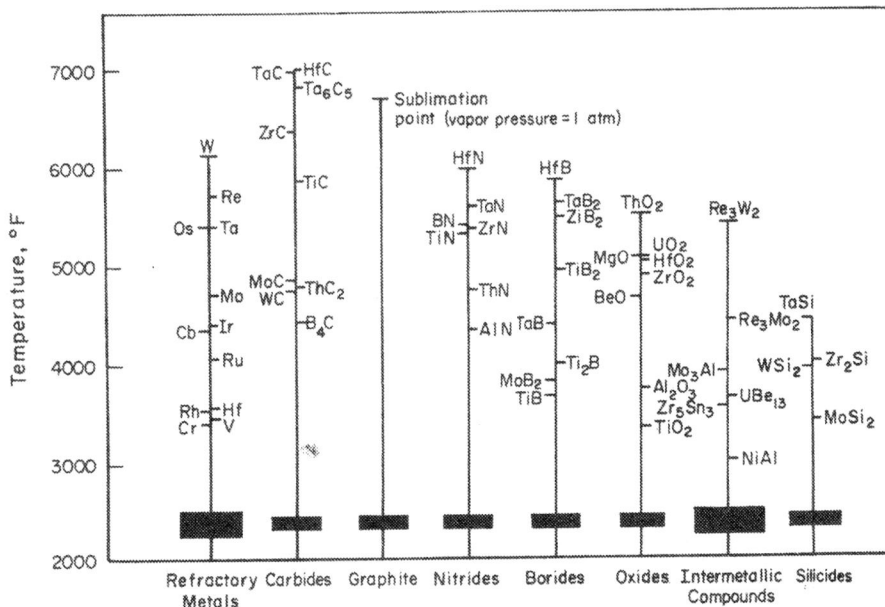

FIGURE 1.—Melting points of refractory materials (ref. 39)

TABLE 2.—*Typical Properties of Selected Refractory Ceramics* [a]

[From ref. 40]

Compound	Melting point, °F	Density, gm/cu cm	Knoop hardness (100 gm)	Coefficient of thermal expansion, 10^{-6} per °F	Thermal conduction, Btu/hr/sq ft/°F/ft	Specific heat, Btu/lb/°F	Electrical resistivity, microhm-cm	Modulus of rupture, 1000 psi	Modulus of elasticity, 10^6 psi	Compressive strength, 1000 psi
OXIDES										
Alumina (Al_2O_3)	3720	3.96	2360.0	8.5 (to 1472)	17.5 (212)	0.36 (1832)	5.0×10^{15} (ohm-cm)	34.0, 24.0 (1832)	54.0, 46.0 (1832)	450, 40 (2552)
Beryllia (BeO)	4620	3.0	9.0 [c]	9.4 (to 2552)	124.0	0.9 (1832)		30.0, 25.0 (1832)	55.0, 49.0 (1832)	110, 30 (2552)
Magnesia (MgO)	5070	3.6	5.5 [c]	13.7 (to 2552)	20.4 (212)	0.54 (212)		23.0, 20.0 (1832)	31.0, 21.0 (1832)	120
Thoria (ThO_2)	5790	9.7	640.0 [e]	9.55 (to 2552)	5.44 (212)	0.11 (1832)		15.0, 15.0 (1832)	35.0, 32.0 (1832)	210, 10 (2552)
Urania (UO_2)	5200	10.02			1.9 (1833)	0.18 (1832)		20.0, 15.0 (1832)	24.0, 16.5 (1832)	290, 20 (2552)
Zirconia (ZrO_2)	4930	5.6	7.0 [c]	5.6 (to 2192)	12.8					
CARBIDES										
Boron carbide (B_4C)	4440	2.5	2800.0	4.5 (to 1472)	15.7		0.3–0.8 (ohm-cm)	44.0, 22.0 (1832)	65.0	414
Silicon carbide (SiC)	5180	3.2	2740.0	4.8 (to 1832)	7.9 (2732)		10–10^5	50.0, 64.0 (2552)	60.0, 49.0 (2896)	120
Hafnium carbide (HfC)	7030	12.7	2900.0	6.3 (to 1202)	12.9		109.0	35.0, 4.8 (3092)		298
Zirconium carbide (ZrC)	5790	6.7	2850.0	6.7 (to 2552)	11.8		70.0	15.0, 2.5 (3632)	69.0, 51.0 (1832)	109
Titanium carbide (TiC)	5700	4.9	3200.0	10.0 (to 3632)	18.2		190.0	33.0, 13.5 (3632)	64.0	
Tungsten carbide (WC)	4800	15.8	2100.0	5.0 (to 2552)			53.0	78.0	102.0	
Tantalum carbide (TaC)	7020	14.5	1800.0	6.5 (to 1832)	12.9		30.0	31.0, 17.5 (3632)	55.0	
BORIDES										
Titanium diboride (TiB_2)	5400	4.5	3370.0	6.39 (to 2462)	15.1 (392)		15.3	19.0	53.0	97
Zirconium boride (ZrB)	5500	6.1	2300.0	7.5 (to 2462)	13.3 (392)		9.0–16.0	29.8	64.0	
Tantalum boride (TaB)	5610	12.6	2600.0	5.1 (to 2192)	6.3		68.0			
Molybdenum boride (MoB)	4080	9.31	8.0 [d]							
Hafnium boride (HfB)	5880	11.2	2900.0	5.8 (to 2192)	35.9 (1832)		8.0–14.0			
NITRIDES										
Boron nitride (BN)	5430 [b]	2.3	230.0	0.77 [f], 7.51 [g] (to 1832)	16.4 [f], 8.9 [g]			16.0 [g], 7.3 [f]	12.4 [g], 4.9 [f] (77)	45 [g], 24 [f]
Aluminum nitride (AlN)	3990 [b]	3.3	1225.0	5.64 (to 1832)	17.3 (392), 11.5 (1472)			38.5, 18.1 (2552)	50.0, 40.0 (2552)	300
Silicon nitride (Si_3N_4)	3450 [b]	3.2	9.0 [c]	2.5 (to 1832)	8.7			16.0 to 20.0 (77 to 2192)	8.0 (to 1832)	72–90
Titanium nitride (TiN)	5325 [b]	5.4	1770.0	9.3 (to 1832)	18.9		130.0, 340.0 (5306)	34.0	11.4	141
SILICIDES										
Molybdenum disilicide ($MoSi_2$)	3840	6.24	1260.0	4.5 (to 2600)	18.1 (490), 8.3 (2200)	0.13 (1800)	75.0–80.0 (2750)	50.0, 13.0 (2750)	58.0, 12.0 (2750)	57–250
Tantalum disilicide ($TaSi_2$)	4350	9.10	1200.0	4.67 (to 2600)	18.0 (1400), 21.0 (2700)	0.008 (1800)		29.0 (2300), 16.0 (2750)	50.0 (2300), 14.0 (2750)	
Tungsten disilicide (WSi_2)	3960	9.25	1100.0	4.68 (to 2750)	21.0 (1400), 21.0 (2700)	0.008 (1800)		41.0, 45.0 (2750)	43.0, 16.0 (2500)	
BERYLLIDES										
Zirconium beryllide ($ZrBe_{13}$)	3500	2.7	1080.0 [d]	9.86 (to 2750)	18.1 (490), 8.3 (2200)	0.41 (1600), 0.47 (2700)	16.1, 100.0 (2300)	25.0, 25.0 (2750)	47.0, 10.0 (2750)	190, 70 (2500)
Tantalum beryllide ($TaBe_{12}$)	3360	4.2	720.0 [d]	8.42 (to 2750)	17.7 (1400), 21.7 (2700)	0.28 (1600), 0.30 (2700)	43.5, 138.5 (2300)	31.0, 26.0 (2750)	45.0, 10.0 (2750)	150, 190 (1600)
Molybdenum beryllide ($MoBe_{12}$)	3000	3.0	950.0 [d]		18.2 (1600), 17.5 (2600)	0.41 (1600), 0.45 (2600)		42.0 (2300), 13.0 (2750)	15.0 (1600), 1.0 (2750)	

[a] Numbers in parentheses are test temperatures in Fahrenheit; values without temperatures are for room temperature tests.   [b] Dissociation temperature   [c] Mohs scale   [d] Vickers scale   [e] 500 gm   [f] Perpendicular to grain   [g] Parallel to grain.

are also tabulated in "Refractory Ceramics for Aerospace" by Hague et al. (ref. 43). Current information on the properties and applications of these ceramics is gathered and disseminated by the Defense Ceramics Information Center.

Although much information on the properties and characteristics of various ceramic materials is available, there are many gaps in our knowledge, particularly in regard to their mechanical, thermal, and electrical properties at very high temperatures. As with many materials, property data on ceramics should be accepted cautiously because of the many variables (composition, flux content, impurity level, firing history, and finishing operations) that can affect them. Close cooperation between the ceramic producer and user is recommended for critical applications.

METALS

As in the case of ceramics, the selection of metals for ceramic-to-metal joints or seals largely depends on the specific application. The metals most likely to be used in ceramic-to-metal seals in electron tubes and other vacuum devices are steels of various types, copper and copper alloys, nickel and nickel alloys, and the refractory metals. Several of the precious metals are used also, not usually as structural members because of their cost, but as coatings on contacts and as constituents in brazing filler metals. For other applications it may be necessary to select metals on the basis of corrosion resistance or nuclear properties. Regardless of the application, the metal and ceramic members of the joint must be compatible with the expected service conditions and with each other.

Because of the early demand by the metal fabrication industries, elemental metals and their alloys were produced with lower impurity levels and greater uniformity of properties than ceramics. Also, metal property data are more complete, more readily available, and more reliable than ceramics data.

We will not attempt to summarize the properties of metals used in ceramic-to-metal seals in this report. Such information is available in countless texts and handbooks such as the *Metals Handbook* (American Society for Metals) and Smithell's *Metals Reference Book* (refs. 44 and 45). Hampel has reviewed the properties of many metals used in vacuum tube construction in his *Rare Metals Handbook* (ref. 46); Kohl has done likewise in his *Handbook of Materials and Techniques for Vacuum Devices* (ref. 29). Property data on new structural alloys are available from the manufacturers.

An excellent source of current information on the properties of metals, their fabrication and processing, and their application is the Defense Metals Information Center. This organization gathers data and disseminates them in the form of reports, memoranda, and news releases. Data on most of the metals of interest to the electron-tube industry, with the exception of copper and some of the precious metals, are filed by DMIC. Information on copper can be obtained from the Copper Development Association.

BRAZING FILLER METALS

Since brazing is the most commonly used method of fabricating ceramic-to-metal seals, the process and the filler metals used to complete the joint should be discussed here. Although there are differences in the mechanism of bonding, soldering and brazing have much in common; in both cases, a filler metal that melts at a lower temperature than that of the base metal is required for joining. Because of these similarities, there is some confusion regarding the terms "soldering" and "brazing." The American Welding Society includes an arbitrary temperature restriction in defining the processes; soldering is done at temperatures below 800° F and brazing at temperatures above 800° F. "Silver soldering" and "hard soldering" are terms often used as synonyms for brazing.

Brazing filler metals are selected on the basis of the materials being joined and the expected service conditions. For ceramic-to-metal joint applications, the selection of filler metals is somewhat restricted because most brazing alloys do not wet ceramics easily, but procedures have been developed to overcome this difficulty. The filler metal must be compatible with the base materials and must be capable of meeting the service requirements. For example, it is senseless to select a metal and a ceramic for service in a cesium atmosphere and braze the joint with a filler metal that does not possess the required corrosion resistance. Similar care must be exercised in selecting joint materials for use in an oxidizing, reducing, or vacuum environment.

Brazing filler metals usually can be grouped according to their major constituent as follows: (1) copper-base alloys, (2) silver-base alloys, (3) nickel-base alloys, (4) alloys based on the noble metals other than silver, and (5) refractory metal-base alloys. The aerospace and nuclear industries require filler metals that maintain their properties at very high temperatures or in very corrosive environments; thus, many experimental alloys based on titanium, zirconium, or the refractory metals have been developed. For the most part, the requirements for ceramic-to-metal joints posed by the electronics industry can be satisfied with filler metals based on copper, silver, or the other noble metals. (The active-metal brazing process, which uses alloys that contain relatively large amounts of titanium or another active metal, is an exception.) Nickel-base alloys are used less frequently than types (1) and (2) because of

TABLE 3.—*Some Commercially Available Noble-Metal Brazing Filler Metals*

[From ref. 47]

Alloy type	Ag	Au	Pd	Cu	Ni	Mn	Cr	Co	Pt	Temperature, °C	
										Solid	Liquid
Ag	100.0									960	960
Ag-Cu	72.0			28.0						780	780
Ag-Cu	50.0			50.0						779	875
Ag-Cu-Ni	77.0			21.0	2.0					779	830
Ag-Cu-Ni	71.5			28.0	0.5					780	795
Ag-Cu-Ni	62.5			32.5	5.0					775	889
Ag-Cu-Ni	71.5			28.0	0.5					780	795
Ag-Cu-Ni-Mn	65.0			28.0	2.0	5				750	850
Ag-Cu-Pd	54.0		25	21.0						901	950
Ag-Cu-Pd	68.0		5	27.0						807	810
Ag-Cu-Pd	58.0		10	32.0						824	854
Ag-Cu-Pd	65.0		15	20.0						850	900
Ag-Pd	95.0		5							970	1010
Ag-Pd	90.0		10							1002	1065
Ag-Pd	80.0		20							1070	1160
Pd-Ni			60		40.0					1238	1238
Ag-Pd-Mn	75.0		20			5				1071	1121
Ag-Pd-Mn	64.0		33			3				1149	1232
Ag-Mn	85.0					15				960	971
Pd-Co			65					35		1230	1235
Au-Cu		94.0		6.0						965	990
Au-Cu		65.0		35.0						1000	1020
Au-Cu		30.0		70.0						1015	1035
Au-Ni		82.0			18.0					950	950
Au-Ag-Cu	20.0	60.0		20.0						835	845
Au-Ag-Cu	5.0	75.0		20.0						885	896
Au-Cu-Ni		81.5		15.5	3.0					900	910
Au-Cu-Ni		35.0		62.0	3.0					990	1025
Au		100.0								1063	1063
Au-Pd		87.0	13							1260	1305
Au-Pd		75.0	25							1380	1410
Pd			100							1552	1552
Pt-Pd-Au		5.0	20						75	1645	1695
Au-Ni-Cr		72.0			22.0		6			974	1065
Pt									100	1769	1769

the reactions that occur between the base metal and filler metal and because of their lower ductility. Table 3 lists filler metals based on the noble metals used in electron-tube construction (ref. 47).

The properties of filler metals and the techniques of brazing are reviewed extensively in the literature (refs. 29, 48 and 49), but several points should be emphasized here. Metals with low-vapor pressures, such as zinc and cadmium, should not be used for ceramic-to-metal joints exposed to a high vacuum. Also, since filler metals are adversely affected by impurities, "vacuum-tube grade" alloys should be specified for critical applications; these are vacuum-processed alloys in which the impurity content is held to a very low level.

The furnace atmosphere must be considered when ceramic-to-metal joints are brazed, because it affects the base-material properties and the wetting properties of the filler metal. Depending on the materials being joined and the metallizing process used, oxidizing, reducing, or vacuum environments can be used. Titanium and the refractory metals, which are adversely affected by gaseous contaminants, should be brazed in a very good vacuum or in a very pure inert-gas atmosphere. Similar precautions are required when the active-metal process is used for brazing ceramic-to-metal joints.

CHAPTER 3

Ceramic-to-Metal Joint Configurations

To realize the benefits of ceramic-metal structures, it is necessary to use the materials and procedures best suited to their fabrication. Thus, all phases of production from selecting the joint materials to inspecting the finished product must be reviewed in light of the expected service conditions. Some of the variables that must be considered to ensure the reliability of joints and seals will be examined below.

JOINT DESIGN

Selection of Materials

Ceramics

The properties of ceramic materials must be carefully considered and matched against the service requirements, because it is unnecessarily costly to overdesign a ceramic-to-metal structure. For example, the ceramics used for electron-tube envelopes must be dense, vitrified bodies; those used as supports inside tubes need not be so dense and may be slightly porous for ease of outgassing. Larsen suggests three areas of application, each of which emphasizes certain properties which are possessed by ceramics (ref. 50).

1. *Refractory ceramics.*—These ceramics retain their strength and structural integrity at temperatures exceeding 2000° F. They are generally insulators, relatively inert to most environments, and resistant to many molten metals and fused salts. Ceramics of this type can withstand high, steady-state temperatures indefinitely; however, the build-up of thermal stresses resulting from excessive thermal shocks can produce cracking. Ceramics with low elastic moduli, low thermal expansion coefficients, and high heat conductivities are usually resistant to thermal shocks; they are used as structural members or metal coatings to provide oxidation resistance.

2. *Electrical ceramics.*—These ceramics are characterized by their high resistivity, low dielectric losses, and high breakdown voltage. They are used for high-temperature applications where structural integrity is important along with excellent electrical properties. Hard glasses and mica can also be used as electrical insulators; however, the usefulness of these materials is limited by their service temperatures—about 1000° F for glasses and 1500° F for mica.

3. *Mechanical ceramics.*—The usefulness of these ceramics is governed by their strength, hardness, creep resistance, and chemical inertness at high temperatures. Included among these materials are the metal-bonded carbides and ceramic-metal combinations, also known as cermets.

Other properties of ceramics that may be of importance, depending on the service requirements, are surface finish, porosity, uniformity of density, vacuum tightness, freedom from internal strains, transparency, and machinability.

Larsen has suggested the use of a chart (table 4) that lists the refractory, electrical, and mechanical properties of many available ceramic bodies. Knowing the service requirements, it is a relatively simple procedure to select the ceramics that most nearly meet these requirements. When two or more ceramic materials appear to be suitable, final selection may be based on other considerations such as cost, availability, and ease of fabrication.

13

TABLE 4.—*Properties of Ceramics*

[From

	REFRACTORY							
	Silicon carbide (ceramic-bonded)	Silicon carbide (silicon nitride-bonded)	Mullite (synthetic) ($3Al_2O_3 \cdot 2SiO_2$)	Thoria (ThO_2)	Magnesia (MgO)	Zirconia (stabilized) ($ZrO_2 \cdot 2\%$ CaO)	Fused silica (SiO_2)	Pyroceram (9605)	Hard glass
PHYSICAL PROPERTIES									
Melting or softening temp., °F	3500	3540	3320	5830	5070	4710	3050	2460	
Max. use temperature, °F	3200	3000	3000	4890	4170	4600	2700	1300	1300
Specific gravity	2.6	2.9	2.6	10.5	3.6	6.1	2.2	2.6	2.4
Thermal conductivity Btu/sq ft/hr/ °F/in.:									
at RT to 212° F				73	240	14	9.5	2.4	4.0
at 1000° to 1100° F				23	84	14			
at 1830° F				20	49	17			
at 2200° F	109	114	15			7			
Expansion coefficient, micron/in./ °F:									
at RT									
RT to 570° F								0.56	7.7
RT to 930° F				5.0					
RT to 1470° F						7.5			
RT to 2600° F	2.4	2.4	2.5			3.1			
Thermal shock resistance	fair	good	good	poor	poor	fair	excellent	excellent	good
ELECTRICAL PROPERTIES									
Dielectric strength, v/mil at RT			300					300	350
Dielectric constant at 1 mc at RT			7.0				3.8	6.1	4.0–14
Resistivity, ohm-cm:									
at RT to 212° F			10^{13}				10^{17}	10^{13}	10^{17}
at 570° F									
at 900° F									
at 1500° F			10^{6}		10^{7}				
at 2200° to 2300° F				2×10^{4}	8×10^{11}	32			
MECHANICAL PROPERTIES									
Flexural strength, psi×1000 at RT	2.2	5.6 (at 2460° F)	0.5 (at 2460° F)			27	15	37	15
Tensile strength, psi×1000 at RT	low	3	18					20	8
Compressive strength, psi×1000:									
at RT	15	20	150	214		300			100
at 590° F				71					
at 2200° F				28		114			
Modulus of elasticity, psi	13.2	17		21	12	32	10.5	19.8	100
Hardness:									
Mohs	9	9.6	6–7		6	7			700
Knoop	2500					1100		1100.0	
Charpy impact strength, in.-lb									1.2
Abrasion resistance	excellent	excellent	fair	fair	poor	fair	good	good	good

a 72° to 390° F.
b 390° to 930° F.
c 930° to 1470° F.
d 1430° to 2190° F.
e 72° to 1830° F.

and Application Classifications

ref. 50]

ELECTRICAL							MECHANICAL				
Alumina (93% Al_2O_3)	Alumina (99.5% Al_2O_3)	Beryllia (dense) (BeO)	Steatite ($MgO\cdot SiO_2$)	Cordierite ($2MgO\cdot 2Al_2O_3\cdot 5SiO_2$)	Forsterite ($2MgO\cdot SiO_2$)	Lithia porcelain	Titanium carbide (cobalt-bonded)	Tungsten carbide	Boron carbide	Zirconium carbide	Silicon nitride
3700	3700	4650	2550	2610	2620	----------	5680	5031	4080	6390	3450
3000	3180	3600	1830	2280	1800	1800	2800	4700	3500	----------	2800
3.7	3.8	3.0	2.7	2.1	2.9	2.3	5.5–6.5	12–16	2.5	6.2	3.4
200	217	1460	17.4	8.7	24	14.4	120	64	----------	145	130
64	----------	325	----------	----------	----------	----------	----------	----------	----------	----------	----------
43	59	174	----------	----------	----------	----------	----------	----------	----------	----------	----------
----------	----------	170	----------	----------	10.3	----------	----------	----------	188	----------	----------
----------	a 3.8	----------	----------	----------	----------	----------	4.3–5.8	----------	----------	----------	3.1
----------	b 4.4	4.0	4.5	1.2	----------	----------	4.5–6.0	2.5–4	----------	3.5	----------
4.0	c 4.9	5.0	5.0	----------	----------	----------	----------	----------	----------	----------	----------
----------	d 5.4	----------	----------	----------	----------	----------	----------	----------	----------	----------	----------
----------	e 4.7	----------	----------	----------	6.1	----------	----------	----------	1.7	----------	----------
good	good	excellent	poor	fair	fair	good	good	fair	poor	poor	good
300	330	300	220	100	240	250					
8.9	9.7	7.0	5.9	5.0	6.2	5.6					
----------	----------	10^{14}	----------	----------	1.8×10^{12}	----------	Carbides generally exhibit electrical semiconductor properties.				
----------	1.5×10^{11}	8.0×10^{13}	----------	----------	----------	----------					
1.0×10^{15}	1.4×10^{9}	----------	12.0×10^{16}	5.0×10^{16}	----------	----------					
----------	----------	----------	----------	----------	3.0×10^{5}	----------					
1.0×10^{5}	----------	3.0×10^{12}	----------	----------	----------	----------					
40	46	26	21	8.0	20	8.0	----------	80	44	5 (at 2730° F)	34
30	----------	18.5	10	3.5	10	1.5–3.5	26–134	130	22.5	14.4	15.8
300	300	114	90	30	85	60.0	265–450	500–800	414	238	----------
185	----------	92	----------	----------	----------	----------	----------	----------	----------	----------	----------
71	----------	34	----------	----------	----------	----------	----------	----------	----------	----------	----------
40	52	45	7.5	7	----------	120.0	42–57	102	62	45	42
9	----------	9	----------	7	7.5	7.5	----------	----------	9.3	8.9	----------
1750	----------	1500	----------	----------	----------	1400.0	1400–1800	1880	2800	----------	2090
7	----------	----------	5.0	2.5	4.0	3.0	18–192	----------	----------	----------	----------
excellent	excellent	good	fair	fair	good	good	excellent	excellent	excellent	excellent	good

Metals

As in the case of ceramics, the metals for ceramic-to-metal joints and seals are selected primarily on the basis of the service requirements. The most commonly used metals are low-alloy, medium-alloy, and stainless steel; copper and copper alloys; nickel and nickel alloys; and refractory metals and alloys. Several of the noble metals are used in ceramic-to-metal joints; however, because of their high cost they are usually used as constituents in brazing filler metals rather than as structural members. Other metals may be used to meet a specific requirement, such as resistance to liquid metals and low nuclear cross section.

A chart similar to table 4 could also be prepared to aid in selecting metals for ceramic-to-metal joints and matching the service requirements to the physical, mechanical, and electrical properties of the respective metals. The number of alloys included in each classification, however, would expand the chart excessively.

The selection of the metal member of a joint is governed somewhat by the physical and electrical service requirements but even more by the ceramic used. Characteristics such as magnetic properties, electrical conductivity, and strength determine the selection of a group of metals; final alloy selection is based on the degree to which the linear-expansion coefficients for the metal and ceramic can be matched.

Joint Design and Configuration

The joint configurations that can be used for ceramic-to-metal seals and joints are more limited than those for joining metals because of the need to join materials with differing properties. This problem is similar to that encountered in joining dissimilar metals not metallurgically compatible. However, in this instance we are primarily concerned with eliminating or minimizing stresses resulting from the different expansion coefficients of the joint members.

Ceramics are inherently brittle materials, much stronger in compression than in tension; in comparison to most metals, they have small expansion coefficients. Ideally, the ceramic and metal selected should have identical expansion coefficients over a wide temperature range and without hysteresis, so that these materials can expand and contract at equal rates throughout the joining and operating cycle. Figures 2 and 3 show that this is not the case. As a result, ceramic-to-metal joints must be carefully designed to emphasize the advantages and minimize the disadvantages of each material.

Different expansion coefficients are primarily responsible for introducing stresses into a ceramic-metal structure, but other problems can be created as well. If, for example, Kovar is brazed to a 96-percent alumina body, both ma-

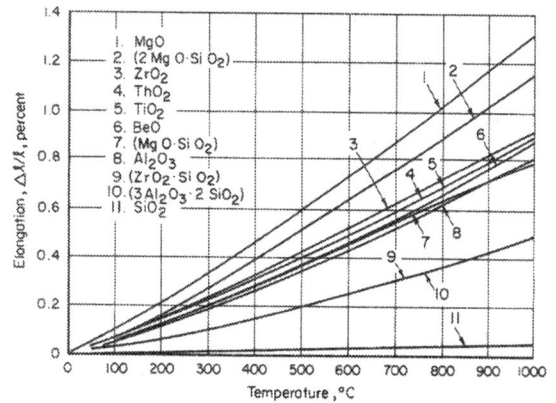

FIGURE 2.—Thermal expansion characteristics of ceramic materials

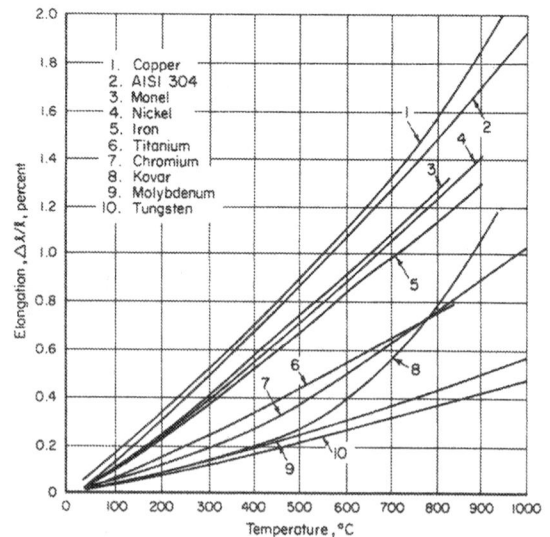

FIGURE 3.—Thermal expansion characteristics of metals used in ceramic-to-metal joints

terials expand at the same rate up to about 450° C, as shown in fig. 4 (ref. 51). Above this temperature, Kovar expands faster than alumina. Thus, a Kovar ring located inside a ceramic ring will produce compressive stresses in the ceramic body at temperatures above 450° C; these stresses can be eliminated if the position of the rings is reversed. However, with the parts reversed, the clearance between the Kovar and ceramic rings increases as the temperature rises, and problems in brazing can occur. When silver-base alloys are used for brazing, the recommended joint clearance is 0.003 to 0.005 inch; however, the clearance for joints brazed with copper ranges from a press fit to about 0.001 inch. A filler metal with sluggish flow characteristics is recommended for brazing joints where large clearances between joint members are expected. Furthermore, the filler metal should be ductile enough to withstand the stresses produced as the joint members contract during the cooling cycle.

Most of the ceramic-to-metal seal configurations have been designed for electron-tube application; however, many of the designs are equally

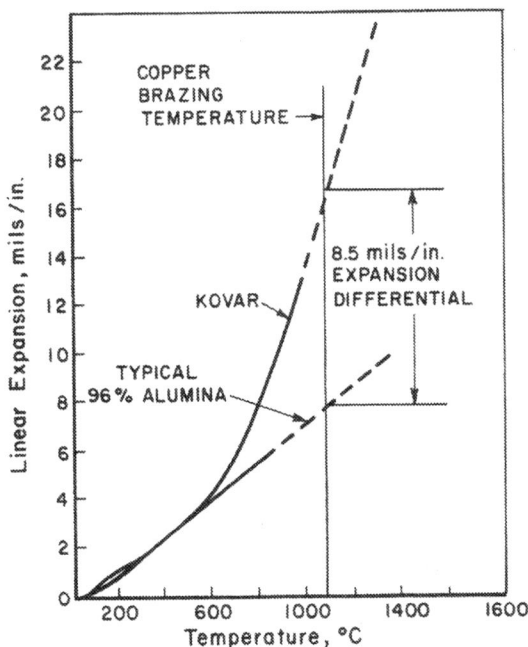

FIGURE 4.—Thermal expansion characteristics of Kovar and alumina (ref. 51)

suitable for other applications. Although some of the seal designs appear complicated, they are composed of basic types, such as the butt seal, lap seal, pin seal, and disk seal. In addition, many types of flexible seals have been developed to accommodate joint members with different expansion coefficients. Special designs have been prepared to join ceramic bodies to massive metal members that cannot flex during the joining or operating cycles; thin metal transition sections have proved useful in such cases. Several typical joint designs are shown in figure 5 (ref. 52); other designs are available in the technical literature.

Several investigators have conducted stress analyses to optimize seal design, indicate areas of high stress concentrations, and provide criteria for designing seals of different sizes (refs. 53 to 55). In a typical study, Cole and Inge derived mathematical expressions to predict the unit stress at the interface of a ceramic-to-metal seal, measured the residual stresses on the outside of the metal member of a ceramic-to-metal seal with strain gauges, and compared calculations of the predicted and measured stresses on the outer surface of the metal member (ref. 53). Using ceramic-to-nickel and ceramic-to-stainless-steel seals, Cole and Inge found the predicted stress to be about 30 percent higher than the measured stress; apparently, the major source of error was to ignore the possible filler-metal effects on the seal stresses. Mark and Lewin analyzed the stresses that were present in ceramic-to-metal butt and disk seals (ref. 55). The theoretical studies were supported by strain gauge measurements and three-dimensional photoelastic model analysis.

SURFACE PREPARATION

Machining (Grinding)

The ceramic bodies for ceramic-to-metal seals may be produced by extrusion, slip casting, dry- or wet-pressing, isostatic pressing, injection molding, and hot compacting. Ceramics designed for metallized assemblies should have smooth contours and simple shapes; sharp edges and corners should be avoided to eliminate concentration of stress. Whenever possible, ceramic bodies should be designed with generous toler-

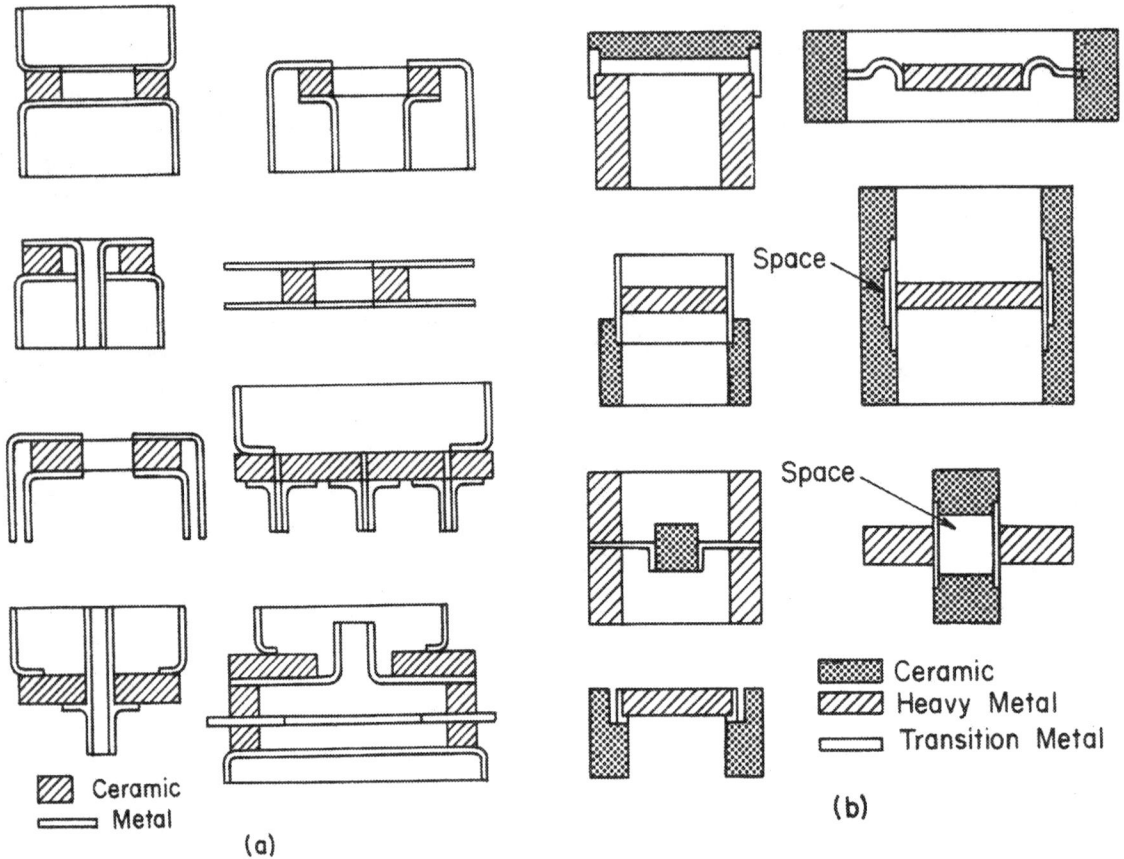

FIGURE 5.—Ceramic-to-metal joint configurations (ref. 52). (a) Butt and lap seal joint designs; (b) joint designs for transitions to thick all-metal members; (c) backup of ductile metal seal with blank ceramic

ances to permit their production without subsequent grinding. Ceramics can be produced in commercial quantities with tolerances of about ±0.005 inch on dimensions up to 0.50 inch and about ±1 percent on larger dimensions. Because diamond-cutting abrasives must be used, grinding is costly even where the surfaces are accessible; costs are prohibitive when less readily accessible surfaces must be finished. In addition, grinding may introduce macrocracks that act as stress risers or notches, as well as stresses that are difficult or impossible to detect. Such stresses can result in excessive brazing clearances and highly stressed joints during the joining or operation cycles.

Despite the disadvantages of grinding, ceramic components used in vacuum-tube construction must often be ground to closer toler-

ances than those to which they can be formed, because the location and spacing of internal structural members are critical to performance. Janssen has discussed the equipment and grinding techniques used to produce precision ceramic bodies (ref. 56).

Cleaning

The cleaning of ceramic surfaces before metallizing and joining is an essential step in the production of reliable ceramic-to-metal seals. As received from the producer, the surfaces of ceramic bodies may be contaminated by organic substances such as traces of oil or grease, fingerprints, and films of adsorbed gases; occasionally, metallic contaminants may also be present. During sintering operations,

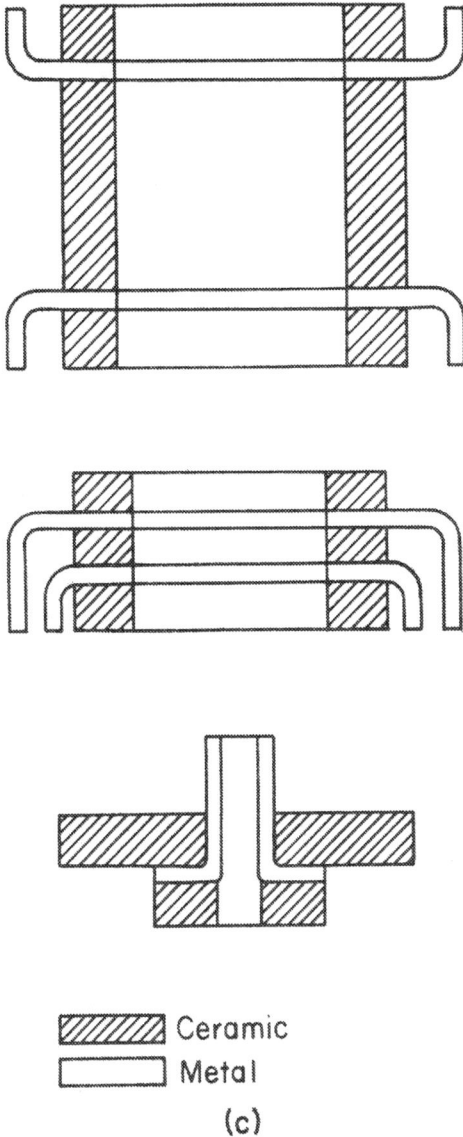

Ceramic

Metal

(c)

mina grit; then, the ceramic body is rinsed in water and acetone and air-dried. Small ceramic parts can also be ball-milled to remove contaminants. As a second step, Johnson and Cheatham recommend that the parts be boiled in clean, concentrated nitric acid for 30 minutes, rinsed in distilled or deionized water, boiled in deionized water for 10 minutes, and rinsed again (ref. 52). Following this treatment, the ceramic parts are given a final rinse in methanol and air-dried at 120° C (248° F).* Then, the parts are fired in air at 1000° ±25° C (1832° ±45° F) for about 10 minutes, cooled, and placed in clean polyethylene bags. Slow heating and cooling rates are used to minimize thermal shocks. Metallizing should be undertaken within 3 days after cleaning.

Other methods using alkaline rinses rather than nitric-acid treatments have also been used to clean ceramic surfaces. The bombardment of ceramic surfaces with gaseous ions to remove surface contaminants has been discussed by Bierlein, Newkirk, and Mastel (ref. 57). This technique has been used to remove oxides and other contaminants from metallic surfaces for critical applications.

After cleaning, the ceramic bodies must be handled carefully with lint-free gloves to prevent recontamination of their surfaces; tweezers, spatulas, and other tools must be plastic-coated or made from nonmetallic materials.

Oxides and other surface contaminants should be removed from the metals used in ceramic-to-metal seals with the same care reserved for ceramics. Johnson and Cheatham recommend the following procedure to clean nickel, Monel, Kovar, and stainless-steel surfaces: Ultrasonically clean the parts in trichloroethylene and rinse in hot water; etch the metals in concentrated hydrochloric acid; and rinse the parts in cold tap water and then in cold, deionized water (ref. 52). The metal parts are then dipped in methanol and air-dried at 120° C. Other

oily substances are reduced to carbon, and metals are converted to oxides; these materials can prevent consistent metallizing and bonding. In addition, they stain the ceramic surface and may cause sparking, electrical leakage, and radio-frequency losses.

Many procedures and techniques have been developed for cleaning ceramic surfaces, depending on the particular ceramic. Loosely adhering particles can be removed by scrubbing the ceramic surface with a water paste of alu-

---

*Both Centigrade and Fahrenheit temperatures are given in this report. The first figure is the temperature cited by the author; the second figure is the converted value of the original data.

etchants for these metals can be found in standard references on metals.

Metallizing

The first ceramic-to-metal joints were successfully produced by brazing a metallized ceramic to a metal member with a silver-base filler metal in the 1930's. Despite the advances in joining technology since then, this is still the most extensively used method of fabricating such joints. Of course, significant improvements in the techniques of metallizing have been made and several new procedures have been developed and evaluated. Also, extensive research on the reactions that occur when a ceramic surface is metallized has contributed to the effectiveness of metallizing.

Metallizing procedures were originally developed to improve the wettability of ceramic surfaces by conventional low-temperature filler metals. Later investigators found that some active metals and their alloys or compounds (e.g., titanium and zirconium) would wet unmetallized ceramic surfaces under certain conditions. Although variations of the so-called active-metal process have been used commercially to produce ceramic-to-metal seals, they have not been accepted to the extent characterized by the metallizing-brazing concept of joining these materials.

In reviewing developments in this area, it should be emphasized that metallizing is a surface preparation for ceramics, not a joining process. While most metallized ceramics are joined to metals by brazing, other processes such as diffusion, pressure, and ultrasonic welding have been used also.

Sintered Metal Powder Processes

These metallizing processes are based on the work of Vatter and Pulfrich in the 1930's and have one common feature, i.e., the metallizing coating is sintered to the ceramic surface. Vatter and Pulfrich mixed finely divided metal powders (molybdenum, tungsten, rhenium, iron, nickel, or chromium) with a suitable binder to form a suspension that was painted on the ceramic surface to be metallized. Then the coating was sintered to the ceramic at a high temperature. Pulfrich used a controlled hydrogen atmosphere for sintering while Vatter used a vacuum. While the desirability of adding certain metal oxides to the final metal powders as a means to improve adhesion between the powder coating and the metal was recognized, Nolte and Spurck are credited with these improvements (refs. 58 and 59).

In 1950, Nolte and Spurck described a procedure to metallize ceramic surfaces with molybdenum at temperatures as low as 1250° C (2282° F). The metallizing mixture was prepared by ball-milling the following ingredients for 24 hours:

160g molybdenum powder (200 mesh)
40g manganese powder (150 mesh)
100cc pyroxylin binder
50cc amylacetate
50cc acetone.

After ball-milling, the mixture was thinned to the proper consistency for painting or spraying, and a 0.001- to 0.002-inch coating was applied to the ceramic surface. Then the coated ceramic was fired in a hydrogen atmosphere for ½ hour at 1350° C (2462° F). Nolte and Spurck recommended a sintering temperature of 1350° C for a metallizing mixture that contained 20-percent manganese; however, satisfactory metallizing was observed when the manganese content was reduced to 10 percent or when the sintering temperature was decreased to 1250° C (2282° F).

The process developed by Nolte and Spurck is known as the "moly-manganese" process. It has been widely accepted by industry as a standard method to metallize ceramic surfaces, and numerous variations have been developed to extend the usefulness of the process. For small production runs and oddly shaped bodies, the metallizing mixture is usually applied with a small brush; care must be observed to apply the coating evenly. Spray-coating, roller-coating, and silk-screening require more elaborate equipment and are thus suitable for large production runs. The metallizing mixture can be supplied on transfer tape for application to the ceramic surfaces (refs. 60 and 61). In producing this tape, the metallizing mixture is spread uniformly on a polyethylene sheet and coated with a pressure-sensitive adhesive that is protected by a paper during shipment and storage. To use, the protective paper is removed and the ceramic bodies are pressed onto the tape; when the ceramic is lifted from the tape, the metallizing coating is transferred to the part. Sintering is done in the usual manner. Standard metallizing coatings can be supplied on transfer tape or proprietary coatings can be ordered. Industry has highly automated tape-transfer operations as well as other methods of applying metallizing mixtures to ceramic bodies.

Engineers at the Sperry Gyroscope Company conducted an extensive program to improve the reliability of the sintered metal powder metallizing process (ref. 53). Starting with a review of the literature, the researchers analyzed the observations and theories of adhesion between the metallizing coating and the ceramic body to determine the classes of materials that appeared to promote adhesion; additional theories on adhesion were also proposed. Based on this analytical approach, over 200 metallizing mixtures were formulated and evaluated. Each mixture was applied to ceramic bodies containing 94.0-, 96.0-, and 99.6-percent alumina; sintering was done at three temperatures between 1250° and 1700° C (2282° and 3092° F). The metallizing powders were mixed with a binder of acetone, amylacetate, and nitrocellulose lacquer and ball-milled for 24 hours. The effectiveness of metallizing was determined by tests that measured adherence, peel strength, compressive strength, and tensile strength; approximately 3200 specimens were prepared and evaluated. The tests indicated that many metals and oxides can be used in place of manganese; at least 16 metallizing mixtures produced seals with equal or greater strength than those produced by the moly-manganese process. The compositions of seven of the most promising mixtures are shown in table 5 along with the processing and test data. The workers at Sperry concluded that:

1. The tensile strength of ceramic-to-metal seals made with the three alumina bodies decreased with increasing alumina content. The maximum tensile strengths were 28 400, 22 000, and 16 100 psi for the 94.0-, 96.0-, and 99.6-percent alumina ceramics, respectively.

2. The optimum sintering temperatures were 1500° to 1600° C (2732° to 2912° F).

3. Metallizing mixtures for the 99.6-percent alumina body invariably required additions of silica or silicate-bearing minerals to produce satisfactory seals.

Other metallizing compositions have been cited by Kohl and Rice (ref. 62) and LaForge (ref. 63). These are basic molybdenum-manganese mixtures with additions of titanium hydride, iron, and various metal oxides to promote adhesion of the metallizing layer to the ceramic surface.

In 1966, the Sperry Electronic Tube Division concluded a program to develop low-temperature metallizing materials (ref. 64). It is known that the microstructure and properties of ceramics change significantly when heated to the high temperatures required to sinter moly-manganese coatings. For example, Cole and Hynes (ref. 33) noted that the seal strength increased at the expense of the modulus of rupture, apparent porosity, and apparent density when the firing temperature exceeded 1550° C (2822° F). Tentarelli, White, and Buck observed sizable increases in the physical dimensions of a 94-percent alumina body when it was heated to temperatures above 1475° C (2687° F). Others have reported increases as high as 1.4 percent in the length of a 96-percent alumina ceramic heated above 1475° C. Such variations in physical dimensions do not occur at lower firing temperatures in the 900° to 1100° C (1652° to 2012° F) range. The stability of ceramics is critical in electronic assemblies where performance depends on close tolerances.

The engineers at Sperry formulated a series of metallizing paints based on metal oxides of molybdenum, manganese, and tungsten. While molybdenum has received the most attention as a metallizing constituent, tungsten has also been investigated to metallize ceramics for high-temperature service (ref. 65). A statistical analysis was made to evaluate two paints, as shown in tables 6 and 7 at 900°, 1000°, and 1100° C (1652°, 1832°, and 2012° F), and coating thicknesses of 0.005, 0.010, 0.015, and 0.020 inch. The metallizing paint was applied to alumina ASTM test pieces (fig. 6) in the desired thickness; then, the coating was sintered to the ce-

TABLE 5.—*Metallizing Compositions*

[From ref. 53]

Composition Number	Composition and weight, g	Sintering temperature, °C	Ceramic, percent Al_2O_3	Peel test values, in.–lb	Compression values, lb	Tensile test values, psi
65	292.5 Mo	1500	94.0	2.5	>4000	28400
	7.5 Ti				>4000	19350
					>4000	16400
91	270 Mo	1500	94.0	2	>4000	15500
	30 LiMnO₃				>4000	15700
					3300	14500
141	291 Mo	1600	94.0	4	3700	12300
	9 Talc				3400	17900
	(MgO–SiO₂)				3900	16100
72	240 Mo	1500	96.0	2	3800	9430
	73.6 CeO₂				3600	22000
					4000	16050
50	255 Mo	1300	96.0	2	2800	15700
	48 SiO₂				3000	13200
	22 Mn				3800	10700
50	255 Mo	1500	99.6	2	2200	14000
	48 SiO₂				2300	16100
	22 Mn				1600	15200
49	255 Mo	1500	99.6	2.75	>4000	11300
	48 SiO₂				3500	11600
	26 MnO				1200	16100

ramic surface in an atmosphere of dissociated ammonia. To evaluate the quality of the metallizing treatment, pairs of the ASTM pieces were brazed with copper and tested in tension. Tentarelli et al. observed that:

1. There were no essential differences in seal strength, regardless of the metallizing temperature or coating thickness.

2. Average tensile strengths of 13 000 to 14 000 psi were obtained with various alumina ceramics, and strengths of 11 300 psi were obtained with beryllia bodies.

3. Thermal-cycling tests were used to evaluate the seal reliability obtained with the MoO_3/MnO_2 low-temperature metallizing system relative to the reliability obtained with conventional high-temperature Mo/Mn systems; no significant differences in the results were noted.

TABLE 6.—*Evaluation of MoO_3/MnO_2 Metallizing Paint**

[From ref. 64]

Metallizing thickness, inches	Tensile strength, psi at indicated reduction temperature		
	900° C	1000° C	1100° C
0.0005	10 300	8 300	10 800
	11 200	10 000	9 900
0.0010	10 900	11 400	7 100
	10 100	9 000	10 100
0.0015	11 100	7 900	5 800
	10 000	11 200	9 400
0.0020	11 600	9 800	9 400
	12 200	13 100	10 500

*Metallized film was 95Mo–5Mn.

41

TABLE 7.—*Evaluation of $WO_3/MnO_2/Fe_2O_3$ Metallizing Paint**

[From ref. 64]

Metallizing thickness, inches	Tensile strength, psi at indicated sintering temperature			
	800° C	900° C	1000° C	1100° C
0.0004	9 700	10 900	10 200	12 900
to	5 500	7 600	14 500	9 800
0.0007	6 600	5 200	17 000	13 300
	9 400	4 700	6 300	3 400
0.0008	5 800	10 400	7 200	11 500
to	15 100	4 700	6 500	10 500
0.0012	11 900	5 700	9 100	8 400
	11 200	6 600	11 100	7 700
0.0013	7 200	3 900	11 100	4 100
to	9 500	13 700	13 200	13 200
0.0017	5 400	6 400	3 300	8 400
	3 700	9 100	3 600	9 600
0.0018	5 600	10 800	7 900	5 100
to	3 700	7 600	6 900	3 700
0.0023	4 600	9 000	6 900	6 900
	3 100	7 300	3 900	8 900

*Metallized film was 94W–5Mn–1Fe.

4. The low-temperature metallizing systems were suited for silk-screen and transfer-tape methods of application. Compared with the moly-manganese process, low-temperature metallizing offers economic advantages associated with lower sintering temperatures and less critical sintering atmospheres.

In a program to develop ceramic-to-metal seals for nuclear thermionic energy converters, Bristow, Grossman, and Kaznoff investigated metallizing systems to prepare essentially pure-alumina ceramic surfaces for brazing (ref. 11). In the conventional moly-manganese process, the manganese is oxidized during the metallizing operation and reacts with (1) the alumina body itself to form manganese alumina compounds, and (2) the fluxing oxides in the ceramic to form a glassy phase that locks the molybdenum coating to the ceramic surface. However, the seal strength tends to decrease as the alumina content of the ceramic increases, presumably because less fluxing oxides are present. For this application, silica-free

alumina ceramics were required for a high-temperature cesium-vapor environment, and the fluxing oxides had to be incorporated in the metallizing mixtures. The composition of the experimental metallizing materials was 60- to 70-volume percent molybdenum powder plus 30- to 40-volume percent fluxing oxides of aluminum, calcium, magnesium, barium, or yttrium (table 8). The powder mixtures were dispersed in an organic binder and ball-milled for 144 hours. After mixing, the metallizing suspensions were thinned to the proper consistency and applied to the ceramic surfaces by silk-screen techniques. The coated ceramics were then sintered at various temperatures in a dry-hydrogen atmosphere. The quality of metallizing was evaluated by preparing and leak-testing ceramic-to-metal seals; these data are summarized also in table 8. Satisfactory seals were obtained with ceramics that were metallized at relatively high temperatures. Similar procedures to metallize high-purity alumina ceramics were developed by

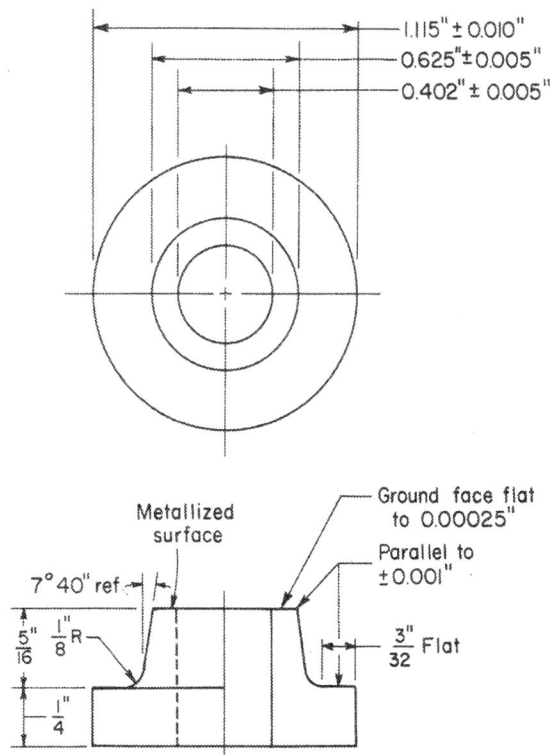

FIGURE 6.—ASTM tensile test piece (ref. 82)

TABLE 8.—*Composition of Metallizing Materials and Results of Sealing Studies*

[From ref. 11]

Metallizing		Oxide constituents, w/o					Sintering temperature, °C	Leak check results	
Mix Number	Oxide composition Number	Al$_2$O$_3$	CaO	MgO	BaO	Y$_2$O$_3$		Vacuum tight	Leakers
1	26	51.8	41.5	6.7	--------	--------	1610	0	6
							1680	0	6
							1750	3	1
							1755	5	1
							1760	3	0
							1800	3	0
							1850	7	0
18	20	80.0	17.0	3.0	--------	--------	1700	0	3
							1750	0	3
							1800	0	6
							1850	3	0
							1900	3	0
21	47	95.0	4.3	7.0	--------	--------	1700	0	2
							1750	0	3
							1800	3	3
							1850	1	2
							1900	0	3
36	48	80.0	20.0	--------	--------	--------	1700	0	3
							1750	0	3
							1800	1	5
							1850	3	0
							1900	3	0
54	44	60.0	--------	--------	40	--------	1750	0	3
55	46	65.0	--------	--------	--------	35	1800	3	0
							1850	2	1
							1900	3	0

Klomp and Botden (ref. 66). Molybdenum powder was added to mixtures of CaO, Al$_2$O$_3$, SiO$_2$, and MgO in the following proportions: 80-weight percent molybdenum plus 20-weight percent oxide powder.

Many investigators have discussed the importance of powder-particle size in metallizing mixtures. It is generally agreed that the size of the particles should be about 1 to 8 microns in diameter; however, fine and coarse particles must be blended to produce metallizing paints with the required viscosity and surface tension. The method used to apply the paint to the ceramic surfaces also has a significant bearing on the particle size and distribution.

Reactive- or Refractory-Metal Salt Coating

Ceramics can also be metallized by painting them with a solution of a refractory-metal salt. The ceramic is then dried and sintered to reduce the metal salt to a metal that bonds to the ceramic, leaving a thin, adherent metal coating on the surface of the ceramic. The ceramic bodies can also be immersed in the solution, but then the sintered metal coating must be ground off areas requiring insulation. In a program to develop low-temperature metallizing procedures, Tentarelli, White, and Buck investigated the use of water-soluble refractory-metal salt for metallizing (ref. 64). The metal salts were dissolved in an appropriate amount of water. Then the solution was painted on alumina ASTM test pieces that were preheated to facilitate drying. After drying, the coated ceramic pieces were heated to 1100° C (2012° F) in a dissociated-ammonia atmosphere whose dew point was 5° C (41° F). The ASTM test pieces were brazed in pairs with copper, and the assemblies were tested in tension to evaluate the metallizing procedure. Table 9 presents the composition of the experimental metallizing solutions along

TABLE 9.—*Evaluation of Low-Temperature Solution-Metallizing Paints*

[From ref. 64]

Composition of metallizing paint	Ceramic, percent Al_2O_3	Average RT tensile strength, psi
4.2g ammonium molybdate 0.2g potassium permanganate	94.0	12 600
18.7g sodium molybdate 0.2g potassium permanganate	94.0	10 050
2.4g ammonium molybdate 0.2g lithium molybdate 0.05g potassium permanganate	94.0	7 300
24.0g ammonium molybdate 0.2g potassium permanganate	94.0	14 000
60.0g ammonium molybdate 3.0g ferrous ammonium sulfate	94.0	3 500
25.0g ammonium molybdate 0.2g ferrous ammonium sulfate	94.0	12 600
30.0g ammonium molybdate 0.5 ml manganese nitrate (50% sol.)	94.0	13 300
40.0g ammonium molybdate 0.4g vanadium pentoxide	94.0	9 500
10.0g molybdenum phosphate 1.0 ml manganese nitrate (50%)	94.0 99.5	9 030 7 790
10.0g molybdenum phosphate 3.0 ml manganese nitrate (50%) 1.0g molybdenum metal powder	94.0 99.5	10 400 9 500
11.0g ammonium molybdate 1.0 ml manganese nitrate (50%) 1.0 ml hydrogen peroxide (30%)	94.0	12 200
10.0g tungstic acid 30.0 ml ammonium hydroxide 0.5g ferrous ammonium sulfate 0.2g manganese tungstate	94.0	12 290
10g tungstic acid 25.0 ml ammonium hydroxide 0.5g manganese tungstate	94.0	13 750
10g tungstic acid 25.0 ml ammonium hydroxide 0.5g ferrous ammonium sulfate	94.0	6 320

with the alumina content of the test pieces and the tensile-test data. The solution of ammonium molybdate and manganese nitrate allowed the least migration into unpainted areas and produced seals with the highest average tensile strength. Tentarelli et al. also observed that:

1. Concentrated metallizing solutions did not produce joints with higher seal strengths, resulted in more loose metallic particles on the ceramic surface after sintering than with a dilute solution, and displayed more migration.

2. Applying the metallizing solution to heated ceramic surfaces appeared to promote uniform coating.

3. The addition of thickening agents to the solutions did not improve the uniformity of the coatings or the seal strengths.

4. The seal strengths obtained by solution-metallizing were comparable to those obtained with low-temperature metallizing mixtures of metal oxides.

Solutions of molten-metal salts have also been used to metallize ceramic surfaces for subsequent joining operations. Straumanis and Schlechten of the University of Missouri School of Mines and Metallurgy, in investigating this process as a means to coat metals with titanium, did some work on coating ceramic surfaces (ref. 67). In this process, a mixture of 90-percent potassium or sodium chloride and 10-percent titanium powder is prepared and heated in a closed vessel under a helium shield until the mixture is molten; titanium powder usually contains 2 to 5 percent oxygen in the form of titanium oxide. After the mixture of metal salts is molten, the metal to be coated is lowered into the bath. The mechanism of coating is not clearly defined and several theories have been advanced to explain it. One states that the titanium oxide and potassium (or sodium) chloride react to form titanium chloride and potassium or sodium oxide; then, the titanium chloride and the metal to be coated react to form a metal chloride and titanium. The titanium is deposited as a film on the metal. Siebert et al. suggested that a reduction mechanism is responsible for depositing titanium on the metal (ref. 68). Of course, other reactions occur during the metallizing of a ceramic surface. The thickness of the coating highly depends on time

and temperature. At 950° C (1742° F), a titanium-rich layer, 0.0002 inch thick, was formed on a steel surface in 2 hours; the layer increased to 0.0004 inch in 6 hours. At 1000° C (1832° F), the layer thicknesses were 0.0004 inch after 2 hours and 0.0007 inch after 6 hours. Quinn and Karlak have simplified the procedures greatly and developed methods to coat ceramic surfaces with zirconium, hafnium, and uranium as well as titanium at temperatures as low as 600° C (1132° F) (ref. 69). To coat a ceramic surface, a thin titanium sheet is coated with a mixture of alkali or earth alkaline halides to a thickness of about 0.030 inch. The ceramic body is placed on the coated titanium sheet and the assembly is heated in an air furnace. The mixture of metal salts melts and flows over the titanium surface, protecting it from oxidation and depositing titanium on the ceramic surface. The thickness of the coating can be varied by regulating the time and temperature of heating.

Vapor-Deposited Coatings

Vapor-deposition processes used to deposit metal coatings on ceramic surfaces can be grouped in two major classifications: chemical vapor deposition and physical vapor deposition. Chemical vapor deposition may be defined as the deposition of elements or compounds in massive form or as a coating process by chemical reaction of the vapors of suitable compounds, usually at a heated surface. Intentional chemical reaction is not involved in physical vapor deposition; the material composing the coating is identical with the source material.

Several variations of the physical vapor deposition process have been used including:

1. *Sublimation and evaporation.*—The coating material is heated until the number of atoms and molecules leaving its surface is sufficient to produce the desired deposit on a ceramic substrate located some distance away. This process is often called "vacuum metallizing," although some nonmetals can also be deposited in this manner.

2. *Sputtering.*—Normal physical sputtering is characterized by the inert-gas ion bombardment of a cathode target, resulting in the ejection of atoms from the target surface into the

surrounding gas atmosphere. Deposition of these target atoms on a nearby substrate produces a thin coating of the target material.

3. *Ion plating.*—Thermally evaporated metallic atoms are ionized and accelerated in an electrical field. The ions impinge on the substrate surface with somewhat higher kinetic energies than in sputtering and with much higher energies than in sublimation or evaporation.

Reed and McRae coated all grades of alumina (up to 99.75-percent alumina) and 96-percent beryllia with molybdenum (ref. 70). The molybdenum was evaporated by a heated filament, and vacuum-tight coatings up to 0.0004 inch thick were deposited in less than 2 minutes. The ceramics were cleaned with water-detergent solutions and air-fired at 1000° C (1832° F). Before evaporation, the ceramic surfaces were further cleaned by a glow discharge. Joints prepared with the molybdenum-coated ceramics had tensile strengths of 20 000 psi for alumina ceramics and 15 000 psi for beryllia ceramics. Holmwood and Glang also used evaporation techniques to coat oxidized silicon wafers with a thin film of molybdenum (ref. 71). The source of the molybdenum vapor was a molybdenum rod heated by an electron beam in a vacuum (2×10^{-7} torr).

Various types of alumina have been metallized with a "particle bombardment" process, developed by Heil and his associates (ref. 72), where ions are sputtered from the target material and accelerated toward the substrate surface by high-frequency energy generated by a magnetron-type discharge. The surfaces of standard ASTM tensile test pieces made from a silica-free alumina ceramic (99.8-percent alumina) were metallized with niobium or vanadium by particle bombardment. The test pieces were joined together in pairs to produce specimens for evaluation. The tensile strength of seals made with niobium-coated ceramic surfaces ranged from about 6600 to 11 000 psi; the strength of seals made with vanadium-coated specimens was about 4500 psi. Alumina test pieces (99.5-percent alumina) were also metallized with tungsten or tantalum. The tantalum-metallized test pieces were copper-coated and brazed in pairs with the silver-copper eutectic

filler metal; the tensile strength of these joints ranged from 12 000 to 13 500 psi. The tungsten-metallized test pieces were nickel-plated and brazed with copper; the tensile strength of these joints was 9000 psi.

The high energy imparted to the ions during particle bombardment causes the particles to bond firmly to the target ceramic substrate; the substrate need not be heated to very high temperatures to ensure good bonding.

Sputtering techniques have also been used by Seeman to metallize alumina surfaces (ref. 73). A primer film of chromium was first deposited on the ceramic surface, followed by an intermediate layer of chromium plus copper, and a final layer of pure copper.

Before using the ion-plating technique to metallize ceramic surfaces (ref. 74), Mattox used it to deposit thin metallic films on metals and noted that this process had the following advantages over other vapor-deposition methods:

1. Since an inert gas is used to establish the glow discharge, the ceramic substrate surface is cleaned by ion bombardment before the film material begins to evaporate. However, after deposition begins, ion bombardment will sputter the deposited film, so that the deposition rate must always exceed the sputtering rate.

2. Ion plating tends to produce more adherent films than other methods.

3. The kinetic energy from ion bombardment is dissipated at the substrate surface as heat and enhances diffusion and reaction rates.

Mattox deposited multi-layer films on lead-zirconate-titanate substrates by ion plating. Aluminum was first deposited on the ceramic surface, because it reacts with the oxygen there and adheres well. To prepare the surface for soldering as well as brazing, a layer of copper or gold was deposited over the aluminum before the aluminum deposition was completed; thus, there was a bottom layer of aluminum, an intermediate layer of aluminum and copper (or gold), and a top layer of copper (or gold).

Metal-Glass Powder Coating

Ceramic insulating materials used in the electronics industry have been metallized for many years with metal-glass powder mixtures. Finely divided metal powders and glass frits are mixed with a suitable inorganic binder to form a paint that is applied to the area of the ceramic to be metallized, and the coated ceramic is fired to promote adhesion of the glass to both the ceramic and metal powder. The thin conductive coatings produced in this manner can be electroplated, or soldered connections can be made directly to them. Noble metals such as gold, palladium, and platinum have been used to metallize ceramics but silver is most common.

In a series of three articles (refs. 75 to 77) on producing thin conductive films on ceramic bodies, Lindquist discusses materials (metal powders, glass fluxes, and binders), coating techniques, and drying and firing methods. Sedenka conducted studies to determine the effect of firing temperature on the adherence of silver to ceramics (ref. 78). He found that adhesion depends on the chemical composition of the glass flux in the silver paint mixture and in the ceramic.

Ceramics have also been metallized by the thermal reduction of aqueous salt solutions of precious metals. The solutions used to deposit thin films of silver, gold, or other metals have been reviewed by Heritage and Balmer (ref. 79).

Electroplating

To ensure the production of reliable ceramic-to-metal seals, most metallized surfaces are coated with nickel, copper, or other metals. The metals are usually deposited by electroplating; however, in some cases, the coatings are produced by reducing oxides of the desired metal. These coatings perform several functions, depending on the method used to produce ceramic-to-metal seal. If the joints are to be brazed with conventional silver- or copper-base filler metals, the coatings serve the following purposes:

1. A metallizing layer is composed of metals and residual oxides not completely reduced during sintering. Such a surface is not conducive to good wetting by the brazing filler metal. Plating with nickel and/or copper eliminates the adverse effects of the surface on the wetting and flow characteristics of the filler metal.

2. When the metals used for metallizing are not wet readily by low-temperature filler metals, plating provides the surface with a metal easily wet by such brazing alloys.

3. To a degree, the plated metal acts as a barrier to the penetration of the metallizing layer by the filler metal. Some filler metals react with the metals used for metallizing, and if the reaction is allowed to proceed too long, the filler metal may penetrate the metallized coating and lift it away from the ceramic. Metallized coatings are usually plated with nickel to retard penetration and copper to provide good wetting.

If the ceramic-to-metal joint is produced by methods other than brazing, the plating may serve other functions. For example, diffusion-welded joints require interface materials that promote diffusion between the metal and the metallized ceramic surface.

Typical electroplating bath compositions and procedures will not be reviewed here, because detailed information is available in standard references on this subject, as well as in the technical literature on ceramic-to-metal joints.

The metals used in ceramic-to-metal joints are frequently plated also. Depending on the metal, plating provides a surface easily wet by the filler metal, protection against oxidation of the metal surface during the brazing cycle, protection against intergranular penetration by the filler metal, and sustained cleanliness of the metal surface during storage.

JOINING PROCESSES

The choice of a joining process to fabricate ceramic-to-metal seals on joints is limited by the physical and mechanical properties of the ceramic materials. The first ceramic-to-metal joints were assembled by mechanical means, and while such techniques are still in use, the high-quality, vacuum-tight joints required by the electronics, nuclear, and aerospace industries are produced by the processes used to join metals. Metal-joining techniques can be divided into three major categories briefly discussed below.

Fusion Joining

The base materials are heated until they melt and fuse. The heat required for fusion is produced by an electric arc, an electron or photon beam, or the resistance of the base materials to an electric current. This category includes the numerous arc-welding and resistance-welding processes. There is little application of fusion joining in producing ceramic-to-metal seals, since ceramic materials cannot withstand the thermal shock associated with welding without cracking. The electron-beam welding of ceramic-to-metal joints has been investigated with limited success.

Solid-Phase Joining

All processes in this category accomplish bonding without changing the solid state of the joint's materials. While most of the processes require heat, the materials in the joint do not become molten. Pressure is usually required to produce joining. Several of the solid-state joining processes have been used to produce ceramic-to-metal seals. Among them are the following: diffusion welding (or bonding), pressure welding, gas pressure bonding, and ultrasonic welding. Electroforming or electrodeposition welding and high-energy welding are also included.

Liquid-Solid Phase Joining

This category includes soldering and brazing—the processes most widely used in fabricating ceramic-to-metal seals. The joint assembly is heated to a temperature below the melting temperature of the base materials but above the melting temperature of the filler metal; thus, a liquid phase is formed at the interface between two solid members, and the molten filler metal is distributed throughout the joint by capillary attraction. Joining is accomplished when the filler metal solidifies.

Adhesive bonding procedures are used to join ceramics to metals for some applications, but not to produce ceramic-to-metal seals, since adhesives have a limited service temperature range (350° F or below for continuous service) and are not suitable for vacuum-tight assemblies. Ceramic adhesives developed for joining metallic materials may be useful in joining ceramic to metals if the joints are not stressed.

JOINT EVALUATION

The soundness of ceramic-to-metal joints and seals must be assured to guarantee the performance of the structure in which they are incorporated. Thus, strict quality control of all phases of the manufacturing process is required from the inspection and testing of all incoming materials to the final inspection and evaluation of the finished product. Tests to determine the properties of ceramic-to-metal joints are an essential part of any quality control program. Such tests fall into the following categories: mechanical tests, electrical tests, leak tests, thermal tests, and tests associated with the service requirements for the joints. These test data should be supplemented by information gathered from visual and metallographic examination of failed joints, so that the processing techniques can be changed to correct apparent deficiencies.

Ideally, it should be possible to determine the properties of a particular ceramic-to-metal joint and extrapolate data for the design of other joints. This is difficult to do because of the current inadequacies of the test procedures and the lack of reproducibility caused by variations in the compositions and properties of the ceramic materials. However, it is possible to evaluate the performance of a ceramic-to-metal joint and establish quality control measures to ensure compliance with the service requirements. Extensive efforts to advance the state-of-the-art in testing ceramics and ceramic-to-metal joints have been initiated by Government, industry, and the technical societies. Shook conducted a survey of the methods used to determine the mechanical properties of brittle materials including ceramics; each test was analyzed to discover its capabilities and limitations (ref. 80). The American Ceramic Manufacturers Association, in an attempt to establish standards for many properties of high-alumina ceramics (ref. 81), has listed about 20 physical, mechanical, and electrical properties of ceramics along with references to test methods and testing conditions. Properties are specified for alumina ceramics on the basis of their alumina content: 80 to 90, 90 to 96, and over 96-percent alumina. Many of the methods used to determine the properties of ceramics and ceramic-to-metal joints are contained in specifications issued by the American Society for Testing and Materials and various Government agencies.

The tests used to evaluate the properties of ceramic-to-metal joints are similar to those used to acquire property data on ceramics and metals. Many tests were derived from those in use by metal fabricators. There are many variations of each test, and despite efforts to standardize there is little industry-wide agreement on the exact testing method, test specimen, or interpretation of results. As a consequence, it is difficult or impossible to correlate data from different manufacturing facilities.

This report will not discuss test procedures in detail, but will indicate the tests used and reference them for further study. Clarke, Ritz, and Girard have prepared an excellent review and discussion of these tests (ref. 28).

Mechanical Tests

Tensile Strength

Tensile strength tests measure the strength of a ceramic-to-metal joint when opposing forces are applied perpendicular to the joint interface and away from the joint. The ASTM tensile test is the nearest equivalent to a standard test that has been accepted and used by industry (ref. 82). The test specimen is prepared by metallizing and brazing together two halves of a ceramic body having a specified size and shape (fig. 6). After being mounted in a standard tensile machine, the specimen is tested in tension to failure. Aligning the specimen so that only pure tension loads are applied has been difficult, and shoulder breaks in the ceramic test pieces are common. These results are frequently interpreted as a positive indication that the joint strength exceeds that of the ceramic, while failure may be caused by improper alignment or stress concentrations induced by the mounting fixtures. While this test is used widely by industry, the cost of the test pieces is relatively high because they must be specially made by a ceramic manufacturer; consequently, it is rare that the number of tests required for a statistical analysis is conducted.

Other tests have been used to measure the tensile properties of ceramic-to-metal joints. Luks and Magee proposed the use of two ceramic hemispheres metallized and brazed together to form a hollow sphere; fluid pressure is then applied until joint failure occurs (ref. 83). Schuck devised a test specimen that was prepared by brazing a metallized ceramic disk inside a Kovar sleeve (ref. 84). Rubber disks were placed inside the Kovar sleeve on either side of the ceramic disk; pressure was applied to the rubber disks by means of steel plugs. The entire assembly was placed in a dynamometer and the pressure was increased in 200-pound increments until a leak developed. As the pressure was increased, the rubber disks were compressed and bulged outward, exerting a radial force on the Kovar sleeve. The pressure at failure was defined as the seal or joint strength. This test is somewhat similar to a peel test.

Peel Strength

Various types of peel tests have been developed to measure the force required to pull a thin metal strip from the ceramic substrate to which it has been brazed. These tests are commonly used by industry, since the specimen is easy to make and test. Equipment for the test is readily available or it can be constructed. To make the test specimen, a thin strip of the metal to which the ceramic will be bonded in the final assembly is brazed to a metallized ceramic; a small tab is left unbrazed so it can be attached to a fixture on the peel test equipment. Using equipment as shown in figure 7, the strip of metal is peeled from the ceramic (ref. 53). A record of the force required for peeling is produced on a strip chart recorder. Sometimes the test is conducted without force measurements and attempts to correlate bond strength with the visual appearance of the joint members may yield inaccurate results.

Flexure Strength

This test measures the strength of a ceramic-to-metal joint when a bending moment is applied. The joint members consist of two round or flat ceramic sections whose length-to-cross-

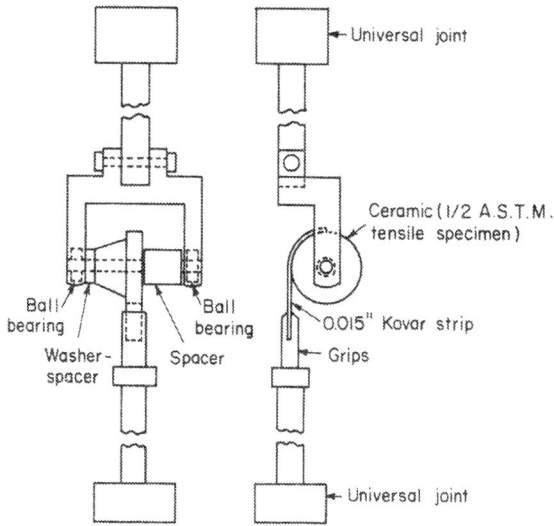

FIGURE 7.—Schematic of drum peel test apparatus (ref. 53)

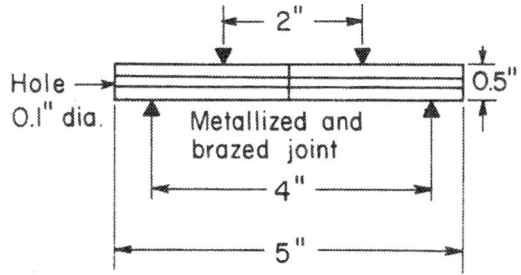

FIGURE 8.—Bending strength test specimen

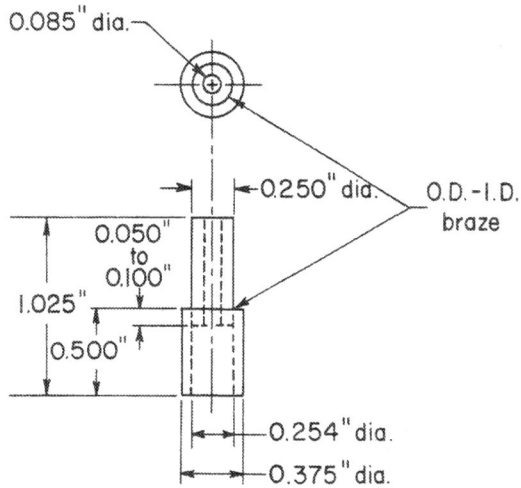

FIGURE 9.—Shear strength test specimen (ref. 86)

section ratio is large. The ends of the sections are metallized and then butt-brazed to a relatively thin sheet of metal. The test can be conducted with three- or four-point loading. In the first case, equal forces are applied in a direction parallel to the joint interface at points equidistant from the joint; a third force is applied in an opposing direction at the center of the specimen. In four-point loading, two equal forces are applied in a direction parallel to the joint interface at points equidistant from the center of the specimen; two equal opposing forces are also applied at points equidistant from the center of the specimen but not directly opposite the points where the original forces were applied. Figure 8 shows four-point loading in the specimen used by Pincus to determine flexure strength (ref. 85).

Shear Strength

This test measures the resistance of the ceramic-to-metal joint to opposing shearing forces. Because of the difficulty in aligning the specimens and obtaining pure shear stresses, this test is not used widely. Reed et al. used the specimen shown in figure 9 to measure the shear strength of a ceramic-to-metal joint (ref. 86). After metallizing the inside diameter (I.D.) of

the large tube and the outside diameter (O.D.) of the small tube, the tubes were brazed together. Reed and his associates claimed the following advantages for this specimen: the joint could be shear-loaded in a conventional compression tester, the specimen could be leak-tested, and residual stresses could be induced in the joint by properly selecting and locating the joint members.

Electrical Tests

Since ceramic-to-metal seals are used for electronic applications, volume resistivity, breakdown voltage, and electrical losses must be measured to determine the effects of the metallizing and joining operations. An accurate determination of electrical losses is particularly important for ceramics used as windows in high-frequency devices for the passage of microwave energy. In studying the factors that contribute to radio-frequency power losses in ceramic-to-metal seals, Reed et al. distinguished between conductive losses and dielectric losses and devised techniques to measure these quantities and determine how they are affected by the metallizing materials, the metallizing process, and the plating materials (ref. 86). Conductive and dielectric losses were determined at low and high radio-frequency power levels. The direct-current resistivity of metallizing was also measured as a function of temperature.

Leak Tests

Leak tests are required when the ceramic-to-metal joint is used as a seal in high-vacuum devices. The helium mass spectrometer leak detector is most commonly used to detect the presence of small leaks. The joint to be evaluated is sealed to the mass spectrometer tube and the area around the joint is flooded with helium. The flow rate of helium through a leak will be detected and measured by the mass spectrometer. The test is extremely sensitive. Considerable experience is required to pinpoint very small leaks. Complete details on the use of this equipment can be found in references on vacuum techniques.

Leaks of larger size can often be detected visually with bubble tests or dye tests.

Thermal Tests

Thermal Conductivity

Thermal conductivity tests measure the quantity of heat that flows across a ceramic-to-metal joint of known dimensions in a specified length of time when a temperature differential across the joint exists. This characteristic has an important bearing on the heat dissipation properties of the structure in which the ceramic-to-metal seal is used. The measurement of heat conductivity is described in ASTM Specification C408–58 (ref. 87).

Thermal Shock and Thermal Cycling Tests

Data on the resistance of the ceramic-to-metal joint to thermal shock and to repeated heating and cooling cycles may be required by the service conditions under which the joint will function. The test variables (maximum and minimum temperature, heating and cooling rates, number of cycles, time between cycles, and test environment, etc.) must be specified in the joint requirements.

The thermal shock resistance is determined by heating the joint to a specified temperature; after equilibrium is established, the joint is cooled rapidly and as uniformly as possible by immersing it in a liquid maintained at a specified temperature. Joints are leak-tested after each thermal-shock cycle.

Thermal cycling tests are performed to determine the behavior of the ceramic-to-metal joint when it is subjected to repeated heating and cooling cycles, such as those required by intermittent operation of a device. The joint is heated until a predetermined temperature is attained. After a specified interval, the joint is cooled to the starting temperature and the cycle is repeated as often as necessary. The heating and cooling rates are usually specified also. Again, leak tests are performed to determine the effect of thermal cycling on the soundness of the ceramic-to-metal joint.

Other Tests

In addition to the tests discussed above, other tests may be required to determine the behavior of the ceramic-to-metal joint under special circumstances. Thus, it may be necessary to measure or determine the resistance of the joints to vibration and shock, to corrosive liquids and vapors, to damage by radiation, and to oxidation at high temperatures.

In addition to standardized tests for determining the properties of ceramic-to-metal seals, there is a pressing need for good nondestructive testing methods to evaluate the overall quality of seals. Ultrasonic and X-ray procedures have been used to some extent, but neither method offers the reliability and versatility required by industry.

CHAPTER 4

Theory of Ceramic-to-Metal Joining

Although ceramic-to-metal joints have been made successfully for many years, a completely acceptable theory of bonding has not yet been developed. Much of our knowledge concerning the formation of ceramic-to-metal joints is based on observations rather than on a basic understanding of the reactions that occur during joining. As a result, ceramic-to-metal joining has developed mainly on an empirical rather than a scientific basis. The same situation prevailed in metals joining; however, progress in this area has been more pronounced because extensive research on the fundamental nature of joining has been conducted in response to the critical requirements of industry and Government.

The lack of understanding regarding bonding is a direct consequence of the number of variables associated with the joining process. The structure of ceramics generally is more complex than that of most structural metals, and, until recently, the importance of knowing the complete history of the fabrication of ceramics from the raw materials stage to the finished product was not realized. The development of the single-phase metal-oxide ceramic for critical assemblies has alleviated this situation somewhat. Added to the uncertainties regarding the base materials used in ceramic-to-metal joints are the variables associated with the preparation of the ceramic for joining and the joining operations themselves.

Many factors contribute to bond formation in a ceramic-to-metal seal or joint, among which are the following: chemical interactions between the joint materials, diffusion across the metal-ceramic interface, mechanical interlocking of one phase with another because of microscopic surface roughness, and the penetration of glassy phases into metallic phases. The extent to which these mechanisms predominate depends on the selected joining technique, the characteristics of the individual joint materials, and on the process variables.

Since Pulfrich and Vatter first produced reliable ceramic-to-metal joints, numerous investigators have studied the mechanisms of bonding and have advanced theories to explain their observations. The theories that have received the most emphasis include the alumina reaction theory, the molybdenum oxide theory, the glass migration theory, and the metal ion-metal theory.

In his early patents on ceramic-to-metal sealing, Pulfrich set forth several recommendations for obtaining sound joints (refs. 1 to 4). He metallized the ceramic surface by coating it with a suspension of finely divided molybdenum particles and then firing the coating in a controlled atmosphere within a closely limited temperature range. Once metallized, the ceramic was electroplated and brazed to the metal with a silver-base filler metal. Pulfrich emphasized that glassy phases should be avoided at the joint interface and further specified that:

1. The metal for metallizing should have a melting temperature about 200° C (392° F) higher than the metallizing temperature, since a porous sintered surface promoted adherence of the filler metal to the metallizing metal.

2. The ceramic should have a eutectic bond that melts about 300° C (572° F) below the softening temperature of the ceramic body and

about 200° C (392° F) below the metallizing temperature.

3. The metal to which the ceramic is ultimately joined should have a melting temperature well above the metallizing temperature and should match the expansion coefficient of the ceramic as closely as possible.

4. The filler metal should melt below the melting point of the ceramic's lowest melting eutectic. Neither the filler metal nor any alloy formed between the filler metal and the base metal should react with the metallizing metal.

Pulfrich speculated that the molybdenum particles were wet by the liquid formed by the eutectic constituent of the ceramic. He noted that the furnace atmosphere should contain enough hydrogen to maintain most of the molybdenum as a metal; however, enough oxygen (about 0.25 percent) should be present to form a trace of molybdenum oxide which subsequently melts and improves bonding. During cooling the eutectic liquid should recrystallize completely to form a crystal structure free from residual glass. Thus, the molybdenum particles would be locked within the crystal rather than within a glass flux.

Although Pulfrich did not advance a formal theory, he was aware of the role of chemical reactions and liquid phases.

ALUMINA REACTION THEORY

In 1953, Pincus attempted to explain the reactions during the molybdenum-manganese method of metallizing (ref. 88). Assuming that manganese is oxidized to manganous oxide in any hydrogen atmosphere whose dewpoint is more than −65° C (−85° F), he traced the chain of events that takes place during sintering. At 1000° C (1832° F), a solid-phase reaction between alumina and manganous oxide forms manganese aluminate spinel at the interface, although no bonding occurs; at 1200° C (2192° F), this compound enters a liquid phase. At 1400° C (2552° F), appreciable sintering of the molybdenum particles occurs, and the liquid spinel locks the hardened layer to the ceramic. Upon cooling the liquid phase forms a glass that appears to extend into the metal layer; the alumina surface appears to be corroded. These

observations were verified by microscopically examining tapered sections of metallized ceramics; the tapered section elongated the ceramic-metal interface for easier study.

Pincus extended his theory to cover the metallizing of forsterite ($2 MgO \cdot SiO_2$), where a reaction at the ceramic-metal interface forms a manganese-containing liquid that again locks the sintered molybdenum particles to the ceramic.

According to this theory the strength of seals made to a 100-percent alumina body should be as strong, or stronger, than seals made to a 90-percent alumina body; experimentally, this is not the case, because the difficulty in making seals increases with increasing alumina content. Workers at Sperry Rand have indicated that other additives, particularly silica, to the metallizing mixture are more effective in promoting bonding than manganese (ref. 53).

MOLYBDENUM-OXIDE THEORY

In a later paper, Pincus suggested that bonds between pure molybdenum and high-alumina ceramics were chemical in nature and depended on a reaction between molybdenum oxide and aluminum oxide (ref. 85). He conducted metallizing studies in a hydrogen atmosphere whose dewpoint was varied between +5° and −68° C. (+40° and −90° F.); joints were examined microscopically to verify the reactions during metallization. The theory depends on the controlled oxidation of the molybdenum particles and an interface reaction between the metal oxide and the ceramic. Even if molybdenum oxide is formed by heating molybdenum in air or a wet hydrogen atmosphere, it volatilizes at 600° to 700° C (1112° to 1292° F), well below normal sintering temperatures. The validity of this theory was not confirmed by later work conducted at Sperry Rand (ref. 53).

GLASS MIGRATION THEORY

The glass migration theory, well substantiated by experimental data, proposes that the strength of ceramic-to-metal joints made by conventional means depends on the glass-phase content of the ceramics. This theory, attributed to Cole and Sommer, is the result of continuing work initiated by Cole and Hynes (refs. 89 and 33). Metallizing studies were conducted using molybdenum-manganese, molybdenum-titanium, and pure molybdenum materials and 94- and 99-percent alumina ceramics. The results of these studies at various sintering temperatures are summarized in table 10. The seal strengths obtained with 94-percent alumina ceramic were greater and more consistent than those obtained with the 99-percent alumina ceramic. Photomicrographs showed that the glassy phase of the 94-percent ceramic had migrated into the metallizing coating and surrounded the molybdenum particles. Cole and Sommer suggested that manganese and titanium oxidized during the heating process and reacted with the ceramic to reduce the viscosity of the glassy phase; then, the glass was free to migrate slightly and lock the ceramic to the molybdenum coating. However, the presence of these metals did not appear to be essential, since excellent seal strengths were obtained with pure molybdenum alone.

TABLE 10.—*Results of Tensile Tests of Ceramic-to-Metal Seals*

[From ref. 89]

Alumina, percent	Metallizing sintering temperature, °C	Tensile strength of seals (lb/sq. in.)		
		80 percent Mo 20 percent Mn	97 percent Mo 3 percent Ti	100 percent Mo
99	1700	2900	6600	1700
	1600	750	2700	600
	1500	3100	2900	220
94	1700	9400	12 000	9600
	1600	9700	*12 000	13000
	1500	8000	10 100	10600

*Average values based on several hundred production control tests.

Since virtually no glass is present in the 99-percent alumina ceramic, the adherence in these ceramic-to-metal seals must be explained by another mechanism. Cole and Sommer suggested that the metallizing components reacted to form a compound that wet both the alumina surface and the molybdenum particles; this explanation is supported by the evidence that stronger seals were obtained with a metallizing mixture containing molybdenum and titanium than with molybdenum alone. The possibility that bonding resulted from an ion exchange was not discounted, however.

In conjunction with this theory, the addition of glass or glass-forming materials to metallizing mixtures has been investigated; such additives have proved useful in promoting bonding.

OTHER THEORIES AND MECHANISMS OF ADHERENCE

Several other theories and suggested mechanisms of adherence have been advanced to explain the phenomenon of ceramic-to-metal bonds.

Helgesson used the electron probe to examine the microstructure produced when high-alumina ceramics (94.0-, 99.0-, and 99.9-percent alumina) were metallized with a standard molybdenum-manganese metallizing mixture (ref. 90). In contrast to other investigators, Helgesson reported greater seal strengths with higher alumina ceramics. He concluded that bonding was caused by chemical forces.

Floyd investigated the metallizing of high-alumina ceramics containing various amounts of a MgO–SiO_2–CaO glassy phase; the metallizing materials were pure molybdenum, pure manganese, and an 80-percent molybdenum 20-percent manganese mixture. Floyd noted that bond strength increased with higher contents of the glassy phase and with the formation of the manganese aluminate spinel. His work appears to support the alumina reaction and glass migration theories (ref. 34).

Reed and Huggins metallized high-alumina ceramics with various mixtures and, on the basis of electron microprobe analyses, reported that molybdenum did not dissolve or diffuse in the reacted metal-oxide phases or the alumina-crystal phase of the ceramic, and the metal oxide compounds of the metallizing layer reacted with the alumina-crystal phase and diffused into the ceramics (ref. 91).

Cowan et al. recently developed procedures using tungsten powder plus small concentrations of yttria to metallize silica-free, high-alumina ceramics for service in a cesium environment (ref. 92). They suggested that WO_3 forms at low temperatures and reacts with yttria in the grain boundaries of the ceramic to form yttrium tungstate. At higher temperatures, the yttrium tungstate is reduced and yttrium migrates into the tungsten metallizing layer to lock the particles in place. Kiwak has discussed several modifications of this metallizing procedure and recommends the use of a $Mo-Y_2O_3-Al_2O_3$ mixture to obtain sound bonds (ref. 93).

In research directed toward the development of low-temperature sealing techniques, Tentarelli et al. advanced the insulator-semiconductor-metal theory to explain bonding. They suggested that a molybdenum nonstoichiometric oxide semiconductor is formed at the ceramic-to-metal interface and that electronic bonding is responsible for adherence (ref. 64).

WETTING AND CERAMIC-METAL INTERFACE REACTIONS

Studies investigating the fundamental nature of wetting phenomena and the reactions that occur along the ceramic-to-metal interface are very similar to studies investigating the wetting of a base metal by a brazing filler metal. Van Vlack, in an excellent discussion of the ceramic-metal interface, states that bonding between ceramic and metallic phases depends on interphase reactions and the atomic structure across the interface (ref. 94). He indicated that the strongest boundaries require primary ionic and covalent bonds and low interfacial energies. Epitaxial coherence between crystalline ceramics and metals is necessary if a glassy phase is absent. Greater adherence between ceramics and metals occurs when there is a large interfacial area.

When a drop of liquid is placed on a flat surface, it may spread or remain in an essentially spherical form. The thermodynamic driving force for spreading or wetting is a decrease in the free energy of the system. Extensive studies have been conducted by Kingery and his associates to develop procedures to measure surface energies, interfacial energies, and contact angles at metal-ceramic interfaces, and to examine metal-ceramic reactions and the effect of certain elements on the formation of bonds (refs. 95 to 99).

Additional research on glass-metal systems by other investigators has resulted in bonding theories (ref. 100) that may be applicable to ceramic-metal systems as well. In working with the adhesion of porcelain enamels (glass) to metals, King et al. set forth the following statements regarding adherence (ref. 101):

1. The enamel at the interface must be saturated with a metal oxide that, in solution with the enamel, is not reduced by the metal.

2. Adherence appears to result from metal-to-metal bonding between the atoms in the base metal and metallic ions in the enamel.

Thus, adherence appears to be chemical in nature. A thorough review and critique of the theories and mechanisms of bonding has been prepared by Clarke et al. (ref. 28).

CHAPTER 5

Ceramic-to-Metal Joining

FUSION JOINING

Electron-Beam Welding

In investigating the direct fusion joining of ceramic materials to themselves and to metals, electron-beam welding is the most promising of the available fusion welding processes, because the electron beam can be precisely controlled to produce an extremely small spot weld or weld bead with minimum melting of the joint materials. Also, reactions between the joint materials and the atmosphere are minimized or eliminated by the high-vacuum nature of this process.

Hokanson, Rogers, and Kern of the Hamilton Standard Division of the United Aircraft Corporation were among the first to evaluate the potentialities of this joining method (ref. 102). Because of the high accelerating voltage, the spot diameter of the Hamilton-Zeiss high-power-density electron-beam unit could be controlled to a diameter less than 0.010 inch. An optical viewing system was provided for precise observation of the welding process. With this equipment, butt- and edge-welds were made with ceramic bodies having alumina contents of 85, 96, and 99.75 percent. Preheating and controlled cooling rates, particularly above 1200° F (649° C), were required to eliminate cracking. The most suitable welds between ceramics were made with a high-voltage, low-current electron beam at about 30 ipm; slower welding speeds produced a weak glassy structure. Flexure tests indicated that a butt-welded joint was about 20 percent as strong as the ceramic. Molybdenum-to-ceramic pin joints were welded as well as crack-free joints between a 96-percent

alumina body and the following metals: molybdenum, tungsten, niobium, and Kovar. While welding these joints, the electron beam was positioned slightly off the joint centerline over the metal; thus, the metal melted and flowed over the ceramic. This welding technique is common in the case of metals with widely differing melting temperatures.

Electron-beam welding was investigated as a method to fabricate ceramic-to-metal seals for use in a thermionic energy conversion system. Engineers at the Los Alamos Scientific Laboratory reported only limited success in welding a ceramic body to molybdenum and a molybdenum-titanium alloy metallized with nickel (ref. 103). In a somewhat similar program conducted by the Bendix Corporation, Dring indicated that strong joints between molybdenum and an alumina ceramic were obtained by electron-beam welding; the ceramic surface was metallized by the moly-manganese process (ref. 104). Joining occurred only when intimate contact between the metal and ceramic was achieved. During welding, the beam was controlled to fuse the molybdenum sheet to the metallized layer. The joints were crack-free and exhibited excellent peel strength; joint failure occurred at the ceramic-to-metallizing interface.

Direct Fusion

Stablein and Araoz of the Argonne National Laboratory have used an arc-image furnace to produce direct fusion between ceramics and metals (ref. 105). In this type of furnace, heat is produced by reflecting and refocusing the light from a high-intensity (10 kilowatt)

37

carbon-arc lamp onto a small area, about ½ inch in diameter, where the specimen is located. The large temperature gradient produced by focusing heats a small area on the specimen while the remainder remains relatively cool. The heat from the furnace can be controlled by an adjustable diaphragm that reduces the cone of light reaching the specimen; also, a screen acts as a filter to reduce the total light energy. The specimen is mounted in a rotating holder so that the specimen is heated uniformly. This equipment has been used to produce butt and tee joints in high-purity alumina tubing (99.75-percent alumina). The tubing was cut to size on a diamond saw without using flux or cement. There was no reduction in the internal diameter of the tubing.

Porembka reported that in directly fusing ceramics to metal at the Columbus Laboratories of the Battelle Memorial Institute (ref. 106), zirconium oxide was joined to tungsten by melting the oxide in contact with the metal in a vacuum. In this approach, the wetting of the solid surface by the molten phase was of primary importance. The absence of an intermediate layer at the joint interface suggested that joining occurred as the result of chemical bonding. However, a slight amount of roughness at the joint interface was noted and indicated that some mechanical locking may have occurred.

Buyers (Hughes Research Laboratories) produced joints between tantalum and a stabilized zirconia ceramic at about 2000° C (3632° F) (ref. 107). The ceramic and tantalum workpieces were placed in contact vertically, with the metal resting on the ceramic. Then, the assembly was heated rapidly to 1955° to 1995° C (3551° to 3623° F) by an induction coil. At this temperature, the interface between the ceramic and metal disappeared as the bond was formed. Intermediate compounds that formed spontaneously at this temperature were identified by X-ray diffraction powder patterns. For seals using zirconia stabilized by oxides, the following compounds were identified: tantalum zirconate, yttrium tantalate, tantalum silicide, aluminum tantalate, and magnesium tantalate. The exact mechanism of joining has not been established as yet.

SOLID-PHASE JOINING

Research in this area of joining has been concentrated on fabricating "graded powder" seals, i.e., a ceramic-metal composite material whose composition varies continuously from a ceramic to a metal. Knecht proposed this method, using the reasoning behind the production of graded seals between hard and soft glasses (ref. 108). In an experimental program at the Air Force Electronic Components Laboratory, the feasibility of the technique was demonstrated by producing a closed-end tube consisting of a metal cylinder with a ceramic body.

From 1954 to 1956, Dunegan developed procedures to produce a ceramic cylinder (0.8 inch O.D., 0.64 inch I.D., and 0.4 inch thick) that was metallized on both ends (ref. 109). In producing the cylinder, a mixture of selected metal powders and a mixture of ceramic and metal powders were used. A thick, smooth layer of metal powder was spread over the die area. Then, a thicker layer of the ceramic-metal powder mixture was spread over the metal powder. Finally, another thin layer of metal powder was spread over the ceramic-metal powder mixture. The die set was assembled and pressure was applied to produce a green powder compact. The compact was prefired at about 250° to 300° C (482° to 572° F) to remove the volatile constituents and then sintered at temperatures exceeding 1000° C (1832° F).

The most successful ceramic-to-metal system consisted of an 85-weight percent alumina and a 30-weight percent spinel $(Fe_{0.5}Ni_{0.5})(Al_{1.0}Cr_{1.0})_2O_4$ that was metallized with a tungsten-copper-nickel alloy plus 25-weight percent chromium. Dunegan stated that the success of the pressed powder seal depended on several factors:

1. The thermal expansion and contraction of the metal and ceramic must be very similar.

2. The firing temperature must be the same for both materials.

3. The firing shrinkage must be the same for both materials and occur at the same rate.

4. There must be a reaction between the ceramic and metal. Bonding appears to occur through an oxide phase.

Dring also investigated the fabrication of a

graded-powder seal for use in thermionic converter applications (ref. 104). Since a closed-end cylinder about 1 inch in diameter was typical of the envelopes used in many converters, this research was to produce an envelope in which a refractory metal formed the closed end and the cylinder wall material was graded from a metal to a pure ceramic at the open end. The most successful composite envelope was fabricated from 12 gradations of alumina and tungsten powders as shown in table 11. The 12 powder mixtures were loaded layer by layer into a die; pure alumina was added at each end of the compact to reduce problems with laminations in the high-tungsten area. The compacts were pressed at 30 000 psi. After removal from the die, the compacts were air-dried and machined to remove the excess alumina from each end. Then, the compacts were sintered at 3365° F (1852° C) for 8 hours. Several envelopes were produced, graded from 95 percent tungsten–5 percent alumina at the closed end to 100 percent alumina at the open end. It was not possible to use pure tungsten at the closed end

of the cylinder because of the laminations produced by the pressing operation. Although the envelopes were not completely vacuum-tight, the leak rate was very low.

The development of a graded-powder ceramic-to-metal seal for high-temperature service in a cesium environment was also undertaken by Bristow et al. (ref. 11). The metallic members of the seal had to be joined to a multi-layered or graded structure whose composition varied from a predominantly oxide core to a predominantly metallic surface. The oxide core had to be electrically insulating. Graded-powder structures were produced by hot-pressing five-layer compacts whose composition varied from 75 percent molybdenum–25 percent alumina at the metal end to 5 percent molybdenum–95 percent alumina at the ceramic end. The powders were placed in an induction-heated die and sintered for 10 minutes at 1600° C (2912° F) under 4000 psi. The necessity to use an alumina-rich composition as the core layer increased the problems of fabricating a dense structure that was vacuum-tight and electrically insulating.

TABLE 11.—*Gradations of Powder Used in Tungsten-Alumina Structure*

[From ref. 104]

Gradation Number	Tungsten [a] weight	Alumina [b] weight	Phosphoric [c] acid solution, cm^3	Percent by weight		Percent by volume	
				Tungsten	Alumina	Tungsten	Alumina
0	0	3.0	0.21	0	100	0	100
1	13.30	0.70	0.10	95	5	80	20
2	8.80	1.2	0.10	88	12	59	41
3	3.08	0.92	0.10	77	23	40	60
4	2.56	1.44	0.10	64	36	26	74
5	1.50	1.50	0.10	50	50	16.5	83.5
6	1.08	1.92	0.13	36	64	10	90
7	0.93	2.07	0.14	31	69	8.15	91.85
8	0.52	1.48	0.10	26	74	6.5	93.5
9	0.40	1.60	0.11	20	80	4.7	95.3
10	0.26	1.74	0.12	13	87	2.86	97.14
11	0.12	1.88	0.13	6	94	1.24	98.76
12	0	21.0	1.45	0	100	0	100

[a]Tungsten—99.75% #3 tungsten powder (table 1) 0.025%–325 nickel powder
[b]Alumina—73.5% B (Table II)
   24.5% A (Table II)
   2.0% magnesium stearate
[c]Phosphoric acid solution—15% aqueous solution of H_3PO_4

Although the results of these studies were encouraging, the need for more research was indicated.

Porembka discussed the fabrication of a graded cermet bond between tungsten and thoria using a series of tungsten-base cermets with decreasing tungsten content and a series of thoria-base cermets with increasing thoria content (ref. 106). Cermets are made by mixing metal powders with the ceramic batch. Densification of the cermet and bonding to the parent materials can be accomplished by sintering or hot-pressing methods. Porembka stated that such a bonding method was particularly useful in joining materials with widely differing expansion coefficients.

GAS-PRESSURE BONDING

Porembka used gas-pressure bonding to fabricate joints between niobium and high-purity ceramic materials (ref. 110). Of the two metal-ceramic systems selected for study, the niobium-chromium-alumina system is a three-component system in which the intermediate material provides the necessary diffusion characteristics with niobium; its oxide is completely soluble in alumina. The niobium-zirconia system represents a two-component system in which the metal oxide has limited solubility in the ceramic. Two types of specimens were used for the study of the niobium-chromium-alumina systems as shown in figure 10; the stainless steel sections were incorporated in each specimen for attachment to the grips in a tensile testing machine. The cermet section was composed of 50-volume percent chromium and 50-volume percent alumina. The specimens were assembled in a capsule that was evacuated and sealed before bonding. Isostatic hot-pressing was conducted for 3 hours with the temperature 2200° to 2300° F (1205° to 1260° C) and pressure 10 000 psi.

Niobium-zirconia specimens were prepared in the same manner as the niobium-chromium-alumina specimens; no intermediate material was present in these joints. Isostatic hot-pressing was conducted under the conditions listed above.

Reduced-section tensile specimens were machined and tested. The test data are summarized

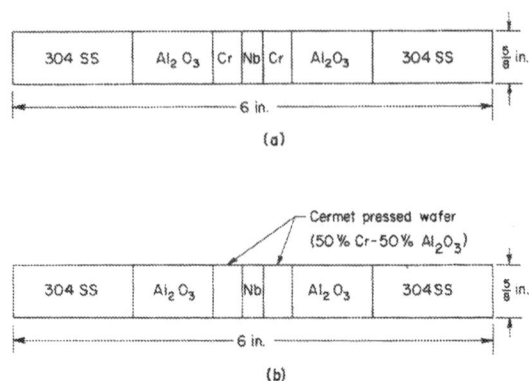

FIGURE 10.—Alumina-chromium-niobium specimen assemblies (ref. 110). (a) Metal wafer composite; (b) cermet composite

in table 12 for the niobium-chromium-alumina system. The specimens prepared with cermet intermediate materials fractured during machining, so that no tensile data were obtained. Similarly, all but one of the niobium-zirconia specimens fractured during machining; a tensile strength of 1370 psi was obtained from this single specimen.

Additional studies conducted during this program included metallographic examinations of niobium-chromium-alumina and niobium-zirconia joints, microprobe analysis of a niobium-chromium-alumina joint, and thermal cycling tests of niobium-chromium-alumina joints.

The following conclusions were advanced:

1. Gas pressure bonded joints between chromium and alumina showed higher strengths than those produced by other joining methods.

2. The infiltration of the metal phase into the asperities and surface pores of the ceramic was a major factor in the strength of pressure-bonded joints.

3. Diffusion in the chromium-alumina and chromium oxide-alumina systems was not measurable by microprobe analysis. No metallographic evidence of diffusion was noted.

4. Niobium-zirconia joints produced by gas-pressure bonding were weak. Joint failure occurred through an intermediate phase formed by diffusion between niobium and zirconium.

5. Niobium-chromium-alumina joints were resistant to limited thermal cycling to 2300° F (1260° C).

TABLE 12.—*Tensile Strength of Pressure-Bonded Niobium-Chromium-Al₂O₃ Assemblies* [a]

[From ref. 110]

Specimen Number	Chromium oxidation treatment [b]	Gage diameter, inches	Breaking load, psi	Breaking stress, psi	Location of fracture
CNA–1	None	0. 345	502	5550	Al₂O₃ [c]
CNA–2	None	. 375	878	10 700	Cr–Nb interface
ANC–10	8 hr at 700° C	. 405	856	6700	Al₂O₃
ANC–11	do	. 465	846	6500	Al₂O₃ [c]
ANC–12	do	. 410	936	7100	Al₂O₃ [c]
ANC–13	2 hr at 700° C	. 441	916	6100	Cr–Nb interface
ANC–14	do	. 476	931	5400	Al₂O₃ [c]
ANC–15	do	. 441	882	5900	Al₂O₃ [c]

[a] All specimens tested at room temperature at strain rate of 0.02 in./min.
[b] Oxidation conducted in 0.1 atm oxygen.
[c] Fractures in shoulder area of reduced section within Al₂O₃ components.

Diffusion Bonding

Only limited research has been conducted on diffusion bonding for fabricating ceramic-to-metal joints, although this process is used extensively in metals joining. In this method of joining, the carefully prepared workpiece surfaces are held in close contact by applying pressure, and joining occurs at temperatures below the melting temperatures of the base materials. An intermediate material may be used to promote or limit diffusion. Diffusion bonding is achieved in a vacuum or a protective atmosphere.

In 1963, Dring discussed the results of research to bond alumina-to-molybdenum and alumina-to-niobium seals by diffusion for service up to 1500° C (ref. 104). Two approaches were used during this study: (1) The formation of seals between a metal and a ceramic, and (2) the formation of seals between a metal and the metallized surface of a ceramic. The initial experimental studies were conducted with sheet specimens; however, the joint design shown in figure 11 was used for final evaluation of a bonding technique.

During the study to bond molybdenum or niobium to an unmetallized ceramic, numerous intermediate materials were applied to the joint surfaces to promote diffusion. These were largely ineffective in producing a strong bond between alumina and molybdenum. However, a 95 percent niobium-5 percent nickel mixture applied to the faying surfaces of the alumina cylinder by plasma-arc spraying techniques was very effective. Dring recommended the following procedures to produce acceptable seals:

1. Dry-lap a circular area on the niobium cap
2. Grit-blast the niobium cap and alumina cylinder
3. Plasma-spray 95 percent niobium-5 percent nickel mixture on the bevelled edge of the alumina cylinder and on the inside of the niobium cap

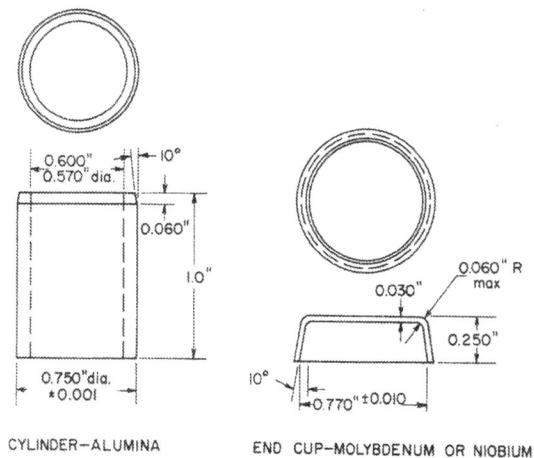

FIGURE 11.—Parts for seal evaluation (ref. 104)

4. Dry-lap cylinder and cap

5. Wet-lap cylinder and cap

6. Assemble joint members

7. Diffusion-bond assembly in a vacuum (10⁻³ torr) for 5 hours at 1524° C (2775° F).

Mechanically strong seals were also produced between molybdenum sheet and the metallized surface of a high-purity alumina body. An alumina cylinder was first metallized with a layer of 85 percent molybdenum-10 percent manganese-5 percent titanium hydride. After this layer was sintered to the alumina cylinder, a second metallizing layer was slurry-coated to the ceramic surface and fired; the composition of this layer was 80 percent molybdenum-20 percent chromium. The faying surfaces of the molybdenum sheet and metallized alumina ceramic were then lapped and the joint members assembled. Joining proceeded in a dry argon atmosphere at 1800° C (3272° F). While joints made in this manner had acceptable strength properties, it was necessary to infiltrate the metallized ceramic with a manganese oxide-alumina eutectic mixture before bonding to obtain hermetic sealing.

Metelkin, Makarkin, and Pavlova, in investigating the diffusion bonding of ceramic-to-metal joints (ref. 111), developed bonds between copper and a variety of ceramic materials such as alumina (72.0, 94.0, and 99.5 percent), sapphire, forsterite, and steatite. The joints were produced in a vacuum as well as a hydrogen atmosphere under the following conditions: pressure 14.4 psi; temperature 1000° C (1832° F); and time 10 minutes. The specimen used for the studies was a copper disk (0.012 to 0.020 inch thick) sandwiched between two ceramic cylinders (0.52-inch O.D., 0.100-inch I.D., and 1.8 inches long). The strongest joints between these materials were obtained in a hydrogen atmosphere; the joints generally failed in the ceramic.

Additional joints were made between a 99.5-percent alumina body and the following metals: stainless steel, Kovar, nickel, palladium, titanium, Nichrome, low-alloy steel, and iron. The bonding conditions were the same as noted above except the temperature was increased to 1250° to 1300° C (2282° to 2372° F). Satisfactory joints were obtained between all of these ceramic-to-metal combinations. When joints between alumina and several refractory metals were attempted, bonding did not occur either in a vacuum or in hydrogen; however, satisfactory joints were obtained when a ductile metal such as copper, nickel, and stainless steel was used as an intermediate material between the metal and ceramic surfaces.

Also studied during this investigation were the effects of hydrogen dewpoint, ceramic grain size, time, pressure, and temperature on the strength of ceramic-to-metal joints. Maximum joint strength was obtained at about 28 psi when the joints were held at temperature for 10 to 15 minutes; however, joint strength increased at high temperatures because the reaction processes are temperature-dependent. The dewpoint of the hydrogen atmosphere also affected joint strength, maximum joint strength was obtained with a dewpoint of about 40° to 45° F. Thus, water vapor provided oxygen for metal-oxide formation.

During an early investigation conducted in the late 1940's, Wellinger (ref. 112) discussed the fabrication of diffusion-bonded joints between copper and steatite and noted that the bonding time required to join these materials at 1000° C (1832° F) with a pressure of 3000 psi was about 2 hours. The bonding time was reduced to 10 minutes if the copper surface was covered with a thin layer of cuprous oxide.

Diffusion bonding may also have taken place in the so-called "ram seal" or "crunch seal" that has been developed by a division of the Radio Corporation of America (ref. 113). However, the mechanism of joining is probably cold pressure welding. High-alumina ceramic cylinders, up to 20 inches in diameter, are ground to a blunt bevel on the end. Then, a copper-plated tool steel cylinder whose inner diameter is smaller than the outer diameter of the alumina cylinder is pushed onto the ceramic cylinder under considerable pressure to produce a room-temperature seal. This metal-to-ceramic seal can be used at temperatures up to 550° C (1022° F) and heat-cycled repeatedly.

Ultrasonic Welding

Ultrasonic welding is a solid-state bonding

process for joining metals by introducing high-frequency vibratory energy into overlapping workpieces. For joining metals ultrasonic welding has several advantages over other joining methods:

1. Since thermal distortion does not occur, close dimensional tolerances in an assembly can be maintained.

2. A variety of dissimilar metal combinations can be joined.

3. Thin sheet stock can be joined to much thicker metallic sections.

Care must be used in selecting and applying this process. Problems have been experienced with microcracking during the ultrasonic welding of heat-resistant alloys and some dissimilar metal combinations. Ultrasonic welding has been used most extensively in the electronics industry to join components made from aluminum and copper alloys.

Scheffer, Liederbach, Pikor, and Miller (ref. 114) used ultrasonic joining techniques to produce a hermetically sealed transistor package. The base of the package was a 94- to 96-percent alumina wafer that measured 0.310 by 0.310 by 0.030 inch. The required seal and terminal patterns on the ceramic wafer were provided by moly-manganese metallizing. After metallizing, the patterns were nickel- and copper-plated and then solder-dipped. The transistor was mounted in a cavity in the ceramic wafer. After assembly, a brass cap was soldered to the wafer to produce a sealed unit. Since a flux could not be used because of the danger of contaminating the ultrasensitive transistor surface, ultrasonic methods were used for joining. A special ultrasonic sealing press was constructed to position the seal components and isolate the soldering tip from any physical connection except to the transducer; the soldering tip was heated by means of an isolated annular heater. Joining was accomplished successfully without a flux.

Dring investigated ultrasonic welding during a study to develop techniques to produce ceramic-to-metal seals for thermionic converter applications (ref. 104). The object of the research was to join 0.001-, 0.005-, and 0.025-inch-thick molybdenum sheet to a moly-manganese metallized alumina ceramic. Although some progress was made in producing single spot welds between the molybdenum sheet and the metallized ceramics, the results were not reproducible enough to proceed with the second-phase operation—the production of overlapping welds. The most suitable joints were made when a tantalum foil was placed between the molybdenum sheet and the ceramic.

High-Energy Welding With Exploding Foils

Vagi and DeSaw conducted an investigation to obtain information on the requirements for joining metals, nonmetals, and metals to nonmetals with exploding foils (ref. 115). In this method, a metal foil is positioned between the two parts to be joined. Then, the foil is exploded by passing a large current through it from a bank of capacitors. Under the proper conditions, joining will occur. Exploding foils behave similar to exploding wires when a large current is passed through them.

Most joining experiments were conducted using a 0.125-inch-wide, 0.001-inch-thick tantalum foil sandwiched between the abutting ends of 0.125-inch-diameter rods; some studies were also made with a reduced center-section type of foil (fig. 12). The rods to be joined were held in place with a modified spot-welding head. Quartz-to-quartz joints were made under the following conditions: preset clamping force 19.5 pounds; charging voltage 5 kilovolts; and capacitance 145 microfarads. The joint efficiency of these bonds was 30 percent. In attempts to join tungsten to quartz under the same welding conditions, the tantalum foil was bonded to the tungsten rod after the explosion but bonding to the quartz rod did not occur. Experiments to join other materials were also conducted, and the authors reported that:

1. The exploding-foil process appears promising for bonding a variety of materials, such as quartz, Fiberglas tape, phosphor-bronze woven tape, and thin-wall zirconium tubing.

2. Metal coatings were produced on the faying surfaces of alumina and graphite.

3. Consistent and continuous edge welds were obtained with metal specimens. This edge-fusion effect was used to join thin-wall tubing.

4. Exploding-foil welding variables must be closely controlled; they include: part material,

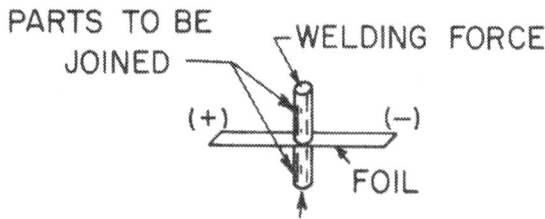

(a) Ribbon-Type Foil

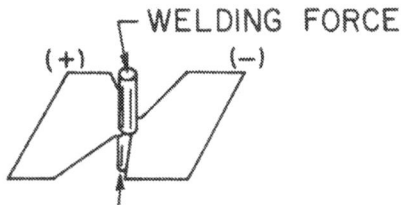

(b) Reduced Center-Section-
Type Foil

FIGURE 12.—Arrangement of foils and parts for bonding with exploding foils

size, and shape; foil geometry and thickness; capacitor voltage and discharge time; welding-force magnitude and alignment of parts; and the reaction of the parts during welding.

Electroforming

Electroforming—a process in which a layer of metal is deposited on a surface or form by electroplating—has been investigated as a method to fabricate ceramic-to-metal joints. After plating, the form is removed to leave a shell of metal whose inside configuration matches that of the form. Hare, of the Stanford Electron Devices Laboratory, studied this joining process in connection with the fabrication of an external-circuit traveling-wave tube that consists of a large number of ceramic-to-metal seals (ref. 116). Hare lists the following steps in producing an electroformed seal:

1. Metallize the ceramic in accordance with the moly-manganese process

2. Plate the metallized surfaces with nickel and sinter

3. Plate the nickel surface with copper and sinter

4. Plate the copper surface with gold

5. Assemble the ceramic-to-metal joint. Make provisions for electrical contact to the metal and metallized surfaces

6. Mask-off any electrically conductive areas not to be plated

7. Clean joint assembly and plate with copper.

Typical joint designs evaluated in this program are shown in figure 13. The initial current density in the electroplating bath was 40 amperes/square foot for 30 seconds; then, the

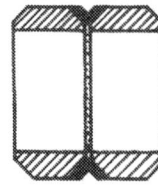

■ Electroform copper
▨ Ceramic
▨ Metal component

(a)

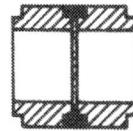

(b)

(c)

(d)

FIGURE 13.—Joint designs for electroformed seals (ref. 116). (a) Vee design of metal-to-ceramic electroform seal; (b) step design of metal-to-ceramic electroform seal; (c) step design using plain ceramic cylinders and metal sleeves; (d) vee design using ceramic disk and metal cylinder

current density was reduced to 10 amperes/square foot. At this level the deposition rate was 0.008 to 0.010 inch in 24 hours. Using the joint design shown in figure 13(b), an assembly of 41 stacked seals was joined by electroplating. The metals used in the electroformed seals included molybdenum, copper, and Kovar (it should be possible to use other structural metals); the ceramic was a high-alumina body.

Hare stated that electroforming is useful where glass windows for electronic devices may be damaged by a high-temperature sealing operation, where sapphire or quartz must be used in seals, and where a large, expensive ceramic-to-metal seal must be made.

In a program to develop ceramic-to-metal seals for thermionic converters, Dring evaluated the electroforming technique (ref. 104). Attempts to join molybdenum to a metallized alumina ceramic by rhodium-electroplating were inconclusive because of the difficulty in plating molybdenum.

Reed and McRae used electroformed iron, overplated with chromium, to produce a seal between niobium or Kovar to metallized alumina (ref. 117) to be used in a liquid-metal environment.

SOLID-LIQUID PHASE JOINING

Soldering

Soldering is a process in which metals are joined below 800° F by using a nonferrous filler metal or solder. The solder is distributed throughout the joint by capillary attraction, provided the proper clearances exist between the joint surfaces. The solder and the base metal are bonded by adhesion and physical attachment; however, sometimes a thin layer of the base metal is dissolved by the solder and an intermetallic compound forms to aid in bonding. The ease with which the solder wets and flows on the base metal surface is a measure of the metal's solderability.

Soldering has long been used in the electrical industry to join glass and ceramic parts to metals. Since it is a low-temperature joining process, there are limitations on the service temperature of soldered joints. While other solders are used occasionally for special joining

applications, most electrical joints are made with tin-lead solders having a tin content of about 60 percent. The solders have excellent wetting and flow characteristics, are quite ductile, and have compositions that approach the eutectic in the tin-lead system, which occurs at about 63 percent tin. Since the solders begin to melt at 361° F, the service temperature of the soldered joint is 200° F or less. Soldered ceramic-to-metal joints are used in electrical insulators, feedthroughs, and contacts on electronic components.

Metallizing is necessary to ensure the solderability of ceramic surfaces. Such surfaces are usually metallized by thermally reducing metal salt solutions and sintering paints composed of precious metal flakes or powders in an organic binder to which a glass frit has been added. (These metallizing procedures were discussed in chapter 3.) Solders are also occasionally used to join a metal to an electroplated ceramic surface metallized by the moly-manganese process. However, this process is best suited for high-temperature applications and its high cost may not be justified for soldering.

Many reports of soldering glass and ceramic-to-metal joints can be found in the literature. In 1949, Jenny discussed the fabrication of soldered ceramic-to-metal seal using the techniques that had been developed for cable terminals, bushings, and connectors (ref. 118). As shown in figure 14, soldered joints should provide flexibility to the joint and minimize the load that must be supported by the solder. To minimize stresses in the ceramic, the metal member should be relatively thin and located on the outside of a cylindrical joint. Jenny recommended the use of tin-lead solders containing silver to minimize dissolution of the thin silver metallizing film; short soldering times were suggested for the same reason. Jenny also provided creep data for common solders.

Bondley discussed the use of low-melting soft solders with the titanium hydride process used to ensure the wettability of ceramic surfaces (ref. 119). However, a dam or well must be provided to hold the solder in place during the interval between melting of the solder and dissociation of the titanium hydride. Bondley also reviewed the characteristics of solders based on

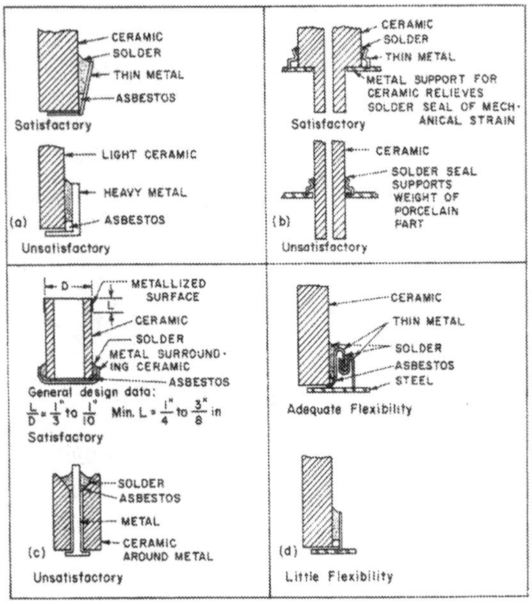

FIGURE 14.—Designs for soldered joints (ref. 118)

tin, lead, and indium. Experimental solders were prepared and evaluated by soldering two ½-inch-diameter ceramic rods together and testing them in flexure by four-point loading; 70-percent joint efficiencies were obtained. The solders were based on lead with additions of 0 to 50 percent silver, 0 to 50 percent indium, and 0 to 5 percent copper.

In 1955, McGuire of the Los Alamos Scientific Laboratory reported on a method to tin the surfaces of metals and nonmetals for subsequent joining operations (ref. 120). He found that an abrasive wheel on a hand grinder could be loaded with a solder by holding a stick of the solder against the rotating wheel. When the loaded abrasive wheel was rotated in contact with a metal surface, the frictional heat melted the solder and, at the same time, friction removed the surface oxides from the metal. The solder immediately wet the cleaned surface and flowed. This technique made it possible to solder a copper electrode to a tinned aluminum surface. The process was further evaluated by tinning other metals (stainless steel, titanium, and many of the refractory metals that are difficult or impossible to solder by conventional means). Soft and hard glasses as well as several ceramics were also tinned in this manner. For example,

a fired magnesium oxide body was tinned with a mixture of Wood's Metal* and 50Sn–50In solder; joints were made later with a 50Sn–50Pb solder.

Brazing

Brazing is uniquely suited to fabricating ceramic-to-metal joints and seals, and despite recent advances in the technology of metals joining, its position as the foremost method of making such joints has not been threatened seriously. In many respects brazing and soldering are similar processes; in both instances, the materials to be joined are heated and joined by a filler metal whose melting temperature is below those of the base materials. The four important differences distinguishing these processes are discussed below.

Brazing proceeds at temperatures above 800° F (430° C), while soldering proceeds below 800° F. Filler metals for soldering are based on such low-melting metals as tin, lead, and indium. Brazing filler metals are based on the noble metals, the heat-resistant metals such as nickel and cobalt, certain reactive metals such as titanium, beryllium, and zirconium, and the refractory metals.

In brazing, as in soldering, the filler metal is distributed throughout the joint area by capillary attraction; however, the clearances between the joint surfaces are more critical because of the reactions occurring during brazing. The recommended joint clearances are determined by the characteristics of the base and filler metals; data on clearances can be found in the *Brazing Handbook* and in the *Welding Handbook*, both published by the American Welding Society. (Certain so-called "wide gap" filler metals have been developed for use where joint clearances cannot be maintained.)

In soldering, the mechanisms responsible for joining are associated with adhesion and physical attraction, although a slight reaction along the joint interface may occur. Since brazing is conducted at high temperatures, the major bonding mechanisms are reactions between the base metals and filler metal, diffusion of the base

*25Pb–50Bi–12.5Cd–12.5Sn.

metal and filler metal constituents across the joint interface, and the formation of intermetallic compounds. The magnitude of these reactions depends on many factors such as the materials being joined, the filler metal, and the brazing cycle.

Soldering is usually done in air using a flux. While fluxes are suitable for low-temperature brazing, they lose their effectiveness at elevated temperatures, requiring other means of protection. The protective environment significantly affects the soundness and properties of the brazed joints. Controlled atmospheres are used to prevent oxidation of the joint materials, reduce surface oxides, or in some cases, produce limited oxidation. Oxide reduction depends on the type of gas, the gas dew point, and the brazing temperature. Vacuum environments also provide oxidation resistance. The filler metals used for vacuum brazing must be selected with care to avoid evaporation. Surface films and oxides are also removed in a vacuum, but the mechanism of removal is not well defined. Finally, the effects of gaseous atmospheres on base metal must be considered. For example, the properties of the reactive and refractory metals are adversely affected by even small traces of gaseous contaminants.

Thus, brazing is a much more critical operation than soldering; extensive knowledge is necessary to select the proper process, filler metal, atmosphere, and brazing cycle.

Although the metallizing of ceramic surfaces is costly and time-consuming, the brazing of metals to such surfaces is a straightforward operation because the metallizing layer ensures wettability of the ceramic by the filler metal. However, certain metals and their hydrides possess the ability to wet bare ceramic surfaces, and "active hydride" and "active metal" processes based on this property have been developed for producing ceramic-to-metal seals and joints.

The joining of ceramics to metals with the active metal or active hydride processes dates back to the middle 1940's when Bondley of the General Electric Company announced the use of titanium hydride for this purpose (ref. 121). Bondley painted the area to be joined with a mixture of fine titanium-hydride powders (300 mesh) suspended in a suitable binder. After drying, the ceramic and metal parts were assembled with a silver-base filler metal in contact with the hydride area. The assembly was heated to 900° to 1000° C (1652° to 1832° F) in a vacuum or in a very pure hydrogen atmosphere. As the titanium hydride dissociated, a residue of pure titanium remained on the ceramic surface; the hydrogen evolved in the atomic state during heating and tended to reduce oxides on the material surfaces. When the filler metal melted, it alloyed with titanium to form a silver-titanium alloy that bonded strongly with the metal and the areas of the ceramic that were coated with titanium hydride.

Pearson and Zingeser, in studying the bonding of ceramics with active metals and their hydrides (ref. 122), extended the work of Bondley and found that hydrides of zirconium, tantalum, and niobium were just as effective as titanium hydride in ceramic-to-metal joints. The effectiveness of various filler metals in making bonds with alumina, synthetic sapphire, beryllia, and thoria (table 13) was evaluated. In addition, Pearson and Zingeser found that titanium and zirconium, produced in reducing titanium and zirconium hydride, could also be used in powder form for ceramic-to-metal joints, thus marking the beginning of the "active metal" joining process. In developing experimental filler metals, Pearson and Zingeser noted that excellent bonds to ceramics, diamonds, sapphires, and other materials were made with an alloy containing 85 percent silver and 15 percent zirconium; aluminum-zirconium, aluminum-silver-zirconium, and silver-titanium alloys were also produced and evaluated. The effect of various brazing environments was also investigated, and a vacuum or a controlled atmosphere of hydrogen or an inert gas were employed to produce ceramic-to-metal joints.

In the early 1950's, research was conducted to further develop the active metal and active hydride processes. In 1951, Kelley received a patent on the use of a titanium or zirconium hydride mixture plus copper, silver, and gold for bonding ceramics (ref. 123). Hume applied the same principles to form a hermetic seal be-

TABLE 13.—*Evaluation of Metal Hydrides for Ceramic-to-Metal Joints*

[From ref. 122]

Material	Hydride	Brazing alloy	Atmosphere	Bond	Results
Aluminum oxide____	ZrH	Pure silver_____	Vacuum_____	Good	
	ZrH	Pure aluminum_____	Vacuum_____	Good	Bonded to tantalum
	ZrH and TiH	None_____	Vacuum_____	Good	Heated to about 1700° C
	TiH	Cobalt_____	Vacuum_____	Good	
Synthetic sapphire__	TiH	Pure silver_____	Vacuum_____	Good	
	ZrH	Pure silver_____	Vacuum_____	Good	
	TaH	Pure aluminum_____	Vacuum_____	Good	
	CbH	Pure aluminum_____	Forepump vacuum___	Good	
	ZrH	Pure silver_____	Dry tank N_2_____	Good	
	None	15 percent zirconium-silver alloy.	Forepump vacuum___	Good	
Beryllium oxide_____	ZrH	Pure silver_____	Vacuum_____	Good	Bonded to molybdenum
	ZrH	Pure aluminum_____	Vacuum_____	Good	
	CbH	Pure aluminum_____	Forepump vacuum___	Good	
	TaH	Pure aluminum_____	Forepump vacuum___	Good	
	None	15 percent zirconium-silver alloy.	Forepump vacuum___	Good	
Thorium oxide_____	ZrH	Pure silver_____	Vacuum_____	Good	
	CbH	Pure aluminum_____	Forepump vacuum___	Good	
	TaH	Pure aluminum_____	Forepump vacuum___	Good	
	None	15 percent zirconium-silver alloy.	Forepump vacuum___	Good	

tween an aluminum bushing and an alumina header (ref. 124). The ceramic header was coated with a suspension of zirconium hydride, a ring of pure silver was placed in the area to be metallized, and the assembly was heated to about 1100° C (2012° F). The zirconium hydride was reduced to elemental zirconium at 500° C (932° F); the zirconium wet the ceramic surface. When the silver melted, it alloyed with the zirconium and coated the ceramic surface. After the assembly cooled, the aluminum bushing was positioned, and the assembly was reheated until the aluminum melted and formed a seal with the metallized ceramic header.

Since these early investigations, joining ceramics to metals by the active metal or active hydride process has advanced significantly. The strengths of joints made by this process are as great as those obtained with joints made by the moly-manganese process. Some difficulty has been experienced in making seals by the active

metal or active hydride process in dry hydrogen. The dew point of hydrogen must be extremely low to prevent oxidation of titanium. According to Chang, the dew point of hydrogen must be about −85° C (−121° F) to reduce titanium oxide at 1000° C (1832° F); such a dew point is difficult to achieve except under laboratory conditions (ref. 125). Producing ceramic-to-metal seals in a vacuum is advantageous in that the parts are outgassed during brazing.

The characteristics of the moly-manganese, active metal, and active hydride processes are summarized in the following paragraphs:

The moly-manganese process is a multi-step sealing process in which the ceramic surface is metallized and plated with one or two metals before brazing can take place. The operations are conducted at a high temperature in a controlled atmosphere of hydrogen; hydrogen firing may discolor some ceramics and produce conductive surfaces. Despite the number of steps

67

required to produce a seal, the moly-manganese process can be automated quite readily, and minor deviations in the process variables can be tolerated.

The active hydride process is essentially a single-step process in which hydride reduction and brazing proceed simultaneously. Joining in a vacuum or in a controlled atmosphere of hydrogen or an inert gas is accomplished at relatively low temperatures, permitting a fast brazing cycle. This process is more difficult to automate than the moly-manganese process, particularly if the joints are produced in a vacuum. Careful control must be exercised in coating the ceramic with the hydride.

The active metal process may be a one-step operation like the active hydride process.* Joining proceeds at high temperatures in a vacuum or in a controlled atmosphere; vacuum joining is not readily automated.

These processes will be discussed further from the application standpoint in the following sections, with emphasis on recent or current research.

Brazing to Metallized Ceramic Surfaces

The technology of joining metals to metallized ceramic surfaces was first applied on a large scale by the electronics industry. Since numerous reports and papers have been prepared on fabricating vacuum-tube components by these methods, only some especially interesting developments will be discussed here.

In the early 1950's, Coykendall of the Machlett Laboratories discussed the procedures used to assemble a UHF power triode designed to operate in the 30- to 2000-megacycle range (ref. 126). Ceramic-to-metal seals as large as 7 inches in diameter were made by brazing nickel-iron rings to cylindrical alumina sections; the ceramic parts were metallized by the moly-manganese process. The joints were brazed with pure silver, so that other, lower melting silver-base alloys could be used for subsequent joining operations. Cronin has discussed the trends in the design of ceramic-to-metal seals for use in

high-power magnetrons (ref. 127). LaForge has prepared an excellent review of the procedures to fabricate seals for high-power pulsed klystrons delivering a peak power output of 30 megawatts at an average power of 30 kilowatts (ref. 63). The authors discussed features of the klystron and the requirements for ceramic windows that provide vacuum sealing for the output wave guide while permitting passage of electromagnetic energy. At first, the high-alumina window sections were metallized with moly-manganese and brazed with either the 82Au–18Ni or 72Ag–28Cu filler metals. The metallized layer was nickel-plated when the silver-copper eutectic alloy was used. The wettability of 95- to 97-percent alumina bodies was improved with an activated metallizing mixture containing molybdenum, manganese, iron, calcium oxide, and silicon oxide.

Two important programs to develop high-temperature seals for vacuum tubes were conducted by the Sperry Gyroscope Company and the Radio Corporation of America (refs. 128 and 129). Engineers at Sperry investigated the effect of process variables (metallizing composition, powder particle size, metallizing layer thickness, method of application, composition of alumina ceramic body, and sintering time and temperature) on the strength and reproducibility of brazed assemblies. The RCA study was concerned with developing ceramic-to-metal sealing techniques that could be used in producing output windows for high-power microwave tubes that would withstand a 700° C (1292° F) bake-out temperature. Procedures were developed to metallize synthetic sapphire (pure alumina), and adherence tests showed that optimum metallizing was achieved with a mixture of 80- to 90-weight percent molybdenum and 10- to 20-weight percent S–641A (a proprietary material that resembles steatite). The studies also included suitable filler metals for ceramic-to-metal seals (table 14). The 62.5 Cu–37.5 Au alloy, which appeared to meet the temperature requirements, was further evaluated to determine the effect of metallizing variables on the strength of synthetic sapphire joints. The joints were made by metallizing sap-

---

*A thin metallized layer of titanium or molybdenum on BeO surfaces has enhanced wetting.

TABLE 14.—*Evaluation of Brazing Filler Metals for Joining Synthetic Sapphire Metallized With Molybdenum and S–641A Mixtures*

[From ref. 129]

Metallizing material, percent	Brazing material, percent	Brazing temperature, °C	Quality	Penetration	Remarks
80 Mo, 20 S–641A*	Copper	1090	Good	Slight	
90 Mo, 10 S–641A	Copper	1090	Good	Very slight	
80 Mo, 20 S–641A	37.5 gold 62.5 copper.	1025	Fair	Slight	
90 Mo, 10 S–641A	37.5 gold 62.5 copper.	1025	Excellent	Very slight	Excellent structure.
80 Mo, 20 S–641A	50.0 gold 50.0 copper.	1005	Fair	Slight	
90 Mo, 10 S–641A	50.0 gold 50.0 copper.	1005	Excellent	None	
90 Mo, 10 S–641A	35.0 gold 62.0 copper. 3.0 nickel.	1035	Fair	Slight	
80 Mo, 20 S–641A	35.0 gold 62.0 copper. 3.0 nickel.	1035	Excellent	Slight	
90 Mo, 10 S–641A	82.0 gold 18.0 nickel.	960	Fair	None	
80 Mo, 20 S–641A	82.0 gold 18.0 nickel.	960	Excellent	None	

*S–641A—a proprietary material that resembles steatite.

phire rods 0.625 inch long and 0.250 inch in diameter and brazing them in pairs. Two small specimens with a cross-sectional area of 0.0049 square inch were machined from each brazed rod and tested; the seal strength data are summarized in table 15. Attempts to metallize synthetic sapphire with tungsten mixtures were discontinued because the seal strength obtained with tungsten-metallized sapphire did not match that obtained with molybdenum metallizing mixtures. Procedures were also developed to metallize and braze beryllia (table 16), and synthetic sapphire and beryllia output windows were fabricated and evaluated. Usually such windows are made by brazing a metallized ceramic disk to a machined ring; however, because voids were encountered, a subassembly consisting of a copper (or copper-gold) ring cast onto the ceramic disk was produced (fig. 15(a)). This assembly was machined to precise dimensions and brazed to the outer copper

sleeve as shown in figure 15(b). Output windows of this type withstood the required 700° C (1292° F) bake-out temperature and passed the thermal cycling test (1000 cycles between 20° and 125° C or 68° and 258° F).

Several research programs have been conducted to develop sealing techniques for constructing energy-converter devices. In 1964, Bristow, Grossman, and Kaznoff developed mixtures composed of 60- to 70-percent molybdenum powder and 30- to 40-percent oxides of aluminum, calcium, magnesium, and yttrium for metallizing high-alumina ceramics (ref. 11). (The results of studies to evaluate these metallizing mixtures were discussed earlier in the section on "Metallizing.") During the course of this program, vacuum-tight seals were produced between alumina and molybdenum, tantalum, and niobium. Goldstein has described the fabrication of ceramic-to-metal seals for spaceborne reactor components (ref. 130). The seals were

TABLE 15.—*Seal Strength Data on Synthetic Sapphire Joints*

[From ref. 129]

Metallizing mixture, percent	Firing time, hours	Firing temp, °C	Plating material	Brazing material, percent	Modulus of rupture, psi			Failure location
					Average	High	Low	
80 Mo. 20 S–641A.*	0. 5	1500	Nickel.	37.5 Au. 62.5 Cu.	36 300	54 000	26 500	Metallizing layer.
90 Mo. 10 S–641A.	0. 5	1700	Nickel.	37.5 Au. 62.5 Cu.	43 700	48 000	39 000	Random: Sapphire. Sapphire-metallizing interface.
95 Mo. 5 S–641A.	0. 5	1800	Nickel.	37.5 Au. 62.5 Cu.	41 900	47 500	36 300	Random: Sapphire-metallizing interface. Sapphire.
90 Mo. 10 S–641A.	0. 5	1600	Nickel.	37.5 Au. 62.5 Cu.	40 400	45 600	39 000	Random: Sapphire-metallizing interface. Sapphire.
90 Mo. 10 S–641A.	0. 5	1750	Nickel.	37.5 Au. 62.5 Cu.	47 000	51 400	37 900	Sapphire.
92.5 Mo. 7.5 S–641A.	0. 5	1750	Nickel.	37.5 Au. 62.5 Cu.	34 500	49 300	12 800	Random: Metallizing and metallizing-sapphire interface.
92.5 Mo. 7.5 S–641A.	0. 5	1800	Nickel.	37.5 Au. 62.5 Cu.	42 300	46 300	39 600	Sapphire.
90 Mo. 10 S–641A.	2. 0	1750	Nickel.	37.5 Au. 62.5 Cu.	47 400	55 900	33 300	Random: Sapphire. Sapphire-metallizing interface.
90 Mo. 10 S–641A (fine ink).	0. 5	1700	Nickel.	37.5 Au. 62.5 Cu.	45 200	50 900	42 300	Sapphire.

* S–641 A—a proprietary material that resembles steatite.

NOTE: Modulus-of-rupture figures are influenced by the strength of the sapphire used, which may vary between 25 000 and 60 000 psi with no visible cause for the variance. All samples with breaks in the sapphire lower than 37 000 psi were deleted from calculations.

required to withstand operating temperatures of 1000° C (1832° F), be resistant to high-temperature cesium vapor, and possess adequate resistance to thermal cycling. To meet these requirements, a metallized 97-percent alumina ceramic was brazed to niobium with vanadium as the filler metal. Dring of the Bendix Corporation also conducted studies to produce ceramic-to-metal seals for a cesium-vapor-filled thermionic converter (ref. 104). A 99.5-percent alumina body, metallized with a molybdenum-titanium mixture, was brazed to molybdenum with the following filler metals: nickel, nickel-iron, cobalt, and iron. Sound joints were obtained with cobalt and iron, but the thermal-expansion coefficient for these metals accentuated the expansion mismatch between the molybdenum metal member and the alumina ceramic, causing gross cracking. Additional seals were made between alumina and niobium using pure palladium as the brazing filler metal. Niobium was used instead of molybdenum to improve the expansion match between the ceramic and metal joint members. For this seal, the alumina body was metallized with a tungsten-yttria mixture. Vacuum-tight seals were produced but precise temperature control during brazing was imperative, since the filler metal tended to erode the

TABLE 16.—*Seal Strength Data on Beryllia Joints*

[From ref. 129]

Metallizing mixture, percent	Metallizing time, hour	Firing temperature, °C.	Plating material	Brazing material, percent	Modulus of rupture, psi			Failure location
					Average	High	Low	
100 Molybdenum (fine).	0.5	1550	Nickel____	37.5 Au____ 62.5 Cu.	26,400	30,600	20,000	Random metallizing layer.
90 Molybdenum, 10 S–641A* (fine).	0.5	1600	Nickel____	37.5 Au____ 62.5 Cu.	33,100	38,000	27,100	90 percent in ceramic; 10 percent random.

*S–641A—a proprietary material that resembles steatite.

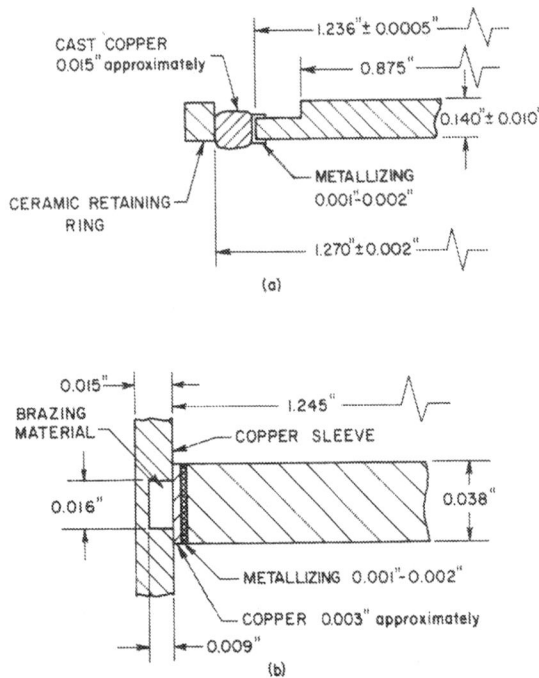

FIGURE 15.—Design and assembly of brazed synthetic sapphire and beryllia output windows (ref. 129). (a) Subassembly for output windows; (b) brazed output windows (enlarged section)

niobium. To overcome the problem, Dring suggested the use of the niobium-palladium eutectic alloy for brazing.

In another program to design and construct equipment for the direct conversion of nuclear to electrical energy, engineers at the Los Alamos Scientific Laboratory developed ceramic-to-metal seals for use at 1000° C (1832° F) in a cesium vapor environment (refs. 103 and 131). The seal materials also had to be resistant to radiation damage. The basic cell for energy conversion consists of a nuclear fuel element that acts as an electron emitter, a metal base section to which the fuel element is attached, and a metal electron collector section; the inner space of the cell is filled with cesium vapor. Ceramic-to-metal seals are required to isolate the cell's base and collector sections. A schematic of the basic thermionic converter cell is shown in figure 16. The base and collector sections were machined from Nb–1Zr; silica-free alumina was selected for isolating the sections. In 1963, Brundige and Hanks discussed the procedures to produce the ceramic-to-metal seals (ref. 103). Several high-alumina ceramics were metallized with tungsten in accordance with procedures developed by Cowan and Stoddard (ref. 92). Two joint designs were used to make the ce-

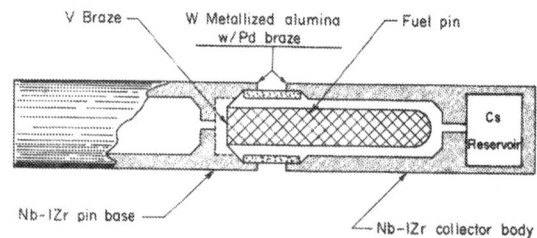

FIGURE 16.—Schematic of basic thermionic converted cell (ref. 103)

ramic-to-metal seal; figure 17(a) shows the seal brazed with the 65Pd–35Co filler metal in the form of washers or rings. In the second design (fig. 17(b)) the filler metal was plated on the metallized ceramic before assembling the joint; palladium and cobalt were electrodeposited in alternate layers to the approximate composition 65Pd–35Co. The resulting joints were leaktight but unreliable. In an extension to this program, Hanks, Kirby, and LaMotte evaluated several filler metals using the joint design shown in figure 17(c); the joints were brazed in a cold-wall vacuum furnace at 10^{-4} torr (ref. 131). These

data are summarized in table 17. While brazing with palladium and vanadium filler metals was satisfactory, some erosion of the base metal occurred. This problem was largely eliminated by using induction heating to produce a very short brazing cycle. Kirby and LaMotte also report on research to evaluate other filler metals, conducted by the Pyromet Company (ref. 132).

Brazing With Active Metals and Hydrides

The concept of fabricating ceramic-to-metal seals by the active metal or active hydride

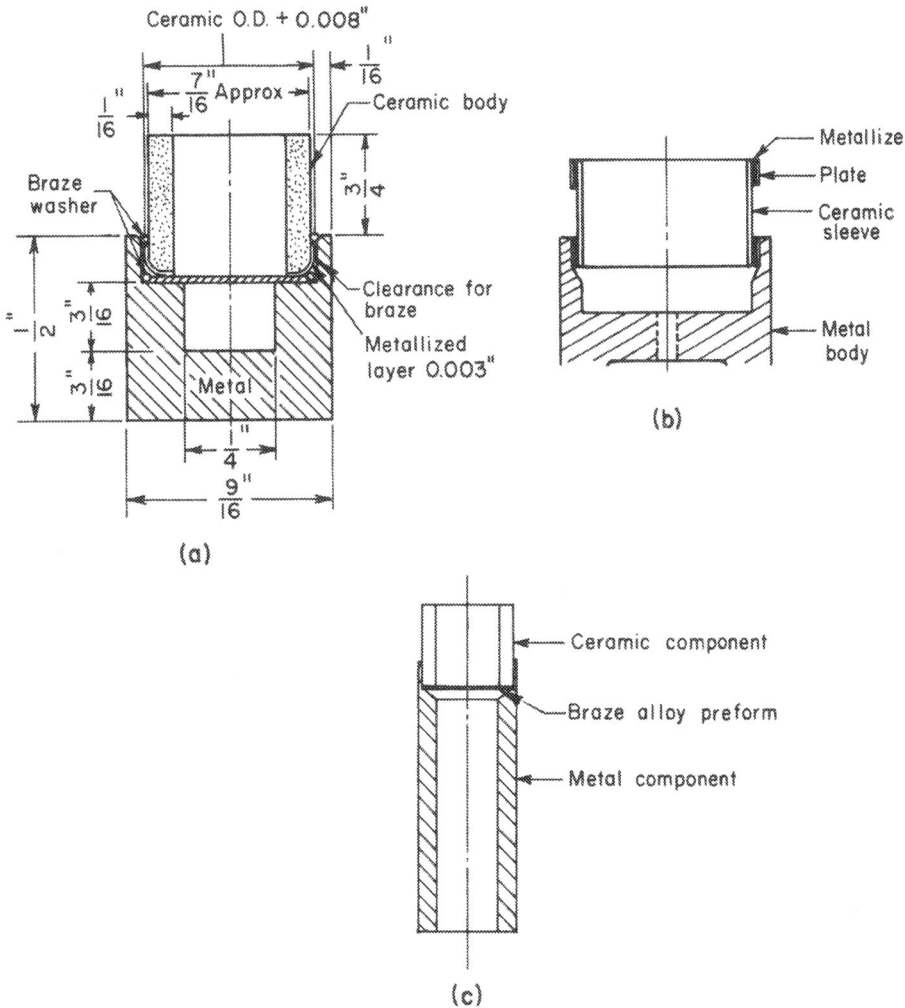

FIGURE 17.—Ceramic-to-metal seals for thermionic converter cells (refs. 103, 131). (a) Straight seal; (b) plated seal; (c) simplified straight seal

TABLE 17.—*Ceramic-to-Metal Wetting Tests on Metallized Ceramic Surfaces*

[From ref. 131]

Brazing alloy composition, w/o	Ceramic [a] type	Brazing temperature, °C	Results
Vanadium	Al–14	1960	Wet metal and ceramic well. Considerable alloying with metal. No evidence of metallizing after brazing.
Vanadium	Lucalox	1960	Same as above.
Palladium	Al–14	1590	Wet metal and metallized ceramic well. Considerable alloying with metal. Metallized coating still intact.
Pd–35Co	Al–14	1260	Wet metal and metallized ceramic well. Considerable alloying with metal. Metallized coating still intact.
Co	Al–14	1540	Wet the metallized ceramic well. Extreme alloying with the metal.

[a] Ceramic type	Density, percent theo	Composition
Al–14 (pressed)	93	Al_2O_3–0.5 w/o Y_2O_3.
Al–14 (slipcast)	96	Al_2O_3–0.5 w/o Y_2O_3.
Lucalox	98	Al_2O_3–0.3 w/o MgO.

process was first applied in the electronics industry. However, in recent years, these joining processes have found other uses to meet the need for high-temperature vacuum-tight seals in the nuclear and aerospace industries.

In 1954, Bender of Sylvania Electric Products, Inc., discussed research to evaluate the active metal process for joining zirconia ceramics (ref. 133). Specifically, he conducted wetting tests to determine the percentage of titanium that should be added to the silver-copper eutectic alloy (72Ag–28Cu) to obtain maximum wetting in a dry hydrogen atmosphere; alloys with 10 percent titanium produced the best results. He also investigated the silver-zirconium system for joining zirconia and alumina ceramics; good wetting and bonding were observed when the tests were conducted in a vacuum. Evans, of Sylvania, published an article in 1954 discussing the feasibility of using the active metal process to produce typical vacuum-tube seals (ref. 134). The filler metal for this work was the silver-copper eutectic alloy on a titanium wire core. Martin (Edgerton, Germeshausen, and Grier) discussed

research on the active hydride process to seal windows in high-power magnetrons (ref. 135). Joints between alumina and Kovar produced in a vacuum of 10^{-3} torr or lower by painting the ceramic with a suspension of fine titanium hydride and brazing the joint with 72Ag–28Cu filler metal withstood sustained bake-out temperatures of 700° C (1298° F). A ceramic-to-stainless steel bellows joint was also produced by this method. A simplified version of the active metal process has been used successfully by a firm engaged in producing electrical feed-through components for vacuum processing equipment (ref. 136). After the parts are assembled, a band of titanium is painted on the ceramic surface, a ring of 72Ag–28Cu is positioned, and the joint is brazed in a vacuum.

Recently, Fox and Slaughter investigated the use of experimentally developed active metal alloys for producing ceramic-to-ceramic and ceramic-to-metal joints, some of which may be potentially useful in nuclear reactor technology (ref. 12). The filler metals, 68Ti–28Ag–4Be and 49Ti–49Cu–2Be, were originally developed for joining graphite to metal. However, studies indicated that good wetting and flow occurred between 49Ti–49Cu–2Be and alumina, beryllia, and uranium oxide; the 68Ti–28Ag–4Be alloy produced good wetting on alumina surfaces but only fair wetting on beryllia and uranium oxide. Several other titanium- or zirconium-base alloys showed promise in wetting oxide ceramics. During the initial studies, sound joints were produced between the following combinations of materials: alumina-to-alumina with 49Ti–49Cu–2Be; alumina-to-Zircaloy 2 with 48Ti–48Zr–4Be; uranium oxide to molybdenum with 49Ti–49Cu–2Be; uranium oxide to niobium with 68Ti–28Ag–4Be; uranium oxide to zirconium with 95Zr–5Be; and beryllia to zirconium with 95Zr–5Be. To demonstrate the possible uses of these active metal alloys for fuel element applications, a compartmented aluminum oxide assembly was vacuum-brazed with the 49Ti–49Cu–2Be alloy; a simulated beryllia fuel bundle was brazed with the same alloy. An alumina-to-titanium bearing assembly and a Hall cell assembly that included joints between synthetic sapphire and zirconium were also fabricated.

In developing ceramic-to-metal sealing techniques for the production of output windows in high-power microwave tubes, engineers at the Radio Corporation of America investigated the active metal and active hydride processes (ref. 129). They evaluated titanium, zirconium, and vanadium foils and hydrides of titanium and zirconium with foils of nickel, copper, and a 37.5Au–62.5Cu used as the brazing materials. Active metal alloys were prepared by combining foils of the active metal with foils of nickel and copper in proportions to produce a eutectic or near-eutectic alloy, wetting tests were conducted, and specimens were prepared to determine the modulus of rupture. The studies indicated that active metal seals can be developed with strengths greater than those currently obtained if an intermediate metal with a low expansion coefficient is introduced into the joint. It was also concluded that ceramics having lower expansion coefficients than alumina probably cannot be joined by the active metal process.

Research to develop a space power alternator capable of delivering 300 watts electrical power has been underway at the Aerospace Electrical Division of the Westinghouse Electric Corporation since 1960. The power system consists of a nuclear reactor as a heat source, a thermodynamic system to convert thermal energy to mechanical energy, and an alternating-current generator to convert mechanical to electrical energy. Since the generator is connected directly to a potassium-vapor-driven turbine, a bore seal is needed in the rotor cavity to isolate the stator electrical windings from the potassium vapor. The tubular bore seal consists of a thin ceramic section hermetically sealed to metal members at each end. The ceramic-to-metal sealing studies undertaken by the Eimac Division of Varian Associates have been reviewed by Hoop in a recent summary report (ref. 10). Major accomplishments are discussed below.

1. *Tensile tests.*—Studies were conducted to select the ceramic and metal members of the seal as well as a suitable brazing alloy. The test specimens were made by brazing a thin metal section between two halves of the ASTM test specimen, CLM–15. Active metal filler metals that contained substantial amounts of zirconium and titanium were used for most joints; a limited number of specimens were metallized and brazed. These specimens were tested in tension after exposure in potassium vapor for 1000 hours at 1100° F (593° C), 5000 hours at 900° F (482° C), and 10 000 hours at 900° F. On the basis of these studies, a ceramic body having the following composition appeared most resistant to corrosion by potassium vapor: 97 percent alumina, 1.5 percent each of calcia and magnesia, and less than 0.1 percent silica. The most suitable metal was an alloy resembling Kovar and Nb–1Zr. Several filler metals appeared promising.

2. *Subscale tubular bore seals.*—Metal end sections were brazed to 2.5-inch-diameter ceramic cylinders with 83Ti–17Cu, 75Zr–19Nb–6Be, and 68Ti–28V–4Be as filler metals. The assemblies were filled with a measured amount of potassium, evacuated, sealed, and tested for 1000 hours at 1100° F (593° C); all but two of the ten specimens were leaktight after exposure to potassium vapor.

3. *Large tubular bore seals.*—Procedures to braze large tubular bore seals were studied, and the ceramic cylinder used for this work had the diameter (11.5 inches) of the full-scale bore seal; however, it had a heavier wall thickness (0.190 inch) and was shorter (4 inches) than the full-size unit. Seven specimens were brazed in accordance with the procedures shown in table 18. As can be noted, it was difficult to obtain leaktight seals. Since leaktight tensile specimens had been produced with the same joint materials, the cause of the difficulty was investigated. Based on metallographic studies of specimens brazed with the rapid brazing cycle used to produce tensile specimens and the slow brazing cycle used to produce the large bore seal configurations, it was concluded that brittle, intermetallic compounds had formed during the slow heating cycle; some cracks were observed in these microstructures. Titanium-base active metals appeared unsuitable for sealing unless the joint could be heated and cooled rapidly. However, a further test indicated that vacuum-tight seals could be produced with the 75Zr–19Nb–6Be filler metal using slow heating and cooling cycles.

Similar research by Westinghouse is under-way to select bore seal materials for advanced space power systems where the seals may be exposed to potassium, NaK, or lithium vapors at temperatures from 1000° to 1600° F (538° to 971° C); the accomplishments of this program have recently been reviewed by Kueser (ref. 137).

1. *Materials selection.*—Extensive screening tests were conducted to select the bore seal materials. The performance of several high-alumina ceramics (94 to 100 percent Al_2O_3) and a high-beryllia ceramic (99.8 percent BeO) was evaluated in room-temperature flexural strength tests conducted before and after test bars were exposed to alkali metal vapors at elevated temperatures. The flexural strength of beryllia was not affected seriously by exposure in potassium vapor at 1600° F for 500 hours. Under the same test conditions, the strength of the alumina ceramics decreased sharply as the Al_2O_3 content decreased from 100 to 94 percent. The strength of a 99.7-percent alumina decreased slightly when the ceramic was exposed to potassium vapors at 1000° F for 500 hours; however, it decreased by one-half when exposure occurred in NaK vapor under the same conditions. Little effect on strength was noted when the 99.8-per-

TABLE 18.—*Trial Brazing of 11.5-Inch-Diameter Tubular Ceramics to 0.030-Inch-Thick Nb–1Zr Sheet*
[From ref. 131]

Brazing alloy	Brazing cycle		Vacuum, torr		Remarks
	Temperature, °F	Hold, minutes	Cold	Hot	
83Ti–17Cu[a]	1830	12	5×10^{-5}	9×10^{-3}	Leaker; metal discolored.
83Ti–17Cu[a]	1850	15	3×10^{-6}	6×10^{-5}	Leaker; braze discolored.
83Ti–17Cu[b]	1980	30	3×10^{-6}	3×10^{-4}	Stress fracture of joint.
80Ti–20Ni	1925	20	7×10^{-7}	1×10^{-4}	Leaker; poor braze to ceramic.
80Ti–20Ni[c]	1925	15	1×10^{-6}	5×10^{-5}	Good braze fillets; backup ring cracked.
72Ti–28Cu[b,d]	1870	15	1×10^{-6}	3×10^{-5}	Columbium alloy cracked.
72Ti–28Cu[b,d,e]	1870	7	5×10^{-6}	4×10^{-5}	Good fillets but porous braze.

[a] Copper washer to form brazing alloy was adjacent to ceramic.
[b] Copper washer to form brazing alloy was adjacent to columbium alloy.
[c] The ceramics were coated with lithium molybdate to improve reaction.
[d] An excess of copper was used to compensate for its evaporation.
[e] A 9-hour hold at 1200° F during cooling to relax residual stresses.

cent beryllia ceramic was exposed in potassium or lithium vapor at 1000° F for 500 hours. Niobium-base alloy Nb–1Zr was selected as the metal member in ceramic-to-metal joints; some joints were also made with D43 (Nb–10W–1Zr–0.1C).

2. *Brazing with metallized ceramic surfaces.*—The effectiveness of tungsten-base metallizing paints in promoting strong, leaktight ceramic-to-metal joints was investigated. The metallizing paints contained 85 to 95 percent tungsten and 5 to 15 percent rare earth oxides, alumina, and/or calcium carbonate. Joints were made using ASTM test specimens made from ceramics containing 99.0 percent alumina, 99.7 percent alumina, and 99.8 percent beryllia. The results of the joining studies with alumina ceramics are shown in table 19; acceptable strength levels

were obtained but problems were encountered with brazing-alloy metallized-coating reactions and cracking in the joint. The tests with metallized beryllia ceramics did not produce acceptable results because the joint strength was too low for the intended application. The studies with metallized ceramic surfaces were discontinued in favor of active metal brazing.

3. *Active metal brazing.*—The use of active metals in brazing ceramic-to-metal joints for service in alkali metal vapors at elevated temperatures was studied extensively. Promising active metal brazing alloys were selected using ASTM test specimens and modulus-of-rupture test bars (table 20). On the basis of these tests the following alloys were selected for further evaluation: 75Zr–19Nb–6Be, 56Zr–28V–16Ti, and 48Ti–48Zr–4Be. Vacuum leak test speci-

TABLE 19.—*Tensile Strength and Leak Testing of Special Metalizing Paints Utilizing ASTM CLM 15 Tensile Test Assembly*

[From ref. 137]

| Paint symbol | Copper braze [a] | | | | | | | Nickel alloy braze [b] | | |
| | Ei3–3W (99.7 percent Al_2O_3) | | | | AD 99 (99 percent Al_2O_3) | | | AD 99 (99 percent Al_2O_3) | | |
	Number of tests	Tensile strength, psi	Number of tests	Leak test	Number of tests	Tensile strength, psi	Number of tests	Leak test	Number of tests	Leak test
W5M	1	>14 850	2	VT						
W8M	1	>12 900	2	VT						
W9M	1	>14 550	2	VT						
W10M	1	>13 200	2	VT						
W11M	1	>14 400	1	VT	2	>12 600	2	VT	1	LKR
W12M	1	[c] 6030	1	LKR	2	>11 250	2	VT	1	VT
W13M	1	>12 900	1	VT	2	>11 160	2	VT	1	VT
W14M	1	[c] 8160	1	VT	2	>13 110	2	VT	1	LKR
W15M	1	[c] 4050	1	VT	2	>13 290	2	VT	1	LKR

[a] Copper brazed in —100° F dewpoint hydrogen at 2040° F with 3-minute hold at temperature. 0.020-inch 70/30 cupro-nickel washer between CLM 15 pieces.

[b] Coast Metals Braze Alloy 52 (3B, 4.5Si, 0.15C, Ni balance) brazed at 1850° F in vacuum at 10^{-5} torr with no hold time at temperature. 0.015-inch Columbium washer between CLM 15 pieces. Columbium washer and metallizing plated with 0.0005 inch Fe; vacuum-sintered for 10 minutes at 1470° F.

[c] Specimens broke in the AD 94 side of the joint.

NOTES:

(1) > indicates a metallizing strength greater than the figure shown. The specimen broke in the ceramic at the stress level noted.

(2) VT indicates a leak rate less than 1×10^{-9} torr-liters/sec as determined in leak-testing procedure.

(3) LKR indicates leaker.

(4) All paints listed were sintered for ½ hour at 3045° F in forming gas, $75N_2$–$25H_2$, 70° F dewpoint.

76

TABLE 20.—*Active Braze Alloy Preliminary Screening Using Cb–1Zr Metal Member with Designated Ceramic*

[From ref. 137]

Braze composition, weight percent	Brazing Temperature, °F	Thermalox 998 99.8% BeO		Ei3–3W Al$_2$O$_3$		Remarks
		VT [b]	Strength (MoR) [a], psi	VT [b]	Tensile Strength [a], psi	
75Zr–19Cb–6Be	1940	2/4	15 260	4/4	9750	Wets columbium alloy D–43 well.
68Ti–28V–4Be	2370	1/4	16 635	1/4	>240	Wets columbium alloy D–43 and tantalum alloy T111 well.
				0/4	>4475*	
56Zr–28V–16Ti	2270	c	15 740	2/4	4850	
48Ti–48Zr–4Be	1940	3/4	16 500	4/4	4900	Wets columbium alloy D–43 well.
				4/4	9575*	
46Ti–46Zr–4V–4Be	1830	--------	16 650	4/4	--------	
				1/4	6938*	
50Zr–30V–20Cb	2415	3/4	15 175	0/4	140	
65V–35Cb	3400	--------	--------	--------	--------	No test.
70Ti–30V	3000	--------	--------	--------	--------	No test.
60Zr–25V–15Cb	2435	--------	14 035	0/4	50	Forms skull on columbium alloy D–43 and tantalum alloy T111.
50Zr–30Ti–20V	2480	0/4	>4 742	0/4	>202	Forms skull on columbium alloy D–43 and tantalum alloy T111, Cb–1Zr alloy.
40Zr–30Ti–30V	2335	4/4	13 165	0/4	245	Wets columbium alloy D–43 and tantalum alloy T111 well.
35Ti–35V–30Zr	2595	0/4	>10 280	0/4	>500	
50Ti–30Zr–20V	2595	--------	--------	--------	--------	No test.
62Ti–30V–8Si	2480	4/4	8 370	0/4	>1175	Wets columbium alloy D–43 and tantalum alloy T111 well.

[a] Modulus of rupture.
[b] Number vacuum tight over total number tested.
[c] Vacuum-tight assemblies fabricated previously on another program (SPUR, Westinghouse).

NOTES:

(1) Brazed in vacuum furnace (10^{-5} torr) at temperature indicated; no hold time. Results shown are from the best braze run for each alloy.

(2) *Made with AD 99 alumina + 99 percent Al$_2$O$_3$.

(3) >Indicates incomplete melting—greater strengths might be expected with increased temperature but were not attempted because of excessive pressure in the furnace at elevated temperature.

(4) Italicized brazes and ceramic-to-metal assemblies were considered most favorable for further evaluation in potassium, potassium-sodium eutectic, and lithium.

mens, modulus-of-rupture test specimens, and tab peel assemblies of high-purity alumina and beryllia were vacuum-brazed to Nb–1Zr or D43 alloy metal members with these three alloys. The ceramic-to-metal joints were tested in the as-brazed condition, after exposure in vacuum for 500 hours at either 1000° or 1600° F, and after exposure in alkali metal vapors for 500 hours at either 1000° or 1600° F. The results are summarized in tables 21 and 22. Metallographic

TABLE 21.—*Effect of 500 Hour 1600° F Potassium Vapor Exposure on the Room Temperature Flexural Strength of Selected Ceramic-Metal Sealing Systems*

[From ref. 137]

Ceramic	Braze, weight percent	Brazing temperature, °F	Key	Flexural strength, psi		
				As brazed	Vacuum exposed 500 hr, 1600° F	Potassium vapor 500 hr, 1600° F
Ei3–3W alumina, 99.7 percent Al_2O_3____	75Z1–19Cb–6Be	1940	$\bar{x}$	25 655	12 965	[a] 0
			s	7 370	[b] 4505	
			n	11	2	4
Ei3–3W alumina, 99.7 percent Al_2O_3____	48Ti–48Zr–4Be	1940	$\bar{x}$	23 342	19 760	[a] 0
			s	4 690	[b] 1440	
			n	12	2	4
Thermalox 998, beryllia 99.8 percent BeO.	75Zr–19Cb–6Be	1940	$\bar{x}$	15 404	17 300	[c] <1000
			s	1 220	[b] 800	
			n	5	2	5
Thermalox 998, beryllia 99.8 percent BeO.	48Ti–48Zr–4Be	1940	$\bar{x}$	16 559	14 250	[c] 10 538
			s	2 500	----------	3740
			n	8	1	5
Thermalox 998, beryllia 99.8 percent BeO.	56Zr–28V–16Ti	2270	$\bar{x}$	13 503	13 985	[c] 11 810
			s	2 870	[b] 1715	850
			n	6	2	5

[a] No MoR structures survived exposure test intact.
[b] Standard deviation has little significance for sample size of two pieces.
[c] Accompanying vacuum leak test assemblies were broken during removal from capsule.

NOTES:
(1) All tests on modulus-of-rupture assemblies (MoR) using columbium–1 percent Zr metal member.
(2) Italicized ceramic-metal sealing systems appear to be the best of those tested.

KEY:
$\bar{x}$—arithmetic mean.
s—standard deviation.
n—number of specimens tested.

examinations of the ceramic-to-metal joints were also made, and all studies were analyzed to select the best ceramic-to-metal seal system for each alkali metal-temperature environment (table 23).

4. Wetting was enhanced by depositing thin layers of titanium and molybdenum on ceramic surfaces by evaporation techniques.

The work on active metal brazing of ceramic-to-metal joints for service in alkali metal vapors at high temperatures has been continued at Westinghouse under a followup program. In a recent quarterly report, Kueser indicated that additional brazing studies had been conducted to select other alloys for joining a 99.8-percent beryllia ceramic to an Nb–1Zr alloy metal member for service in potassium vapor at temperatures up to 1600° F. (ref. 138). The results of these studies are summarized in table 24. The following three alloys were selected for further evaluation: 46Ti–46Zr–4Be–4V, 60Zr–25V–15 Cb, and 35 Ti–35Zr. A topical summary report on this program is expected to be published later in 1968. It has been reported that a 4-inch-diameter BeO bore seal has been operated for 6000 hours in potassium metal vapor at 1300° F.

The active metal process has been investigated in a number of programs as a means to fabricate ceramic-to-metal seals and composite structures for use in cesium-plasma thermionic converters. In addition to producing joints be-

TABLE 22.—*Effect of 500-Hour 1000° F Potassium, NaK, or Lithium Vapor Exposure on the Room Temperature Flexural Strength of Selected Ceramic-Metal Sealing Systems*

[From ref. 137]

Ceramic	Metal member	Braze alloy, weight percent	Brazing temperature, °F	Alkali metal	Key	Room temperature flexural strength, psi			Alkali metal exposed leak test[e]	Remarks room temperature flexural strength and leak tests
						As brazed	Vacuum exposed 500 hr, 1000° F	Alkali metal exposed 500 hr, 1000° F		
Thermalox 998, beryllia 99.8 percent, BeO.	Columbium D-43	75Zr-19Cb-6Be	1940	K[a]	$\bar{x}$ s n	14 343 3740 7	14 450 [d]250 2	14 340 [d]490 2	2/2 VT	Very good results.
Thermalox 998, beryllia 99.8 percent, BeO.	Columbium D-43	48Ti-48Zr-4Be	1940	K[a]	$\bar{x}$ s n	14 130 4090 4	Not tested	9710 [d]1940 2	2/2 VT	Fair results.
Thermalox 998, beryllia 99.8 percent, BeO.	Cb-1Zr	56Zr-28V-16Ti	2270	Li[c]	$\bar{x}$ s n	13 503 2870 6	12 895 [d]555 2	[f]12 000 1	4/4 VT	Good results.
Ei3-3W, alumina 99.7 Al₂O₃.	Cb-1Zr	75Zr-19Cb-6Be	1940	K[a]	$\bar{x}$ s n	25 655 7370 11	21 100 [d]3300 2	21 432 2250 3	3/3 VT	Good results.
Ei3-3W, alumina 99.7 percent Al₂O₃.	Cb-1Zr	48Ti-48Zr-4Be	1940	K[a]	$\bar{x}$ s n	23 342 4690 12	16 335 2140 4	6120 3500 3	4/4 VT	Fair results.
Ei3-3W, alumina 99.7 percent Al₂O₃.	Cb-1Zr	75Zr-19Cb-6Be	1940	NaK[b]	$\bar{x}$ s n	25 655 7370 11	22 100 [d]3300 2	9587 1320 4	4/4 VT	Fair results.
Ei3-3W, alumina 99.7 percent Al₂O₃.	Cb-1Zr	48Ti-48Zr-4Be	1940	NaK[b]	$\bar{x}$ s n	23 342 4690 12	16 335 2140 4	10 390 1820 4	4/4 VT	Fair results.

[a] Oxygen level in associated purity test capsule was less than 10 ppm.
[b] Oxygen levels in two associated purity test capsules were less than 10 ppm.
[c] Associated purity test capsule leaked; no meaningful determination.
[d] Standard deviation has little significance for a sample size of two pieces.
[e] VT indicates helium leak rate of less than 1×10^{-9} torr-liter/second.
[f] One sample only.

NOTE: All tests on modulus-of-rupture assemblies (MoR) using columbium–1 percent zirconium metal member.
KEY:
$\bar{x}$—arithmetic mean.
s—standard deviation.
n—number of specimens tested.

TABLE 23.—*Best Ceramic-to-Metal Sealing System Tested in Each Alkali Metal-Temperature Environment*

[From ref. 137]

Environment	Ceramic	Brazing alloy	Average room temperature flexural strength,[b] psi
Potassium 1600° F	Thermalox 998, beryllia 99.8 percent BeO.	56Zr–28V–16Ti	11 810
Potassium 1000° F	Ei3-3W, alumina 99.7 percent Al₂O₃.	75Zr–19Cb–6Be	21 432
NaK 1000° F [a]	Ei3-3W, alumina 99.7 percent Al₂O₃.	48Zr–48Ti–4Be	10 390
Lithium 1000° F	Thermalox 998, beryllia 99.8 percent BeO.	56Zr–28V–16Ti	>12 000

[a] Marginal usefulness after 500 hours.

[b] Postexposure flexural strength.

NOTE: All seal systems were made with 0.015-inch-thick columbium–1 percent zirconium alloy metal member.

tween metals and metallized ceramic bodies, engineers at the Los Alamos Scientific Laboratory used active metal alloys to braze unmetallized high-alumina ceramics to Nb–1Zr (ref. 131). Using the joint design shown earlier in figure 17(c), seals were made in a cold-wall resistance furnace at 10^{-4} torr with 58V–32Nb–10Ti, 57Ti–19Nb–12V–9.5Cr–2.5Al, and 99.5Zr–0.5Ni. The results of these tests are shown in table 25. Research to join alumina to Nb–1Zr, conducted for the Los Alamos Scientific Laboratory by the Pyromet Company, has been reviewed by Kirby and LaMotte (ref. 132). Several filler metals were evaluated for joining metallized and unmetallized alumina sleeves to an Nb–1Zr base section; subsequently, research was concentrated on 62Ti–4Zr–8Mo–26Fe and 48Ti–48Zr–4Be as filler metals for joining uncoated alumina (Al_2O_3–$0.5Y_2O_3$) to Nb–1Zr. The joint design was similar to that shown in figure 17(c). The results of the brazing tests are shown in table 26. In 1962, Bristow (General Electric Company) reported on an extensive investigation of the active metal process to fabricate ceramic-to-ceramic and ceramic-to-metal joints for service in a cesium vapor environment up to about 900° C (1652°F) (ref. 139). All joints were made with titanium-nickel filler metals. To make ceramic-to-ceramic joints, two ceramic cylinders (0.690 inch O.D., 0.480 I.D., and 0.200 inch long) were brazed together; joints were made by brazing two of the ceramic cylinders to either side of a 0.010-inch-thick metal washer. Of the two ceramics used during these studies, one was a silica-free, high-purity alumina body suitable for a cesium vapor environment and the other a 97-percent alumina ceramic containing 3 percent CaO, MgO, and SiO_2 as fluxing oxides. The metals studied included titanium, nickel, tantalum, Kovar, and Types 304 and 430 stainless steel. Nickel and titanium shim stock of various thicknesses was used as the filler metal. The alloy composition was controlled by stacking up nickel and titanium washers of different thicknesses and placing them in the joint area. For example, an alloy with the composition 71.8Ti–28.2Ni was produced by stacking a washer of 0.001-inch-thick titanium between two washers of 0.0003-inch-thick nickel; the composition of this filler metal is very similar to that of the nickel-titanium eutectic, 71.5Ti–28.5Ni. This investigation emphasized the microstructures that form during brazing, their properties, their behavior in high-temperature environments, and the means that can be used to alter their com-

TABLE 24.—*Summary of Brazing Data for Selecting Brazing Alloys to Join Thermalox 998 Beryllia Ceramic to Columbium–1 Percent Zirconium Metal*

[From ref. 138]

Braze alloy number	Nominal alloy composition, weight percent	Brazing conditions		Strength of brazed specimens								Leak test [e]
		Time, minutes	Temperature, °F	Modulus of rupture				Tab peel strength				
				Number of specimens	Remarks	Average strength, psi	Standard deviation,[d] psi	Number of specimens	Remarks	Average strength, pounds/inch	Standard deviation,[d] pounds/inch	
2	68Ti–28V–4Be	5	2372	14	(a)	17 465	1900	6	(c)	9	2	6/8 VT
5	46Ti–46Zr–4Be–4V	5	1832	17	(a)	16 800	3220	8	(a)	19	17	5/6 VT
6	50Zr–30V–20Cb	5	2516	6	(a)	15 150	2160	4	(b)	21	3	2/2 VT
9	60Zr–25V–15Cb	5	2436	12	(a)	15 000	875	6	(b)	34	9	6/6 VT
10	50Zr–30Ti–20V	10	2732	4	(b)	15 225	3140	2	(b)	21	[f] ±1	0/2 VT
12	35Ti–35V–30Zr	1	2804	3	(a)	24 433	1380	2	(c)	16	[f] ±0	2/2 VT

[a] Good wetting, good fillet.
[b] Fair wetting, incomplete fillet.
[c] Poor wetting, incomplete, granular fillet.

[d] $$S = \sqrt{\frac{\sum (x - \bar{w})^2}{(n - l)}}$$

[e] VT indicates a leak rate $< 1 \times 10^{-9}$ std. cc/sec., 2/3 indicates 2 of 3 samples vacuum tight, etc.
[f] When there were less than three specimens, the standard deviation was not determined; the value shown indicates the spread of values from the average value.

NOTE: Ceramic parts were Brush Beryllia's 99.8 percent BeO body containing 70 ppm SiO_2 (Lot No. 1) or 150 ppm SiO_2 (Lot No. 2). The parts were fabricated by dry-pressing slabs, firing in electric kiln, and subsequent cutting and grinding to shape. The ceramic bar size was 0.1 by 0.1 by 1 inch. Density was between 2.85 and 2.94 g/cc, as specified by vendor. Average postbraze modulus-of-rupture strength of the ceramic was 25 000 psi. Ten bars from above assemblies were tested. The standard deviation was 3230 psi. The metal member was 0.015-inch-thick Cb–1 percent Zr sheet.

TABLE 25.—*Evaluation of Brazing Alloys With Unmetallized Alumina Ceramics*

[From ref. 131]

Brazing alloy composition, w/o	Ceramic type [a]	Brazing temp., °C	Remarks
V–32Co–10Ti	Al–300	1800	Good wetting of ceramic and metal. Limited alloying with metal (0.003 in.) and limited intergranular attack of ceramic.
V–32Co–10Ti	Lucalox	1800	Fair wetting of ceramic, good wetting of metal.
Ti–19Cb–12V–9.5Cr–2.5Al.	Al–14	1700	Good wetting of metal, extensive intergranular attack of ceramic.
Ti–19Cb–12V–9.5Cr–2.5Al.	Lucalox	1700	Good wetting of metal, fair wetting of ceramic.
Zr–0.5Ni	Al–14 and lucalox	1830	Good wetting of metal, but severe attack of the ceramics.
Co	Al–14	1540	Extreme alloying with the metal. Good wetting of metallized ceramic.

[a]	Ceramic type	Density, percent theo	Composition
	Al–300	95	97.6 w/o Al_2O_3
	Al–14 (pressed)	93	Al_2O_3–0.5 w/o Y_2O_3
	Al–14 (slipcast)	96	Al_2O_3–0.5 w/o Y_2O_3
	Lucalox	98	Al_2O_3–0.3 w/o MgO

TABLE 26.—*Evaluation of Active Metal Brazing Alloys With Unmetallized Alumina (Al_2O_3–$0.5Y_2O_3$)*

[From ref. 132]

Spec. number	Filler alloy	Method of placement	Brazing temperature °F	°C	Holding time, min	Results Visual	Leak test
1	Ti–4 Zr–8 Mo–26 Fe	Organic binder.	2280	1249	5	Fillet 90 percent complete.	Gross leak.
2	Ti–4 Zr–8 Mo–26 Fe	Organic binder.	2440	1338	None	Fillet 50 percent complete.	Gross leak.
3	Ti–4 Zr–8 Mo–26 Fe	Organic binder.	2300	1260	1	Fillet 50 percent complete.	Gross leak.
4	Ti–4 Zr–8 Mo–26 Fe	Organic binder.	2290	1254	5	Fillet 100 percent complete.	Gross leak.
5	Ti–4 Zr–8 Mo–26 Fe	Organic binder.	2280	1249	5	Fillet 100 percent complete.	Gross leak.
6	Ti–4 Zr–8 Mo–26 Fe	Organic binder.	2280	1249	5	Fillet 100 percent complete.	Gross leak.
			Rebrazed at 2290.	1254	2	Fillet 100 percent complete.	Gross leak.
7	Ti–48 Zr–4 Be	Organic binder.	2000	1093	5	No fillet	Gross leak.
8	Ti–48 Zr–4 Be	Organic binder.	1970	1077	5	No fillet	Gross leak.
			Rebrazed at 2000.	1093	5	Fillet 90 percent complete.	Leaktight.

TABLE 26.—*Evaluation of Active Metal Brazing Alloys With Unmetallized Alumina (Al_2O_3–$0.5Y_2O_2$)*— Continued

Spec. number	Filler alloy	Method of placement	Brazing temperature		Holding time, min	Results	
			°F	°C		Visual	Leak test
9	Ti–48 Zr–4 Be	Organic binder.	1970	1079	7	No fillet	Gross leak.
			Rebrazed at 2350.	1288	5	Fillet 90 percent complete.	Leaked at 5×10^{-5} cc/sec.
10	Ti–4 Zr–8 Mo–26Fe	Organic binder.	1950	1065	5	Fillet 70 percent complete.	Gross leak.
11	Ti–48 Zr–4 Be	Organic binder.	2140	1171	5	Fillet 50 percent complete.	Leaked at 1×10^{-7} cc/sec.
			Rebrazed at 2350.	1288	5	Fillet 50 percent complete.	Leaked at 1×10^{-7} cc/sec.
12	Ti–48 Zr–4 Be	Zr foil ledge.	1950	1065	5	Fillet 80 percent complete.	Gross leak.
13	Ti–48 Zr–4 Be	Zr foil ledge.	2100	1149	None	Fillet 60 percent complete.	Leaked at 1×10^{-9} cc/sec.
14	Ti–4 Zr–8 Mo–26 Fe	Zr foil ledge.	2290	1254	15	Fillet 80 percent complete.	Leaked at 1×10^{-6} cc/sec.
15	Ti–48 Zr–4 Be	Zr foil ledge.	2300	1260	None	Fillet 50 percent complete.	Ceramic cracked.
16	Ti–48 Zr–4 Be	Zr foil ledge.	2100	1149	15	Fillet 30 percent complete.	Leaked at 5×10^{-7} cc/sec.

position. A brittle intermetallic compound Ti_2Ni forms when the eutectic composition is exceeded toward the nickel-rich side of the nickel-titanium phase diagram. Wisser and Hagadorn have studied nickel-titanium microstructures in brazed ceramic-to-ceramic joints and noted the presence of small cracks in an alloy containing 32.1 percent nickel and large cracks in an alloy containing 35 percent nickel (ref. 140). On the basis of these studies, Bristow concluded that:

1. Seals or sealing alloys containing a titanium phase in contact with alumina were unsuitable for long-time service at 700° C (1292° F) or for short-time service at 900° C (1652° F) because the active metal alloy continued to react with the ceramic at the service temperature and became embrittled.

2. Seals to thick titanium members minimized the formation of Ti_2Ni.

3. The ceramic composition had no apparent effect on the reaction rate between the ceramic and an alloy containing a titanium phase at 900° C (1652° F). At 700° C (1292° F), the reaction rate appeared lower with the high-purity alumina ceramic than with the ceramic containing a glassy phase.

4. Service lives up to 2000 hours were recorded at 700° C (1292° F) for alumina-to-titanium and alumina-to-tantalum seals.

5. At 900° C (1652° F), the only seals that had lives exceeding a few hundred hours were

alumina-to-nickel joints made with titanium shim stock under conditions that precluded the formation of α-titanium or Ti_2N.

During further research on the active metal process, conducted by Bristow, Grossman, and Kaznoff (ref. 11), seals were made between a high-purity alumina body and nickel, molybdenum, tantalum, and niobium. The alumina-to-niobium seals exhibited the best high-temperature properties, being vacuum-tight after 2000 hours exposure at 900° C (1652° F) and after 240 hours exposure at 1075° C (1967° F). The alumina-to-molybdenum joints also had good high-temperature behavior; however, poor properties were observed with the alumina-to-nickel and alumina-to-tantalum joints.

CHAPTER 6

Joining Graphite

PROPERTIES AND USES OF GRAPHITE

The industrial uses of graphite seem almost limitless. Because of its high melting (sublimation) temperature and resistance to thermal shock, much of the graphite produced in the United States is used for electrodes in electric steel-making furnaces and for anodes in electrolytic processing equipment. Graphite is also used for anodes in high-power electronic tubes, for brushes in rotating electrical machinery, as a mold material for casting, as a refractory for lining furnaces, and as a constituent in powder metallurgy products. These applications require little or no graphite-to-metal joining.

Graphite also has become exceedingly important, however, in nuclear and aerospace applications that require graphite to be joined to itself and metallic components. Because of its low absorption cross section for fast neutrons, graphite is used in nuclear reactors as a material for moderators, reflectors, and thermal columns. Also, composites of graphite and materials with high neutron cross sections can be used for control rods and shielding.

Because graphite has exceptional resistance to thermal shock and a favorable strength-to-weight ratio at high temperatures, it is used widely as a nozzle material for solid-fuel rocket engines. The rocket exhaust environment is characterized by temperatures of 600° F (315° C) or higher, supersonic gas flow, reactive and abrasive combustion products, and steep temperature gradients. For such an application, the ideal structural material should have the following properties: low specific weight; high melting temperature; good thermal conductivity; superior resistance to thermal shock; resistance to oxidation, corrosion, and erosion; and adequate strength and ductility over the service temperature range. While graphite does not satisfy all of these requirements, it does possess many of the desired properties. For some rocket nozzle applications, the graphite nozzle is backed up by a metal liner. If the firing time is long enough to cause erosion of the graphite, a tungsten nozzle backed up by a graphite liner is used. In either case, a sound graphite-to-metal joint is required to obtain the desired performance. Graphite has also been used in the fabrication of jet vanes for controlling the yaw and pitch attitudes of rockets.

Selected properties of graphite are shown in table 27; the properties of metals used for high-temperature service are included for comparison (refs. 29 and 46). Not only do the physical and mechanical properties of graphite vary widely in accordance with the method used to produce the graphite and the type and purity of the starting materials, but the requirements of industry vary also. For example, nuclear reactor grade graphite used as a moderator must have a high degree of purity, because impure elements affect the magnitude of the absorption cross section. The mechanical properties of graphite are here of secondary importance, but they are of primary importance when graphite is used as a structural material in the production of aerospace hardware.

Significant progress in producing graphite with consistent properties has been made in response to the requirements of the aerospace and

TABLE 27.—*Selected Properties of Some Structural Materials*

[From refs. 29 and 46]

Metal	Density, g/cc at 20° C	Melting point, °C	Thermal conductivity, cal/cm /°C/cm	Coefficient of linear thermal expansion, cm/cm/°C $\times 10^{-7}$	Vapor pressure, torr 1.316×10^{-3} atm	Young's modulus 20° C, psi
Nickel	8.90	1453	0.198/100° C 0.175/200° C 0.152/300° C 0.142/400° C 0.148/500° C	133.0/25 to 100° C 144.0/25 to 300° C 155.0/25 to 600° C 163.0/25 to 900° C	10^{-8}/912° C 10^{-4}/1247° C 10^{-2}/1407° C 10^{-1}/1667° C	30×10^{-6}.
Iron	7.87	1537	0.163/100° C 0.147/200° C 0.081/800° C	121.0/20 to 100° C 134.0/20 to 300° C 147.0/20 to 600° C 150.0/20 to 900° C	10^{-8}/877° C 10^{-4}/1207° C 10^{-2}/1437° C 10^{-1}/1637° C	28.5×10^{-6}.
Columbium	8.57	2468 ±10	0.125/0° C 0.135/200° C 0.145/400° C 0.156/600° C	73.9/0 to 400° C 75.6/0 to 600° C 77.2/0 to 800° C 78.8/0 to 1000° C	2.00×10^{-4}/2031° C 1.85×10^{-1}/2323° C	12.4×10^{-8}.
Molybdenum	10.22	2620 ±10	0.298/204° C 0.289/427° C 0.272/649° C 0.254/871° C 0.215/1649° C 0.206/2204° C	54.3/20 to 149° C 51.9/20 to 482° C 53.6/20 to 649° C 58.0/20 to 982° C 62.8/20 to 1316° C 66.5/20 to 1593° C	10^{-8}/1582° C 10^{-4}/2167° C 10^{-2}/2627° C 10^{-1}/2927° C 1/3297° C	47×10^{-6}.
Tantalum	16.6	2996	0.130/20° C 0.162/568° C 0.171/828° C 0.179/1106° C 0.188/1416° C	65.0/0 to 1000° C 66.0/20 to 500° C 73.0/27 to 1400° C 78.0/27 to 2400° C	10^{-8}/1957° C 10^{-4}/2587° C 10^{-2}/3067° C 10^{-1}/3372° C 1/3737° C	27×10^{-6}.
Tungsten	19.3	3395 ±15	0.31/20° C 0.275/927° C 0.268/1127° C 0.260/1327° C 0.253/1527° C 0.245/1727° C	44.4/27° C 51.9/1027° C 72.6/2027° C	10^{-8}/2077° C 10^{-4}/2957° C 10^{-2}/3297° C 10^{-1}/3642° C	50×10^{-6}.
Graphite	1.7	3652 to 3697	0.02 to 0.5/20° C	10.0 to 50.0/20° C (varies widely).	10^{-8}/1950° C 10^{-4}/2380° C 10^{-2}/2700° C 10^{-1}/2900° C 10^{0}/3140° C 10^{2}/3800° C	0.8 to 1.4×10^{-6} (varies widely).

nuclear industries. Among the most important advances is the development of pyrolytic graphite, which consists of highly oriented planes of graphite molecules closely stacked in a laminated structure as shown in figure 18 (ref. 140). The bonding forces between atoms in a given plane (a and b direction) are much stronger than the bonding forces between planes. As a result, the properties of pyrolytic graphite are highly directional as shown in table 28 and figure 19. These differences in properties must be considered in designing a structure that incorporates graphite; similarly, the dependence of properties on the crystal orientation of graph-

ite has a profound effect on the design of a graphite-to-metal joint.

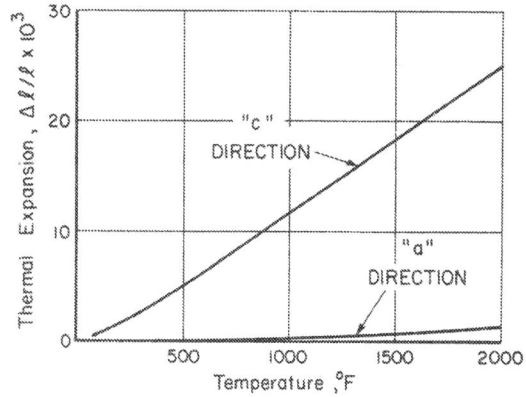

FIGURE 19.—Thermal expansion characteristics of pyrolytic graphite (ref. 140)

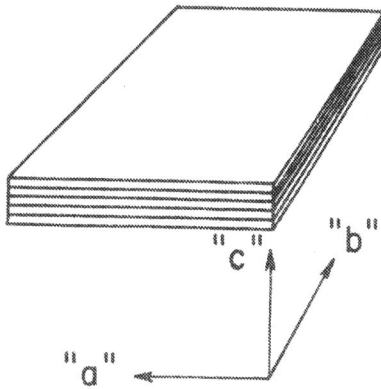

FIGURE 18.—Schematic representation of pyrolytic graphite structure (ref. 140)

TABLE 28.—*Typical Room-Temperature Properties of Pyrolytic Graphite*

[From ref. 141]

	Pyrolytic graphite	
	a or b direction	c direction
Tensile strength, psi	15 000	750
Young's modulus, psi $\times 10^6$	4	
Thermal conductivity (1000° F), Btu–ft/ft² sec F$\times 10^4$	280 to 370	0.8 to 2.9
Specific heat	0.18	0.18
Specific gravity	2.21	2.21

JOINING CONSIDERATIONS

Graphite-to-metal joints are difficult to design and produce because graphite is not wet readily by most conventional filler metals, the physical and mechanical properties of graphite differ significantly from those of most structural metals, and only a few processes can be used for joining. The problems of joining graphite are very similar to those of joining ceramics to other materials, and the same care must be observed in selecting the filler metal, joining process, and joint design.

Most filler metals do not wet graphite well and do not exhibit their usual excellent flow properties. Filler metals used to join graphite to graphite or graphite to metals should contain constituents that have a strong tendency to form carbides, since the bonding mechanism depends on carbide formation. Thus, filler metals containing substantial amounts of titanium, zirconium, chromium, silicon, or niobium are most successful; however, they are not commercially available, although several experimental alloys have been patented.

The differences between the linear thermal-expansion coefficients of graphite, the filler metal, and the base metal cause the difficulty in designing and producing graphite-to-metal joints; graphite, in comparison with metallic elements, has a small expansion coefficient. The graphite and metal workpieces expand and contract at different rates during the heating and cooling portions of the joining cycle; as a result, stresses that can cause cracking are established in the joint area. Graphite-to-metal joints must be carefully designed to minimize such stresses. Because of graphite's low tensile strength, the joint should be designed so that the graphite workpiece is compressed. Particular care must be observed when joining pyrolytic graphite, because the expansion coefficient varies widely as a function of crystal orientation.

FUSION JOINING OF GRAPHITE

The methods that can be used to join graphite to itself and to metals are limited to the nonfusion joining techniques, i.e., solid-phase joining and liquid-solid-phase joining. Most research has been concentrated on the diffusion welding and brazing processes. Kareta and Nefedov reported on experimental studies to arc-weld graphite-to-graphite joints (ref. 142). The joints were produced with a straight-polarity, direct-current arc; a consumable graphite electrode was used as the filler metal. Although a joint was produced, its strength was too low for practical use. No reports on arc-welding graphite to metal were found.

NONFUSION JOINING OF GRAPHITE

Solid-Phase Joining

Only limited research has been conducted on the solid-phase or diffusion welding of graphite to itself or to metals, because satisfactory joints cannot be produced without a diffusion aid. Kareta and Nefedov reported that anode-grade, electrode-grade, and pyrolytic graphite plates were diffusion-welded in an inert atmosphere using titanium, zirconium, niobium, tantalum, or hafnium foil inserts between the graphite workpieces (ref. 142). The joining conditions were as follows: temperature 2300° to 3000° C (4172° to 5432° F); pressure 1422 psi; and time 3 to 6 minutes. The tensile strength of joints made with a 0.004-inch-thick zirconium insert varied from 2247 psi for anode- and electrode-grade graphite to 7082 psi for pyrolytic graphite in the a or b direction.

According to a British patent, graphite has been diffusion-welded to mild steel in a pure carbon dioxide atmosphere (ref. 143). The workpieces were bonded at 4480 psi for 200 hours; the bonding temperature was not given.

Solid-phase joining was investigated in a program directed toward joining tungsten to graphite for aerospace applications at temperatures up to 5000° F (2760° C) (ref. 144). Various materials were used to promote diffusion. This research is summarized below:

1. For joining graphite to tungsten in a vacuum at 3600° F (1982° C), the joint interface was coated with a mixture of molybdenum, carbon, and ruthenium powders to promote bonding; however, little or no bonding occurred.

2. A titanium-carbide layer was produced on a graphite substrate by coating the substrate with titanium hydride and firing the specimen at 4700° F (2593° C). A similar procedure was used to produce a titanium coating on the tungsten substrate. Then, the coated surfaces were butted together and the assembly diffusion-welded at 4700° F (2593° C) in a vacuum. Joining did not occur. Attempts to join graphite coated with titanium carbide to tungsten coated with colloidal graphite were also unsuccessful.

3. The use of a titanium foil insert between the graphite and tungsten surfaces was also unsuccessful. The solid-phase joining approach was eliminated during this program, because a technique involving the formation of a liquid phase during bonding appeared more successful. (This work will be discussed in the next section.)

Bondarev reported that a variation of diffusion-welding—eutectic-diffusion welding—produced sound joints between graphite and several refractory metals (ref. 145). A titanium foil insert, 0.004 inch thick, was electroplated with a 0.0008- to 0.001-inch-thick layer of copper and placed between the workpieces. Joints between graphite and molybdenum, niobium, or tantalum were produced in a vacuum diffusion-welding unit under the following conditions: temperature 890° to 1050° C (1634° to 1922° F); pressure 43 to 100 psi; time 5 to 10 minutes. An eutectic occurs in the Ti–Cu system at 885° C (1607° F) where the weight percent of copper is 72. Under the welding conditions cited above, the Ti–Cu eutectic alloy formed and wet the graphite and the refractory metal. The maximum joint strength occurred when the welding temperature was between 960° and 980° C (1760° and 1786° F). The graphite-to-tantalum joint had the highest strength 1735 psi; the strengths of the graphite-to-molybdenum and graphite-to-niobium joints were slightly lower.

Eutectic-diffusion welding was also used to join graphite to Zircaloy-2 sheet stock, 0.035 inch thick, as a means of stabilizing advanced sodium graphite reactor (ASGR) moderator cans (ref. 146). After depositing a controlled amount of copper on the metal surface by plasma-arc spraying, the graphite and copper-coated Zircaloy-2 parts were held in intimate contact in a vacuum furnace and heated to a temperature somewhat higher than the temperature at which the zirconium-copper eutectic forms. The joining temperature of 1800° F (982° C) ensured rapid and complete formation of the Zr–Cu eutectic when using the recommended thickness of copper (0.001 inch).

Liquid-Solid-Phase Joining

Brazing, a liquid-solid-phase joining method, has proved to be the most suitable process for joining graphite to itself and metals. Most of the research on the development of filler metals and brazing procedures discussed in the following sections has been conducted by firms and Government agencies active in the nuclear and aerospace fields.

Graphite-to-Graphite Joining

Graphite-to-graphite joints using zirconium as a filler metal for nuclear applications have been investigated by Burnett and Marengo (ref. 147). Zirconium wets graphite well, but a layer of zirconium carbide is formed during brazing. At high operating temperatures, differential thermal expansion between the graphite and the carbide causes failure because of the brittleness of the zirconium carbide. Molybdenum disilicide has also been investigated for use in producing graphite-to-graphite joints (ref. 148). Little or no penetration of the graphite occurred, and thermal cycling tests indicated

suitability of these joints for some nuclear applications.

In another program Lindgren investigated the use of titanium, molybdenum, molybdenum disilicide, silicon, zirconium, and an experimental Ni–Cu–Mo alloy as filler metals for brazing graphite and caps to a graphite sleeve (ref. 149). The most satisfactory joints were produced with silicon or zirconium.

The Japanese Atomic Energy Research Institute has been active in developing methods to join graphite to graphite with a brazing filler metal containing (by weight percent) 30 to 50 Ni, 0 to 20 Ti, and Bal. Fe (refs. 150 and 151). This alloy, used also to produce graphite-to-metal joints, is based on Invar (Fe–36Ni), which has a small expansion coefficient comparable to that of graphite; titanium was added to improve the wetting properties of this alloy. Graphite-to-graphite joints have been evaluated by sustained high-temperature tests at 600° C (1112° F); the joints did not appear to be brittle.

Davidson and Ryde received a patent on a method to join graphite to itself or to metals (ref. 152). Nickel is deposited on the graphite surfaces when the graphite workpieces are heated to 200° to 500° C (392° to 932° F) in an atmosphere of nickel carbonyl; then, the parts are joined with conventional brazing alloys.

Several additional filler metals, developed primarily to braze graphite-to-metal joints but also used to produce graphite-to-graphite joints, are discussed in the section below.

Graphite-to-Metal Joining

The most extensive research on brazing graphite to itself and metals has been conducted by the Oak Ridge National Laboratory (ORNL) in connection with the molten salt reactor program for the last 7 or 8 years (refs. 12, 153, and 154). Initially, the wetting characteristics of selected commercial and experimental filler metals on graphite were determined; later, filler metals were developed that would wet graphite readily and be resistant to corrosion by molten fluoride salts. Commercial filler metals used to braze stainless steels and nickel-base alloys were screened early in the program.

Neither the nickel-base filler metals nor the nickel-gold eutectic alloy wet graphite; wetting was achieved with a titanium-cored silver-copper alloy that has been used on other occasions to braze graphite to metal. Two experimental filler metals—48Ti–48Zr–4Be and 49Ti–49Cu–2Be—also wet graphite well (ref. 12). Experimental alloys based on the nickel-gold alloy system were then prepared and evaluated. (This alloy system was selected because the nickel-gold eutectic, Au–18Ni, was ductile and very resistant to corrosion by molten salts.) Tantalum or molybdenum were added to the basic alloys; these metals are strong carbide formers and, unlike zirconium and titanium, possess the required corrosion resistance. Numerous ternary alloys in the nickel-gold-tantalum (Ni–Au–Ta) and nickel-gold-molybdenum (Ni–Au–Mo) alloy systems were prepared and evaluated by wetting tests conducted at 1300° C (2372° F). In the Ni–Au–Ta alloy system, 60Au–10Ni–30Ta proved to be most effective in brazing graphite-to-graphite and graphite-to-molybdenum joints; however, this alloy had limited ductility and displayed fillet cracking. Filler metals that contained less tantalum were suitable for brazing graphite-to-molybdenum but not graphite-to-graphite joints. More promising filler metals were discovered in the Ni–Au–Mo alloy system. The 35Au–35Ni–30Mo alloy wet graphite readily and possessed excellent flow properties; it could be used to join graphite to itself or to molybdenum. Similar alloys containing lower concentrations of molybdenum were quite suitable for joining graphite to metal; an alloy containing 15 percent molybdenum was more ductile than the alloy mentioned above. The 60Au–10Ni–30Ta and 35Au–35Ni–30Mo alloys were used to fabricate leak test specimens that consisted of a graphite tube with a brazed molybdenum cap at each end (ref. 154).

Recent research at ORNL has been concerned with joining graphite to INOR–8 (Hastelloy N), a nickel-base alloy with the composition 71Ni–17Mo–7Cr–5Fe (refs. 155, 156). The large difference between the expansion coefficients of graphite and INOR–8 results in cracking of the brazed graphite-to-INOR–8 joints upon cooling from the brazing temperature. Insert materials, placed between the faying surfaces of the graphite and INOR–8 parts, minimize this problem; the expansion coefficient of the insert material should be between that of graphite and INOR–8. To investigate the suitability of refractory metals for this application, joints between the following metal combinations were brazed with a nickel-base alloy and the gold-nickel eutectic: INOR–8 to tungsten, INOR–8 to molybdenum, molybdenum to niobium, and tungsten to niobium. It was found that crack-free joints could be made with a ductile filler metal such as Au–18Ni, but not with the brittle nickel-base alloy. Joints between graphite and the refractory metals were also brazed with 70Au–20Ni–10Mo (very ductile) and 35Au–35Ni–30Mo (slightly ductile); with these joints, the ductility of the filler metal had little influence on cracking. Crack-free joints were produced between graphite and molybdenum and graphite and tungsten; the expansion coefficients of these metals are similar to that of graphite. Cracking was observed when graphite was brazed to niobium or tantalum, which have significantly larger expansion coefficients than graphite. On the basis of these studies, further work to join graphite to INOR–8 using molybdenum or tungsten as an intermediate material is currently underway.

In another investigation directed toward nuclear reactor applications, graphite was brazed to a structural alloy, Nilo–K; leaktight specimens were prepared to determine the permeability of graphite at high temperatures (900° C or 1652° F) and high pressures (ref. 157). Satisfactory tube-to-end cap joints were brazed with a titanium-cored silver-copper eutectic alloy. The graphite and Nilo–K workpieces were assembled and brazed in an evacuated silver envelope.

Several investigations to develop procedures to braze graphite to other materials have been associated with aerospace applications. In 1962, Ikeuye and Grow reported on developing filler metals to braze pyrolytic graphite to beryllium oxide (ref. 141). The three titanium-base alloys—93Ti–7Ni, 93Ti–7Cr, and 53Ti–47Cr—developed and evaluated during this program wet the graphite and beryllium oxide base materials in a vacuum or inert-gas atmosphere.

Difficulties were experienced because of the differing expansion coefficients of pyrolytic graphite and beryllium oxide; special joint designs were prepared to minimize the stresses established during the brazing cycle (fig. 20).

Two types of graphite have been brazed to the niobium-base alloy Nb–1Zr or Type 316 stainless steel for a space radiator (refs. 158 and 159). The graphite-metal combinations, the compositions of the brazing alloys, and the brazing temperature are shown in table 29; brazing proceeded in a vacuum of 10^{-5} torr. The brazing alloys, developed by the Oak Ridge National Laboratory, were deposited on the metal sub-

strates by plasma-arc spraying procedures; about 50 percent of the titanium in the 66Ag–26Cu–8Ti alloy and 25 percent of the beryllium in the 48Zr–48Ti–4Be alloy were lost during spraying; the alloy compositions were adjusted to compensate for these losses. The thermal stability and joint integrity were evaluated by aging and thermal cycling tests conducted in a vacuum of 10^{-5} torr; the test results are summarized in table 30. In additional aging tests, conducted to determine the long-time stability of the brazed joints, graphite-to-Type 316 stainless steel joints, brazed with 66Ag–26Cu–8Ti, were encapsulated, evacuated to 10^{-3} torr or below, and aged at 1350° F (732° C) for 250, 1000, and 4000 hours. Graphite-to-alloy Nb–1Zr joints, brazed with 48Zr–48Ti–4Be, were aged under the same conditions at 1500° F (815° C) for similar periods. The results of the aging tests indicated that the Ag–Cu–Ti alloy was not compatible with Type 316 stainless steel at 1350° F (732° C) for periods longer than 1000 hours; no difficulties were experienced with the graphite-to-alloy Nb–1Zr joint after the 4000-hour aging period.

Tungsten-to-graphite joints for service at temperatures up to 5000° F (2760° C) have been studied at Narmco Research and Develop-

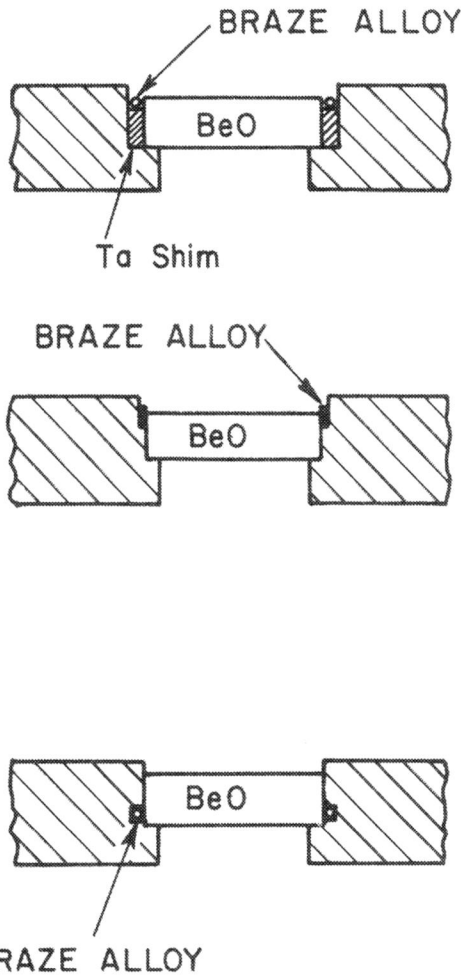

FIGURE 20.—"Frame" specimens modified to partially absorb BeO expansion as strain in the pyrolytic graphite (ref. 140)

TABLE 29.—*Graphite-to-Metal Joint Combinations*

[From ref. 159]

Material combinations	Brazing alloy, weight percent	Brazing temperature, °F
Graphite G to Cb–1Zr alloy.	66Ag–26Cu–8Ti.	1750.
	48Zr–48Ti–4Be.	1925.
	85Au–10Ni–5Fe.	2025.
Graphite G to Type 316 stainless steel.	Same as above.	Same as above.
Expanded pyrolytic graphite to Cb–1Zr alloy.	Same as above.	Same as above.
Expanded pyrolytic graphite to stainless steel Type 316.	Same as above.	Same as above.

TABLE 30.—*Thermal Stability Test Results of Graphite-to-Metal Joints*

[From ref. 159]

Material combination	Brazing alloy	Metallographic analyses	
		After aging for 500 hours at 1350° F	After 500 thermal cycles at 350° to 1350° F
316 S.S.–graphitite G	66Ag–26Cu–8Ti	Crack near braze	No cracking.
316 S.S.–expanded pyrographite.	66Ag–26Cu–8Ti	No cracking	No cracking.
Nb–1Zr–graphitite G	66Ag–26Cu–8Ti	No cracking	No cracking.
Nb–1Zr–expanded pyrographite.	66Ag–26Cu–8Ti	No cracking	No cracking.
316 S.S.–graphitite G	48Zr–48Ti–4Be	Crack near braze	Crack near braze.
316 S.S.–expanded pyrographite.	48Zr–48Ti–4Be	No cracking	No cracking.
Nb–1Zr–graphitite G	48Zr–48Ti–4Be	No cracking	No cracking.
Nb–1Zr–expanded pyrographite.	48Zr–48Ti–4Be	No cracking	No cracking.
316 S.S.–graphitite G	85Au–10Ni–5Fe	Crack near braze	Crack near braze.
316 S.S.–expanded pyrographite.	85Au–10Ni–5Fe	No cracking	No cracking.
Nb–1Zr–graphitite G	85Au–10Ni–5Fe	No cracking	No cracking.

ment (refs. 144 and 160). Much of the research was directed toward developing a suitable coating for the graphite and/or tungsten surfaces to enhance joining. The most successful coating, 60TaC–30WC–10ZrC, resulted from a mixture of tungsten and tantalum powders plus zirconium-hydride powder. The graphite surface was machined to a flatness of 200 microinch or less, coated with a wash coat of rhenium powder, painted with a slurry of tantalum, tungsten, and ZrH_2 powders, and dried in an oven at 150° F (65° C). The specimen was then fired at 5400° to 5500° F (2982° to 3038° C) in an argon atmosphere. The brazing alloy, 21W–79V, in the form of a mixture of the individual metal powders suspended in a suitable vehicle, was slurry-coated on the tungsten surface. After the coating was dry, the carbide-coated graphite and the filler-metal-coated tungsten parts were assembled and brazed in an argon atmosphere at 4200° to 4400° F (2315° to 2427° C); the parts were weighted to ensure contact during brazing. Some brazed specimens were also made with a 25W–75Re filler metal in the same manner as discussed

above. Thermal shock tests, the severest of which consisted of heating the specimen to a minimum temperature of 4425° F (2440° C) in slightly more than 10 seconds, followed by cooling to below red heat in 15 seconds, were conducted in the heat flux generated by a plasma-arc torch; no joint failures resulted. Joint tensile strengths of 530 psi at room temperature and 193 psi at 4000° F (2204° C) were obtained. In using the techniques discussed above to braze a simulated rocket nozzle configuration, some difficulty was experienced in maintaining good contact between the graphite backup structure and the tungsten liner; this problem was overcome by segmenting the graphite part.

Graphite was brazed to tungsten using pure titanium as the filler metal during an earlier investigation (ref. 161). Since titanium flowed sluggishly on tungsten and reacted quickly with graphite, it was necessary to preplace the filler metal in the joint and provide a reservoir of filler metal. The joints were brazed in an argon atmosphere at 3240° F (1782° C) minimum; the brazing time was minimized to retain maximum joint ductility.

References

1. PULFRICH, H.: Ceramic-to-Metal Seal. U.S. Patent 2 163 407, June 20, 1939.

2. PULFRICH, H.: Vacuum-Tight Seal. U.S. Patent 2 163 408, June 20, 1939.

3. PULFRICH, H.: Ceramic-to-Metal Seal. U.S. Patent 2 163 409, June 20, 1939.

4. PULFRICH, H.; AND MAGNER, R.: Ceramic-to-Metal Seal. U.S. Patent 2 163 410. June 20, 1939.

5. VATTER, H.: Ceramic Seals. German Patent 645 871, Apr. 1935.

6. SIEMENS; AND HALSKE, R. G.: Uniting Metal to Ceramic Materials. British Patent 475 878, Nov. 1937.

7. SIEMENS; AND HALSKE, R. G.: Producing Evacuated Vessels for Electrical Apparatus. British Patent 487 679, June 1938.

8. SIEMENS; AND HALSKE, R. G.: Evacuating and Sealing Vacuum Vessels. British Patent 492 480, Sept. 1938.

9. SIEMENS; AND HALSKE, R. G.: Tight Joints Between Ceramics and Metals. German Patent 734 274, Mar. 1943.

10. HOOP, J. G.: Generator Development, Spur Program, Part II, Generator Stator Bore-Seal. Westinghouse Elec. Corp., Tech. Rept. TDR–63–677–PT–2 (AF 33(657)–10922, AF 33(615)–1551), Feb. 1967.

11. BRISTOW, R. H.; GROSSMAN, L.; AND KAZNOFF, A. J.: Research and Development Program of Thermionic Conversion of Heat to Electricity. Vallecitos Atomic Lab. Gen. Elec. Co., vol. 1, Final Tech. Summary Rept. GEST–2035 (NOBs 88578), June 1964.

12. FOX, C. W.; AND SLAUGHTER, G. M.: Brazing of Ceramics. Welding J. vol. 43, No. 7, July 1964, pp. 591–597.

13. CHORNE, J.; BRUCH, C.; FEINGOLD, E.; SUTTON, W. H.: Development of Composite Structural Materials for High Temperature Applications. Gen. Elec. Co., Missile and Space Div., Final Rept. (NOW–65–0176), Nov. 1965.

14. BARR, F. A.; AND JONES, R. A.: Composite Ceramic Radome Manufacture by Mosaic Techniques. Whittaker Corp., Final Rept. AFML–TR–65–418, (AF 33/657/–10111), Jan. 1966.

15. HERRON, R. H.: Friction Materials—A New Field for Ceramics and Cermets. Am. Ceram. Soc. Bull., vol 34, No. 12, 1955, pp. 395–398.

16. SMALLEY, A. K.; RILEY, W. C.; AND DUCKWORTH, W. H.: Al_2O_3—Clad UO_2 Ceramics for Nuclear-Fuel Applications. Am. Ceram. Soc. Bull., vol. 39, No. 7, July 1960, pp. 359–361.

17. BLOCHER, J. M.; AND OXLEY, J. H.: Chemical Vapor Deposition Opens New Horizons in Ceramic Technology. Am. Ceram. Soc. Bull., vol. 41, No. 2, Feb. 1962, pp. 81–84.

18. THORNTON, H. R.: Bond Strength and Elastic Properties of Ceramic Adhesives. J. of the Am. Ceram. Soc., vol. 45, No. 5, May 1962, pp. 201–209.

19. BENZEL, J. F.: Ceramic-Metal Adhesive Combinations. Am. Ceram. Soc. Bull., vol. 46, No. 12, Dec. 1963, pp. 748–751.

20. BROWN, D. A.: Application of Ceramic Adhesive-Braze Alloy Combination Technique to the Bonding of Assemblies. Boeing Co., Military Aircraft Systems Div., Final Rept. ASD–TDR–62–631 (AF 33/616/–8002), June 1962.

21. CASTNER, S. V.; KALLUP, C., JR.; AND SKLAREW, S.: Application and Evaluation of Reinforced Refractory Ceramic Coatings. Marquardt Corp., Tech. Doc. Rept. ML–TDR–64–81 (AF 33/616/–8209), June 1964.

22. NIEHAUS, W. R.; SHROUT, J. E.; AND ANDERSON, R. G.: Structural Heat Shield for Re-entry and Hypersonic Lift Vehicles (High Temperature Composite Structure) Test Evaluation. Aeronca Manufacturing Corp., Final Rept. TDR–64–267–PT–2 (AF 33(657)–7151), Mar. 1966.

23. KUMMER, D. L.; ROSENTHAL, J. J.; AND LUM, D. W.: Shielded Ceramic Composite Structure. McDonnell Aircraft Corp., Final Rept. TR–65–331 (AF 33(657)–10996), Oct. 1965.

24. ANON.: Development of Ablative Thrust Chambers and Throat Inserts Suitable for Use on the Gemini and Apollo Vehicles. Thompson Ramo Wooldridge, Inc., Final Rept. NASA–CR–65055 (NAS9–1619), May 1965.

25. VATTER, H.: History of Ceramic-to-Metal Sealing Techniques. Vacuum-Technik, vol. 4, No. 2, 1956, pp. 180–192.

75

26. JENKINS, D. E.: Ceramic-to-Metal Sealing, Its Development and Use in the American Radio Valve Industry. Electronic Eng., vol. 27, No. 6, June 1955, pp. 290–294.

27. VAN HOUTEN, G. R.: A Survey of Ceramic-to-Metal Bonding. Ceram. Bull., vol. 38, No. 6, 1959, pp. 301–307.

28. CLARKE, J. F.; RITZ, J. W.; AND GIRARD, E. H.: State-of-the-Art Review of Ceramic-to-Metal Joining. Texas Instruments, Inc., Summary Tech. Rept. AFML TR 65–143 (AF 33(615)–1477), May 1965.

29. KOHL, W. H.: Handbook of Materials and Techniques for Vacuum Devices. Reinhold Publishing Corp., 1967.

30. PULFRICH, H.: The Technology of Metal-Ceramic Tubes. Bur. of Ships, Library Translation No. 410, Apr. 1951.

31. RIGTERINK, M. D.: Ceramic Electrical Insulating Materials. J. of the Am. Ceram. Soc., vol. 41, No. 11, 1958, pp. 501–506.

32. LaFORGE, H.: Ceramic Tube Structure Quality. Ceram. Age, vol. 63, No. 2, Feb. 1954, pp. 13–21.

33. COLE, S. S.; AND HYNES, F. J.: Some Parameters Affecting Ceramic-to-Metal Seal Strength of a High-Alumina Body. Am. Ceram. Soc. Bull., vol. 37, No. 3, 1958, pp. 135–138.

34. FLOYD, JAMES R.: Effect of Composition and Crystal Size of Alumina Ceramics on Metal-to-Ceramic Bond Strength. Am. Ceram. Soc. Bull., vol. 42, No. 2, Feb. 1963, pp. 65–70.

35. KINGERY, W. D.: Oxides for High-Temperature Applications. Proceedings of International Symposium on High-Temperature Technology. Oct. 6–9, 1959, McGraw-Hill Book Co., Inc., 1960, pp. 76–89.

36. McCLELLAND, J. D.: The Refractory Oxides. Ceram. News, Sept. 1963, pp. 16–17.

37. SCHNEIDER, S. J.: Compilation of the Melting Points of the Metal Oxides. National Bureau of Standards, Monograph 68, Oct. 1963.

38. RYSHKEWITCH, E.: Metal-Oxide Ceramics. International Science and Technology, No. 2, 1962, pp. 54–61.

39. JAFFEE, R. I.; AND MAYKUTH, D. J.: Refractory Materials. Battelle Memorial Institute, Defense Metals Information Center, Memo 44, 1960.

40. HAUCK, J. E.: Guide to Refractory Ceramics. Materials in Design Eng., vol. 58, No. 1, July 1963, pp. 85–96.

41. LITZ, L. M.: Graphite, Carbide, Nitride, and Sulfide Refractories. Proceedings of International Symposium on High-Temperature Technology, Oct. 6–9, 1959, McGraw-Hill Co., Inc., 1960, pp. 90–112.

42. HIESTER, N. K.; ED.: High-Temperature Technology. Proceedings of International Symposium, Sept. 8–11, 1965.

43. HAGUE, J. R.; LYNCH, J. F.; RUDNIK, A.; HOLDEN, F. C.; AND DUCKWORTH, W. H., EDS.: Refractory Ceramics for Aerospace, A Materials Selection Handbook. Battelle Memorial Institute, 1964.

44. ANON.: Metals Handbook. Eighth ed., Am. Soc. for Metals, 1961.

45. SMITHELLS, C. J.: Metals Reference Book. Third ed., Butterworth, Inc., 1962.

46. HAMPEL, C. A.: Rare Metals Handbook. Reinhold Publishing Corp., 1961.

47. PATTEE, H. E.; AND EVANS, R. M.: Brazing for High-Temperature Service. Battelle Memorial Institute, Defense Metals Information Center, Rept. No. 149, Feb. 21, 1961.

48. ANON.: AWS Committee on Brazing and Soldering: Brazing Handbook. Am. Welding Soc., Inc., 1962.

49. ANON.: Welding Handbook Committee: Welding Handbook. Fifth ed., sec. 3, Am. Welding Soc., Inc., 1965.

50. LARSEN, L. R.: When Designing With Ceramic Materials. Product Eng., vol. 33, No. 22, May 29, 1961, pp. 47–51.

51. KUTZER, L. G.: Joining Ceramics and Glass to Metals. Materials in Design Eng., vol. 61, No. 1, Jan. 1965, pp. 106–110.

52. JOHNSON, C. W.: Manual of Metal-to-Ceramic Sealing Techniques. Sperry Gyroscope Co., Electronic Tube Div., Sperry Pub. NA–27–0001 (AF 30 (602)–2371), May 1963.

53. CHEATHAM, E.; COLE, S. S., JR.; INGE, J. E.; LARISCH, H. W.; STYHR, K. H., JR.: Metal-to-Ceramic Seal Technology Study. Sperry Gyroscope Co., Electronic Tube Div., Final Tech. Rept. RADC–TR–60–236 (AF 30(602)–2047), Oct. 1960.

54. KESSLER, S. W.: Stress Analysis of Butt-Type Ceramic-to-Metal Seals. RCA Review, vol. 26, No. 6, 1965, pp. 262–275.

55. MARK, R.; AND LEWIN, G.: Stress Analysis in the Design of Ceramic Seals. Am. Ceram. Soc. Bull. vol. 44, No. 8, 1964, pp. 599–604.

56. JANSSEN, W. F.: Tailor-Made Precision Ceramic Parts. Materials and Methods, vol. 44, No. 9, 1956, pp. 107–109.

57. BIERLEIN, T. K.; NEWKIRK, H. W., JR.; AND MASTEL, B.: Etching of Refractories and Cermets by Ion Bombardment. J. of the Am. Ceram. Soc., vol. 41, No. 6, 1959, pp. 196–200.

58. NOLTE, H.; AND SPURCK, R.: Metal Ceramic Sealing With Manganese. Television Eng., vol. 1, No. 11, Nov. 1950, pp. 14–16, 18, 39.

59. NOLTE, H. J.: Metallized Ceramic. U.S. Patent 2 667 432, Jan. 26, 1954.

60. SPURCK, R. F.; DOOLITTLE, H. D.; ETTRE, K.; AND VARADI, P. F.: Use Metallizing Tape for High-Quality Ceramic-to-Metal Seals. Ceram. Industry, vol. 79, No. 3, 1962, pp. 88–91, 94.

61. ETTRE, K.: Automatic Metallizing. Ceram. Age, vol. 81, No. 6, June 1965, pp. 57–60.

62. KOHL, W. H.; AND RICE, P.: Electron Tubes for Critical Environments. Stanford Research Center, WADC Tech. Report 57–434 (AF 33(616)–3460), Mar. 1958.

63. LaFORGE, L. H.: Application of Ceramic Sections in High-Power Pulsed Klystrons. Am. Ceram. Soc. Bull., vol. 35 No. 3, Mar. 1956, pp. 117–122.

64. TENTARELLI, L. A.; WHITE, J. M.; AND BUCK, R. W.: Low-Temperature Refractory Metal-to-Ceramic Seals. Sperry Rand Corp., Final Rept. ECOM–03734–F, U.S. Army Electronics Command, DA–36–039–AMC–03734(E), Apr. 1966.

65. VARADI, P. F.; AND DOMINIQUEZ, R.: Tungsten Metalizing of Ceramics. Am. Ceram. Soc. Bull., vol. 45. No. 9, Sept. 1966, pp. 789–791.

66. KLOMP, J. T.; AND BOTDEN, T. P. J.: A New Sealing Method for High-Purity Alumina. Paper presented at British Ceram. Research Assoc. Symposium on the Use of Ceramics in Valves (Stoke-on-Trent, England), Mar. 29–30, 1965.

67. STRAUMANIS, M. E.; AND SCHLECHTEN, A. W.: Titanium Coatings on Metals and Ceramics. Metall., vol. 10, No. 10. 1956. pp. 901–909.

68. SIEBERT, M. E.; McKENNA, Q. H.; STEINBERG, M. A.; AND WAINER, E.: Electrolytic Reduction of Titanium Monoxide. J. of the Electrochem. Soc., vol. 102, No. 5, 1955, pp. 252–262.

69. QUINN, R. A.; AND KARLAK, R. F.: Method of Coating a Body with Titanium and Related Metals. U.S. Patent 3 022 201, Feb. 20, 1962.

70. REED, L.; AND McRAE, R. C.: Evaporated Metallizing on Ceramics. Am. Ceram. Soc. Bull. vol. 44 No. 1, Jan. 1965, pp. 12–13.

71. HOLMWOOD, R. A.; AND GLANG, R.: Vacuum-Deposited Molybdenum Films. J. of the Electrochem. Soc., vol. 112, No. 8, 1965, pp. 827–831.

72. HEIL, O.; AND IMOROZOVSKY, B.: Particle Bombardment Bonding and Welding Investigation. Heil Scientific Labs., Inc., Tech. Rept. ECOM–00429–F (DA–28–043–AMC–00429(E), Task IP6–22001–A–005–01–12), July 1967.

73. SEEMAN, J. M.: Ion-Sputtered Thin Films. Paper presented at the 6th Annual Structures and Materials Conference of the Am. Inst. of Aeron. and Astronaut. (Palm Springs, Calif.), Apr. 5–7, 1965.

74. MATTOX, D. M.: Metallizing Ceramics Using a Gas Discharge. J. of the Am. Ceram. Soc., vol. 48, No. 7, 1965, pp. 385–386.

75. LINDQUIST, C.: Here are Materials and Techniques for Coating Ceramics. Ceram. Industry, vol. 69, No. 1, 1957, pp. 85–88.

76. LINDQUIST, C.: How to Apply Conductive Coatings. Ceram. Industry, vol. 69, No. 2, 1957, pp. 101–103.

77. LINDQUIST, C.: How to Dry and Fire Conductive Coatings. Ceram. Industry, vol. 69, No. 3, 1957, pp. 144–150.

78. SEDENKA, A.: Studies of the Effect of Firing Temperature on the Adherence of Silver to Ceramics. Am. Ceram. Soc. Bull., vol. 38, No. 4, 1959, pp. 139–141.

79. HERITAGE, R. J.; AND BALMER, J. R.: Metallizing of Glass, Ceramic and Plastic Surfaces. Metallurgia, vol. 47, No. 4, 1953, pp. 171–174.

80. SHOOK, W. B.: Critical Survey of Mechanical Property Test Methods for Brittle Materials. Wright-Patterson Air Force Base, Air Force Materials Lab., Aeron. Systems Div., Tech. Doc. Rept. ASD–TDR–63–491 (AF 33(657)–8064), July 1963.

81. ANON.: Standards of the Alumina Ceramic-Manufacturers Association for High Alumina Ceramics. Second ed., Am. Ceram. Manufacturers' Association, 1964.

82. ASTM Specification F19–64: Tension and Vacuum Testing Metallized Ceramic Seals. American Society for Testing and Materials Standards, part 8, 1964, p. 336.

83. LUKS, D. W.; AND MAGEE, J.: Hollow Sphere Method for Determining Tensile Strength of Ceramics. Am. Ceram. Soc. Bull., vol. 41, No. 12, 1962, p. 816.

84. SCHUCK, J. F., JR.: Ceramic-to-Metal Seals—How They Are Made. Ceram. Industry, vol. 65, No. 6, 1955, pp. 79–81, 110–111.

85. PINCUS, A. G.: Mechanism of Ceramic-to-Metal Adherence. Ceram. Age, vol. 63, No. 3, March, 1954, pp. 16–20, 30–33.

86. BARNES, C., ET AL: Metallurgical Research and Development for Ceramic Electron Devices. Eitel-McCullough, Inc., Final Rept. TR–66–1, R/E–66–115 AD–636950 CFSTI–HC (DA–36–039–SC–90903), Jan. 1966.

87. ASTM Specification C408–58: Standard Method of Test for Thermal Conductivity of Whiteware Ceramics. Am. Soc. for Testing and Materials Standards, part 13, 1964, pp. 367–371.

88. PINCUS, A. G.: Metallographic Examination of Ceramic-Metal Seals. J. of the Am. Ceram. Soc., vol. 36, No. 5, May, 1953, pp. 152–158.

89. COLE, S. S.; AND SOMMER, G.: Glass-Migration Mechanism of Ceramic-to-Metal Seal Adherence. J. of the Am. Ceram. Soc., vol. 44, No. 6, June 1961, pp. 265–271.

90. HELGESSON, C. I.: Bonding Mechanism in Molybdenum-Manganese Ceramic-to-Metal Seals. Trans. of Chalmers U. of Technology, Gothenburg, Sweden, Nr. 292, 1964, pp. 3–8.

91. REED, L.; AND HUGGINS, R. A.: Electron Probe Microanalysis of Ceramic-to-Metal Seals. J. of the Am. Ceram. Soc., vol. 48, No. 8, Aug. 1965, pp. 421–426.

92. COWAN, R. E.; HERRICK, C. C.; AND STODDARD, S. D.: Mechanism of Adherence of Tungsten to Bodies Containing Yttria. Paper presented at the Fall Meeting of the Electronics Division of the Am. Ceram. Soc. (Phila.), Sept. 1964.

93. KIWAK, R. S.: Metal-to-Ceramic Seals for High-Temperature Service. Am. Soc. for Metals, Tech. Rept. P10–3–64, 1964.

94. VAN VLACK, L. H.: The Metal-Ceramic Boundary. Metals Eng. Quarterly, vol. 5, No. 4, Nov. 1965, pp. 7–12.

95. HUMERICK, M.; AND KINGERY, W. D.: Metal-Ceramic Interactions, III. Surface Tension and Wettability of Metal-Ceramic Systems J. of the Am. Ceram. Soc., vol. 37, No. 1, Jan. 1954, pp. 18–23.

96. KINGERY, W. D.: Metal-Ceramic Interactions, IV. Absolute Measurement of Metal-Ceramic Interfacial Energy and the Interfacial Adsorption of Silicon from Iron-Silicon Alloys. J. of the Am. Ceram. Soc., vol. 37, No. 2, 1954, pp. 42–45.

97. KINGERY, W. D.; AND HALDEN, F. A.: Metal-Ceramic Interactions, V. Note on Reactions of Metals With Titanium Carbide and Titanium Nitride. A. Ceram. Soc. Bull., vol. 34, No. 4, 1955, pp. 117–119.

98. KINGERY, W. D.: Role of Surface Energies and Wetting in Metal-Ceramic Sealing. Am. Ceram. Soc. Bull., vol. 35, No. 3, 1956, pp. 108–112.

99. ALLEN, B. C.; AND KINGERY, W. D.: Surface Tension and Contact Angles in Some Liquid Metal-Solid Ceramic Systems at Elevated Temperatures. Trans. of the Metallurgical Soc. of AIME, vol. 215, Feb. 1959, pp. 30–36.

100. PASK, J. A.; AND FULRATH, R. M.: Fundamentals of Glass-to-Metal Bonding, VIII. Nature of Wetting and Adherence. J. of the Am. Ceram. Soc., vol. 45, No. 12, 1962, pp. 592–596.

101. KING, B. W.; TRIPP, H. P.; AND DUCKWORTH, W. H.: Nature of Adherence of Porcelain Enamels to Metals. J. of the Am. Ceram. Soc., vol. 42, No. 11, 1959, pp. 505–525.

102. HOKANSON, H. A.; ROGERS, S. L.; AND KERN, W. I.: Electron Beam Welding of Alumina. Ceram. Industry, vol. 81, No. 2, 1963, pp. 44–47.

103. BRUNDIGE, E. L.; AND HANKS, G. S.: Ceramic-to-Metal Seals for High Temperature Operation. Los Alamos Scientific Lab., Rept. N63–20486 (W–7405–Eng–36/LAMS–2917/OTS), Apr. 1, 1963.

104. DRING, M. L.: Ceramic to Metal Seals for High-Temperature Thermionic Converters. Bendix Corp., Red Bank Div., Tech. Rept. RTD–TDR–63–4109 (AF 33(657)–10038), Oct. 1963.

105. STABLEIN, P. F.; AND ARAOZ, C.: Technique for Fusion Bonding Ceramics. Review of Scientific Instruments, vol. 34, No. 11, 1963, pp. 1275–1276.

106. POREMBKA, S. W.: Joining Ceramics to Metals for High Temperature Service. Battelle Technical Review, Sept. 1964, pp. 2–7.

107. BUYERS, A. F.: Ceramic-Metal Bonding Stable in Excess of 2248° K. J. of the Am. Ceram. Soc., vol. 46, No. 5, 1963, pp. 244–245.

108. KNECHT, W.: Application of Pressed Powder Technique for Production of Metal-to-Ceramic Seals. Ceram. Age, vol. 63, No. 2, 1954, pp. 12–13.

109. DUNEGAN, H. C.: A Study and Evaluation of Methods of Producing a Ceramic-to-Metal Seal by Pressed Powder Techniques for Automatic Mass Production. American Lava Corp., Final Rept. (AF 33(600)–27329), Aug. 16 to Nov. 30, 1956.

110. POREMBKA, S. W., JR.: Nonglassy Phase Ceramic-Metal Bonding. Final Rept. to the Battelle Development Corp., Battelle Memorial Institute, Nov. 29, 1963.

111. METELKIN, J. J.; MAKARKIN, A. Y.; AND PAVLOVA, M. A.: Welding Ceramic Materials to Metals. Welding Production, vol. 14, No. 6, 1967, pp. 10–12.

112. WELLINGER, R.: A Critical Survey of Methods of Making Metal-to-Ceramic Seals. Electron Tube Research Lab., Univ. of Illinois, Contract W33–038–ac–14742, Apr. 30, 1949.

113. MARTIN, J. E.; AND TUNIS, A. C.: Development of the Radial Compression Seal. RCA Eng., vol. 3, No. 3, Dec. 1957–Jan. 1958, pp. 9–11.

114. SCHEFFER, H.; LIEDERBACK, W.; PIKOR, A.; AND MILLER, W.: How to Ultrasonically Seal Hermetic Transistor Packages. Ceram. Industry, vol. 79, No. 6, 1962, pp. 50–52, 64.

115. VAGI, J. J.; AND DESAW, F. A.: Investigation of Bonding With Exploding Foils. Welding J., vol. 43, No. 12, 1964, pp. 521s–525s.

116. HARE, M. D.; KELLER, R. F.; AND MENESES, H. A.: Electroformed Ceramic-to-Metal Seal for Vacuum Tubes. Stanford U., Stanford Electronics Lab., Tech. Rept. 453–3, (DA 36(039)–SC–73178), Nov. 17, 1958.

117. REED, L.; AND MCRAE, R. D.: Ceramic-Metal Seals for Liquid Metal Environments. Paper presented at the SAMPE Symposium (Seattle, Wash.), Nov. 17–19, 1963.

118. JENNY, A. L.: Soldered Ceramic-to-Metal Seals. Prodct. Eng., vol. 18, No. 12, 1947, pp. 154, 157.

119. BONDLEY, R. J.: Low-Melting Solders in Metal-Ceramic Seals. Ceram. Age, vol. 58, No. 6, 1951, pp. 15–18.

120. MCGUIRE, J. C.: New Method for Soft Soldering Metals and Ceramics. Rev. Sci. Instr., vol. 29, No. 9, 1955, pp. 893.

121. BONDLEY, R.: Metal-Ceramic Brazed Seals. Electronics. vol. 20, No. 7, 1947, pp. 97–99.

122. PEARSALL, C. S.; AND ZINGESER, P. K.: Metal to Nonmetallic Brazing. Tech. Rept. 104, MIT, Research Laboratory of Electronics, Apr. 5, 1949.

123. KELLEY, F. C.: Metallizing and Bonding Non-metallic Bodies. U.S. Patent 2 570 248, Oct. 9, 1951.

124. HUME, G. W.: New Aluminum-Ceramic Bond Produces Hermetic Seal. Materials and Methods, vol. 41, No. 4, 1955, pp. 110–111.

125. CHANG, W. H.: A Dew Point-Temperature Diagram for Metal—Metal Oxide Equilibria in Hydrogen Atmospheres. Welding J., vol. 35, No. 12, 1956, pp. 662s–624s.

126. COYKENDALL, W. E.: Annular Ceramic Tube. Ceram. Age, vol. 63, No. 3, 1954, pp. 33–36.

127. CRONIN, L. F.: Trends in Design of Ceramic-to-Metal Seals for Magnetrons. Am. Ceram. Soc. Bull., vol. 35, No. 3, 1956, pp. 113–116.

128. JOHNSON, C.; WARASKA, I.; COLE, S.; AND STYHR, K.: Ceramic-Metal Seals for High-Power Tubes, Final Tech. Report No. X63–12994. Sperry Gyroscope Company, Great Neck, New York. Contract AF 30/602/–2371/RADC–TDR–63–43, NA–8250–8331, Jan., 1963.

129. GRIMM, A. C.; AND STRUBHAR, P. D.: Dielectric to Metal Seal Technology Study. RCA, Electron Tube Division, Industrial Tube Products, Power Tube Operations, Tech. Rept. RADC–TDR 63–472 (AF 30 (602)–2652), Oct. 1963.

130. GOLDSTEIN, M. B.: 1000° Plus C Ceramic-Metal Seal for Spaceborne Reactors. Varian Associates, Eimac Div., Space/Aeronautics, Rept. A65–35191, vol. 44, No. 9, 1965, pp. 100, 102, 104, 106, 108.

131. HANKS, G. S.; KIRBY, R. S.; AND LaMOTTE, J. D.: Ceramic to Metal Brazes for Use at 1100° C. Los Alamos Scientific Lab., Rept. LA–DC 7159, July 25, 1965.

132. KIRBY, R. S.; LaMOTTE, J. D.: Feasibility of Brazed Joints Between CB–1 ZR Alloy and Alumina Ceramics. Los Alamos Scientific Lab., Rept. LA–3302–MS (W–7405–Eng–36), Aug. 25, 1965.

133. BENDER, H.: High-Temperature Metal-Ceramic Seals. Ceram. Age, vol. 63, No. 4, 1954, pp. 15–17, 20, 21, 46–50.

134. EVANS, T. L.: Ceramic-to-Metal Seals for Vacuum Tubes. Ceram. Age, vol. 64, No. 2, 1954, pp. 9–13.

135. MARTIN, S. T.: Ceramic-to-Metal Sealing Problems Encountered in Magnetron Construction. Ceram. Age, vol. 63, No. 5, 1954, p. 27.

136. ANON.: The Specialization in Ceramic-to-Metal Bonding to Create High Vacuum Seals. Welding J., vol. 43, No. 6, 1964, pp. 517–518.

137. KUESER, P. E.; et al: Bore Seal Technology Topical Report. Research and Development Program on Magnetic Electrical Conductor, Electrical Insulation, and Bore Seal Materials, Westinghouse Elec. Corp., Aerospace Electrical Div., Rept. NASA–CR–54093 (NAS 3–4162), Dec. 1964.

138. KUESER, P. E.; et al: Development and Evaluation of Magnetic and Electrical Materials Capable of Operating in the 800 to 1600° F Temperature Range, Report No. NASA CR–54357, Westinghouse Electric Corporation, Aerospace Electrical Division.

139. SLIVKA, M. J.; BRISTOW, R. H.; AND GIBBONS, M. D.: Vapor Filled Thermionic Converter Materials and Joining Problems. Plasma Research Pertinent to Thermionic Converter Operation, Final Tech. Summary Gen. Elec. Co., Rept. R–572F (Contract NOBs–86220), Dec. 1962.

140. WISSER, G. R.; AND HAGADORN, M. W.: An Improved Nickel-Titanium Ceramic-to-Metal Seal. General Telephone and Electronics Research and Development J., vol. 1, No. 1, 1961, pp. 43–46.

141. IKEUYE, K. K.; AND GROW, G. R.: Brazing Beryllium Oxide to Pyrolitic Graphite. Welding J. Research Supplement, vol. 41, No. 8, Aug. 1962, pp. 346s–349s.

142. KARETA, N. L.; AND NEFEDOV, N. N.: Welding of Graphite Materials. Automatic Welding, vol. 20, No. 1, 1967, pp. 56–58.

143. NORVEYS, J. J.: Manufacture of Composite Bodies Comprising a Metallic Component and a Component of Porous Material. British Patent 802 086, 1960.

144. ARMSTRONG, J. R.; NORTHRUP, J. B.; AND LONG, R. A.: Tungsten to Graphite Bonding. Narmco Research and Development, Final Summary Rept. (DA–04–495–ORD–3028), Dec. 13, 1961.

145. BONDAREV, V. V.: On the Problem of Brazing Graphite and Some Other Materials. Welding Production, vol. 14, No. 6, 1967, pp. 17–19.

146. LEE, S. K.: Process Development for Bonding Zircaloy–2 Cladding to Graphite Moderator. North American Aviation, Atomics International Div., NAA–SR–9604, Aug. 1, 1964.

147. BURNETT, R. C.; AND MARENGO, G.: An Appraisal of Zirconium as a Brazing Medium for Fuel Boxes. United Kingdom Atomic Energy Authority, D.P. Rept. 27, Apr. 1961.

148. BURNETT, R. C.; AND MARENGO, G.: The Use of Molybdenum Disilicide as a Brazing Medium for Fuel Boxes. United Kingdom Atomic Energy Authority, D.P. Rept. 67, Nov. 1961.

149. LINDGREN, J. R.: Development of Brazed and Cemented Joints for the HTGR Fuel-Element Assemblies. General Atomic Division of General Dynamics Corp., GA–2105, May 1961.

150. ANDO, Y.; TOBITA, S.; AND FUJIMURA, T.: Development of Bonding Methods for Graphite Materials. Japanese Atomic Energy Research Institute, JAERI–1071, Oct. 1964.

151. FUJIMURA, T.; AND ANDO, Y.: Method of Bonding Graphite Articles With Iron-Base Brazing Alloys. U.S. Patent 3 177 577, Apr. 13, 1965.

152. DAVIDSON, H. W.; AND RYDE, J. W.: Improvements in or Relating to Methods of Joining Graphite to Graphite and Graphite to Metal Surfaces. British Patent 865 592, Apr. 19, 1961.

153. McPHERSON, R. G.: Molten-Salt Reactor Program. Quarterly Progress Report for Periods Ending January 31 and April 30, 1960, Oak Ridge National Lab. ORNL–2973, 1960.

154. DONNELLY, R. G.; GILLILAND, R. G.; FOX, C. W.; AND SLAUGHTER, G. M.: The Development of Alloys and Techniques for Brazing Graphite. Paper presented at Fourth National SAMPE Symposium (Hollywood), Nov. 13–15, 1962.

155. BRIGGS, R. B.: Molten Salt Reactor Program. Semiannual Progress Report for Period Ending February 28, 1964, Oak Ridge National Lab., ORNL–3812, June 1965.

156. ANON: Molten Salt Reactor Program. Semiannual Progress Report for Period Ending February 28, 1966, Oak Ridge National Lab., ORNL–3936, June 1966.

157. STAPLETON, B.: Brazing Graphite to Nilo–K Tubes. United Kingdom Atomic Energy Authority, IGR–140 (RD/CA), 1959.

158. DEL GROSSO, A.: Brazed Graphite-Metal Composites for Space Radiators. Trans. Am. Nuclear Soc., vol. 8, Nov. 1956, pp. 408–409.

159. MADSEN, J.; KING, P. P.; AND JOHNSON, R. P.: Experimental Evaluation of Expanded Pyrolytic Graphite for Use in Space Radiators. Douglas Aircraft Co., Inc., Final Tech. Rept., AFAPL TR–67–59, May 1967.

160. ARMSTRONG, J. R.; AND LONG, R. A.: Tungsten to Graphite Bonding. Narmco Research and Development, Final Tech. Rept., Contract DA–04–495–ORD–3028, Sept. 30, 1962.

161. McPHERSON, R. F.: Develop Brazing Parameters for Joining Tungsten to Graphite. Aerojet Gen. Corp., Rept. M–2078, May 3, 1960.

Bibliography

ANTON, N.: Fused Vacuum-Tight Metal-to-Ceramic, Ceramic-to-Glass, Metal-to-Glass, and Metal-to-Mica Sealing by Powdered Glass Techniques. Ceram. Age, vol. 63, No. 6, 1954, pp. 15–16, 18–19.

BAUM, E. A.: Development of High-Temperature Ceramic Rectifiers, Thyratrons, and Voltage Regulator Tubes. Gen. Elec. Co., Final Rept. NASA–CR–54303 (NAS3–2548), Oct. 1964.

BERNETT, E. C.: Metallurgical Examination of Development-Type Thermionic Power Converters. Jet Propulsion Lab. Calif. Institute of Technology, Tech. Rept. 32–548 (NAS7–100), Dec. 15, 1963.

BLOOMQUIST, T. V.: Solder-Glass Sealing of Microwave Antenna Windows. Harry Diamond Labs., Rept. HDL–TM–64–28, Nov. 1964.

BOGOWITZ, R. W.; METCALF, A. G.; AND LONG, J. V.: Development of Brazing Processes for Pyrolytic Graphite. Solar Research Labs., NASA–CR–71768 (NAS7–100), Mar. 1966.

BROWN, D.: Ceramic Adhesive-Braze Alloy, Combination Technique. Paper presented at the ASM Western Metal Congress (Los Angeles), Mar. 18–22, 1963.

BRUSH, E. F.; AND ADAMS, C. M., JR.: Vapor Coated Surfaces for Brazing Ceramics. Welding J., vol. 47, No. 3, 1968, pp. 106s–114s.

CHANDLER, H. H.: Manufacturing Research-Bonding-Ceramic to Metal. Convair Div. of General Dynamics Corp., Rept. MR 59–2 (AF33(657)–7248), Aug. 1962.

DALTON, R. H.: How to Design Glass-to-Metal Joints. Product Eng., vol. 36, No. 9, 1965, pp. 62–71.

DENTON, E. P.; AND RAWSON, H.: Metallizing of High Al₂O₃ Ceramics. Trans. of the British Ceram. Soc., vol. 59, No. 2, 1960, pp. 25–37.

ESPE, W.; HIX, P.; AND KEJHAR, J.: A Vacuum-Tight, Reliable Ceramic-to-Metal Seal. Silikattechnik, vol. 15, No. 7, pp. 205–211, 1964.

FEITH, K. E.; AND LAWRIE, W. E.: Ultrasonic Methods for Nondestructive Evaluation of Ceramic Coatings. Armour Research Foundation, Tech. Doc. Rept. WADD–TR–61–91, Part II. (AF 33(616)–6396), Mar. 1961.

GIBBONS, W. F.: Ceramic Seals Applied to Engineering Problems, Ceramics, vol. 9, No. 117, 1958, p. 8.

HICKOX, G. K.; AND MOTAL, D.: Diffusion Brazing of Tungsten to Graphite. Aerojet-General Corp., Phase I Rept., Contract AF 33(657)–8890, Apr. 1963.

HYNES, F. J.: Alumina Bodies for Ceramic-to-Metal Seals. Ceramic Industry, vol. 66, No. 2, 1956, pp. 87–91, 102.

INGE, J. E.; AND SWANSON, J. E.: Experimental Investigation of Stresses in a Ceramic-to-Metal Brazed Joint. Experimental Mechanics, vol. 2, No. 10, 1962, pp. 289–295.

ISKOLDSKIY, J. J.; RUBASHOV, M. A.; GORBUNOV, A. E.; KOSYREVA, V. I.; AND STROGANOVA, V. V.: Molybdenum-Manganese Base Paste for Joining Metals to Ceramics. Wright-Patterson Air Force Base, Foreign Technology Div., Rept. FTD–TT–65–493, Aug. 1965.

KIRCHNER, K. F.: A 19-Inch-Diameter Bakeable Metal Ceramic Seal. Princeton U., Plasma Physics Lab., Rept. MATT–228 (AT/30–1/–1238), Oct. 1963.

LAWRIE, W. E.; AND OESTRICH, M. D.: Nondestructive Methods for the Evalution of Ceramic Coatings. IIT Research Institute, Tech. Doc. Rept. WADD–TR–61–91, Part IV (AF 33(657)–8938), May 1964.

LERMAN, L.: Brazing Ceramics to Metals. Metal Progress, vol. 79, No. 3, 1961, pp. 126–128.

MANFREDI, R. E.; AND NOLTE, H. J.: Applications of Ceramics to Vacuum Tubes. Am. Ceram. Soc. Bull., vol. 35, No. 3, 1956, pp. 105–107.

METELKIN, I. I.; SCHMELEV, A. E.; AND POSDEEVA, N. V.: Method of Obtaining Vacuum-Tight Joints With Mismatched Soldered Parts of Metal to Ceramic. Wright-Patterson Air Force Base, Foreign Technology Div., Rept. FTD–TT–64–484, July 1964.

PILLIAR, R. M.; CARRUTHERS, T. G.; AND NUTTING, J.: Metal-Ceramic Interactions. Leeds U., Department of Metallurgy, England, Final Rept., Feb. 1965.

RECHIN, F. J.: Manufacturing Ablative Nozzles and Nose Cones. Thompson, Ramo, Wooldridge, Inc. Paper presented at SAE National Aeronautics and Space Engineering and Manufacturing Meeting (Los Angeles), Oct. 4–8, 1965.

SAUNDERS, R. D.: Investigation of the Metallizing of Beryllia Ceramics. English Electric Valve Co., Ltd., Chelmsford, England, Rept. 18/15–1, Dec. 1965.

STERRY, J. P.: Testing Ceramic-Metal Braze. Metal Progress, vol. 79, No. 6, 1961, pp. 109–111.

THORTON, R.: Bond Strength and Elastic Properties of Ceramic Adhesives. J. of the Am. Ceram. Soc., vol. 45, No. 5, 1962, pp. 201–209.

81

100

VOGAN, J. W.; AND TRUMBULL, J. L.: Metal-Ceramic Composite Materials. Boeing Co., Tech. Documentary Rept. ML–TDR–64–83 (AF 33(616)–7815), June 1964.

WALLIS, D. R.: Metal-Ceramic Seal Technology. M-O Valve Co., Ltd., Wembley, England, Rept. 14702C, ACSIL/64/4986, Nov. 1964.

WALLIS, D. R.: Metal-Ceramic Seal Technology. M-O Valve Co., Ltd., Wembley, England, Rept. 14871C, NSTIC/04418/65, Nov. 1965.

WILLIAMS, J. C.; AND NIELSEN, J. W.: Wetting of Original and Metallized High-Alumina Surfaces by Molten Brazing Alloys. J. Am. Ceram. Soc., vol. 45, No. 5, 1959, pp. 229–235.

WILLIAMSON, M. H.: Eutectic Bonding Semiconductor Dice to Metallized Ceramic Substrate. Boeing Co., Aero-Space Div., Molecular Concepts in Microelectronics, Proceedings of the 4th Annual Microelectronics Symposium (St. Louis, Mo.), May 24–26, 1965.

Subject Index

83

Report on

ASPECTS OF

METAL-TO-CERAMIC

BRAZING

by

Dr. Glen R. Edwards
Dr. Stephen Liu

Center for Welding and Joining Research

Project Summary

Research has continued to develop fundamental understanding of wetting and spreading of reactive metals on alumina and to model the metal-to-ceramic bonding process. A study was completed on the modeling of spreading kinetics of reactive brazing alloys on ceramic substrates. Based on nucleation and growth of islands of reaction product at the ceramic-metal interface, in particular, the Avrami type surface nucleation and growth kinetics, Dr. Alan Meyer developed a non-empirical theoretical spreading model for reactive metals. Dr. Paulo Camargo focused on the effect of minor contamination in the metal-to-ceramic bonding system. Residual oxygen in the system plays a fundamental role in the wetting and bonding process. Too low an oxygen content will result in non-spreading of the liquid filler metal. Too high an oxygen content will result in excess growth of the interfacial product layer which jeopardizes the mechanical integrity of the joint. These two research programs are complementary since they both expanded the understanding and modeling of the bonding mechanism. The concept of joining ceramic to metal using ductile, multilayer reactive metal coatings is also being developed. Mr. Don Bucholz, a Ph.D. student, is carrying out a program to explore the technological barriers of using multilayer coatings of ductile and reactive metals in brazing. Both diffusion bonding (solid state) and diffusion brazing (transient liquid phase bonding) are being examined. Satisfactory bonds have been obtained using the diffusion bonding process and detailed chemical and structural analysis of the bond region are in progress.

A short description of each of the three programs are given in the following.

1. Modeling of the Spreading Kinetics of Reactive Brazing Alloys on Ceramic Substrates: Alan Meyer (Concluded)

The objective of this study was to fundamentally model the spreading kinetics of liquid metal alloys on ceramic substrates without utilizing empirical curve fitting. A non-empirical theoretical spreading model for reactive metals on ceramics was derived based on the progress of the nucleation and growth of islands of reaction product at the ceramic-metal interface. The spreading radius as a function of time was predicted utilizing a computer program to numerically solve for the fraction of surface covered by reaction product with time based on Avrami-type surface nucleation and growth kinetics. Reaction rate constants (k) of between 1.10^{-2} and $5x10^{-2}$ s^{-1} were obtained for copper-titanium alloys on alumina. Within the uncertainty in the measurements, the k values were not a function of composition and temperature (between 1120° and 1200°C, and 3 to 20 weight percent titanium). The rate constants values for silver-titanium alloys were between $1x10^{-3}$ and $3x10^{-3}$ s^{-1} which are an order of magnitude lower that the k values obtained for copper-titanium and alloys on alumina.

2. Role of Oxygen in the Cu-O-Ti/Sapphire Interfacial-Region Formation: Paulo Camargo (Concluded)

Most reactive metal brazing studies using sapphire as the substrate assumed stoichiometric dissociation of Al_2O_3 and that the oxygen released as the source of oxygen, ignoring the oxygen concentrations of the precursor alloy and gas phase, which could have a dramatic influence on the thermodynamic activity of the active element in the melt. Sessile drop experiments using sapphire single crystals were conducted with three Cu-Ti-O alloys containing 4 atomic percent titanium and 1200, 860, and 2400 ppm of oxygen. The oxygen partial pressure in the reactor was controlled to between 10^{-7} and 10^{-30} atm, with ppm-level additions of hydrogen to the argon gas carrier. Equilibrium between the hydrogen-oxygen-water vapor in the gas phase controlled the oxygen potential. Detailed sapphire/metal and metal/gas/interfacial regions were analyzed using SEM, EDS and XPS (X-ray photoelectron spectroscopy). Measurements and thermodynamic modeling showed that Ti-O coadsorbed to the surfaces and that $TiO_{(g)}$ gas phase assisted in the reactive metal brazing process. The thickness of the Ti-O co-adsorption layer (interfacial reaction product) increased dramatically with increases in the bulk oxygen concentration. By lowering the oxygen content to 100 ppm, flow of the molten filler metal was inhibited. The complex oxide layer (Cu_3Ti_3O, Cu_2Ti_4O) reported by other researchers was also observed to form as a result of high oxygen content (2400 ppm), high temperature, and long reaction time.

3. Diffusion Bonding Ceramic to Metal Using Ductile, Multilayer reactive Metal Coatings: Donald Bucholz (Research in Progress)

The main objectives of this study are to investigate and model the behavior of ductile, multilayer reactive metal coatings as filler metal for bonding ceramics to metals. Activities for the past year were directed towards characterization of the bond layer, understanding the interface formation, variation of the initial ductile and reactive metal deposition conditions and their relation to process parameters. Detailed chemical analysis of the bond layer cross section was also necessary for the development of a multilayer interdiffusion model. Multilayer Ni/Ti coatings of 1 and 25μ total thickness, with variations in the total number of layers, were produced for joining specimens. They undergo heat treatment where time at temperature, temperature and bonding pressure are the process variables. The composition and structure of the multilayer coatings and bonded cross section were analyzed using x-ray diffraction, SEM/EDS. TEM work is also planned for the bond region characterization. Using literature data for comparison, the measured composition profiles are being modeled to understand the interdiffusion in the Ni-Ti multilayers.

I.   MODELING OF THE SPREADING KINETICS OF REACTIVE BRAZING ALLOYS ON CERAMIC SUBSTRATES: COPPER-TITANIUM AND SILVER-TITANIUM ALLOYS ON POLYCRYSTALLINE ALUMINA

The objective of this study was to model the spreading kinetics of liquid reactive metals on ceramic substrates without utilizing empirical curve fitting. Most of the current reactive spreading models assume that complete reaction product coverage is always obtained at the substrate-liquid interface while the model developed in this study assumes that the spreading rate is controlled by the rate of reaction product nucleation and growth. The spreading kinetics of a metal brazing alloy liquid droplet on a ceramic substrate were modeled based on interfacial energy considerations and Avrami-type nucleation and growth reaction kinetics. The results of the model were compared to spreading data generated for the copper-titanium/alumina and silver-titanium/alumina systems.

Advanced ceramics are finding new applications in many areas including the aerospace, automotive, nuclear and electronics industries. However, cost considerations or additional mechanical, electrical or thermal property requirements frequently favor ceramic-metal or dissimilar ceramic component combinations. Joining these different materials by brazing with metal interlayers is a commercially viable technique. A fundamental understanding of the kinetics and thermodynamics of the processes occurring at the ceramic-metal interface will be useful in tailoring the interface to maximize strength and to control other physical properties (e.g. electrical or thermal conductivity).

A. EXPERIMENTAL PROCEDURE

Two types of experimental testing incorporating a sessile drop testing configuration, Figure 1, were undertaken. First, a procedure was developed to generate isothermal spreading data for copper-titanium and silver-titanium alloys on alumina. This data was necessary to test the validity of the derived spreading model. Second, a methodology was developed for measuring the fractional coverage with time of the alumina surface by reaction product. This information can be used to calculate the surface reaction rate constant, k, for input into the computer model.

1. Materials Used

Two systems were chosen for this investigation: copper-titanium/alumina and silver-titanium/alumina. All of the studies were be performed on alumina because thermodynamic and surface energy values were available in the literature; relatively high purity polycrystalline material was easy to obtain and relatively inexpensive. The metal systems chosen were both simple binary systems with a non-reactive base metal (copper or silver) and a reactive metal addition (titanium), instead of the ternary and quaternary systems used in commercial brazing, so that complications due to the interactions between the different components could be avoided. First, the copper-titanium system was chosen because many of the currently used brazing alloys contain these two materials and because the furnace used is limited to approximately 1500°C which precludes the use of pure titanium. Several copper-titanium intermetallics exist for this system, but above 1000°C only a single liquid exists for approximately 45 to 95 weight percent titanium, Figure 2. The silver-titanium system was chosen for comparison to the copper-titanium system. Only low titanium compositions can be studied for this system due to a large solid-liquid two phase field for this system, Figure 3.

The substrates used for all of the tests were 0.64-mm-thick Coors AD-996 polycrystalline alumina substrates. The Coors AD-996 electronic substrate has a surface roughness of 75 to 125 micrometers (CLA), an average grain size of approximately 1.2 micrometers, and an impurity concentration of 0.4 percent [1]. The substrates were cleaned with nitric acid and rinsed with ethanol before testing.

In the majority of the sessile drop tests, a piece of pure titanium (99.99% purity) was introduced to a piece of pure copper (99.99% purity) or pure silver (99.9% purity) blocks during testing to obtain the desired test composition. For small number of the spreading tests, a piece of pre-alloyed copper- 20 w/o titanium (greater than 99% purity) was used.

2. Measurement of Spreading Radius versus Time

The spreading radius versus time was determined for liquid copper-titanium and silver-titanium alloys on alumina using a sessile drop configuration. All of the metal samples had a nominal mass of 1.20 ± 0.02 g and the titanium concentration was always within ± 0.2 weight percent of the desired composition.

Initially, spreading area and spreading radius data for liquid copper-titanium alloys on alumina were generated by heating samples in a vacuum furnace under an argon overpressure. The samples were held for variable times and temperatures, and the final drop contact diameters were measured after the samples were cooled. However, the results were inconclusive due to a large variation between samples as seen in Figure 4.

Therefore, it was concluded that an in situ measurement technique was needed. A tube furnace with a viewing window was used to obtain in situ diameter measurements of the sessile drop Figure 5. This system recirculated argon through an oxygen scavenger to reduce the oxygen partial pressure to approximately 0.006 Pa (50 ppb). The tests were performed at an argon overpressure of approximately 10 kPa. The nitrogen partial pressure was not measured an a nitrogen scavenger was not used. The nitrogen partial pressure was determined by the purity of the argon. Thus the nitrogen partial pressure was estimated to be approximately 1 Pa (10 ppm). The spreading of the drops was videotaped and the spreading diameter was measured from the videotape by freezing frames on a Leco Image Analyzer. These measurements, accurate to ± 0.1 mm, were corrected to account for the non-spherical contact area. The average radius was calculated by multiplying the measured radius by a proportionality constant. This constant was the ratio of the final radius measured perpendicular to the viewing direction divided by the final radius measured in the viewing direction. The assumption of an elliptical contact area was consistent with experimental observations. A preferred spreading direction, probably related to the processing of the square plates, was observed; the maximum spreading radius was always oriented perpendicular to one of the plate edges.

In the second stage of testing, sessile drop tests were performed using three different initial metal configurations, Figure 6. The first configuration was a pre-alloyed block of copper-20 w/o titanium placed directly on an alumina plate (Figure 6a). The second configuration consisted of stacked titanium, copper and alumina (Figure 6b). In the third configuration, titanium was stacked on the alumina substrate and the copper was placed between two plates above the titanium (Figure 6c). At the desired test temperature, the liquid copper fell onto the titanium and formed a copper-20 w/o titanium alloy. In one special case, the copper and the titanium were placed side by side and the furnace was tilted at the test temperature to induce copper-titanium contact. This test technique resulted in the same test configuration as Figure 6c.

First, tests were performed at 1000, 1100, 1200 and 1300°C using the stacked configuration (Figure 6b) . Then, two tests were performed at 1200°C for all three sample configurations to evaluate the effect of the starting configuration on the spreading kinetics. Each test lasted approximately 26 hours (90 ks). Long testing times were used to obtain the equilibrium spreading radii.

Based on the results of the second phase of testing, the configuration shown in Figure 6c was the only one of the three configurations which resulted in "true isothermal" data. In the third phase of testing, this same configuration was used to generate spreading radius versus time data as a function of composition and temperature for copper-titanium and silver-titanium alloys on alumina. In some cases, the same sample configuration was attained by placing the titanium next to the silver or copper, and then tilting the furnace to achieve contact between the liquid metal and the solid titanium. Similar to the configuration in Figure 6c, tilting the furnace at the testing temperature prohibits contact between a reactive metal and a solid substrate until after the isothermal holding temperature is reached. In this phase of testing, copper-titanium/alumina samples were tested in a composition range of one to twenty weight percent titanium and a temperature range of 1120 to 1200°C; while silver-titanium/alumina samples were tested in a composition range of one to five weight percent titanium and a temperature range of 1000 to 1100°C. A total of seventeen copper-titanium tests and four silver-titanium spreading tests were performed. These tests lasted for between ten and twelve hours (36 to 43 ks).

3. Fractional Coverage Measurements

The fractional coverage as a function of time of the alumina surface by reaction product was determined using short duration sessile drop tests. The same furnace set-up used for the spreading radius measurements was used for these tests (Figure 5). The following procedure was be used. The sample was held for a given time at a constant temperature using the sample configuration which resulted in "true isothermal" data (Figure 6c). In all of the tests, a 2.0-g liquid metal drop was allowed to spread on an alumina substrate for between 20 and 600 seconds under an argon atmosphere. The furnace was then shut down and the sample was exposed to air which oxidized the titanium and stopped spreading. After cooling to room temperature, the metal drops were dissolved by successively placing the samples in concentrated nitric and hydrochloric acids. Then the fraction of the alumina substrate covered by reaction product was measured using a Leco image analyzer. The surface were magnified 200 to 1000 times under a light microscope and the image was transferred to an image analyzer where the fractional coverage was determined using the Leco 2001 software package. Different coverage regions were correlated with their respective coverage times based on analyses of the videotapes of the spreading. The relatively large drops (2.0 g) were used to maximize the interfacial area. This precluded the extraction of other spreading information was from the videotapes because gravity is a factor in the determination of the drop geometry for drops of this size. Three sessile drop tests of varying times were performed for each composition and temperature to minimize surface roughness and surface cleanliness effects.

In the copper-titanium/alumina system, large spreading radius changes occurred even for very short times (less than 100 seconds). Rings of varying fractional coverage were observed on these samples. Thus, from five to eleven coverage regions were quantified for the three samples at each test condition. Ten area fraction measurements were made on each coverage region, and the average fractional coverage and the standard deviation of the mean were determined. Fractional coverage measurements were made for times ranging from 30 to 390 seconds.

Only three fractional coverage measurements were obtained for each silver-titanium/alumina test condition because the relatively small radius changes limited the fractional coverage measurements to the initially covered region for each sample. Outside of the initially covered region, it was difficult to match the fractional coverage rings with time. For this system, fifteen area fraction measurements were made in each coverage region and then averaged. The tests lasted for 40 to 600 seconds.

4. Test Matrices

The final testing configuration (Figure 6c) was used to generate isothermal spreading radius versus time data for both the copper-titanium/alumina and silver-titanium/alumina systems. For both systems, the changes in spreading kinetics as a function of temperature and composition were quantified. Multiple tests at two temperatures (1120 and 1200°C) and one composition (Cu-20 w/o Ti) were performed to check the reproducibility of the measurements. The long time spreading data was also be used to calculate the reaction product/liquid interfacial energy (γ_{SL}^{II}). The fractional coverage versus time data was used to calculate the surface reaction rate constant, k. The test matrix for the copper-titanium/alumina system is presented in Table 1 and the test matrix for the silver-titanium/alumina system is presented in Table 2. At temperature and compositions where r(t) and k were both measured, the results were used to compare the theoretical model and the experimental data. This resulted in a total of thirteen comparisons because two of the silver-titanium/alumina spreading radius versus time runs revealed no measurable radius change with time.

Table 1:   Copper-titanium/alumina test matrix.

w/o Ti	Temperature (°C)		
	1120	1160	1200
1	r		
3	r,k		
5	r,k	r,k	r,k
7	r,k		
8.5	r		
10	r,k	r,k	r,k
15	r		
20	r(2),k	r,k	r(2),k

$r = r(t)$ and γ_{SL}^{II}
k = reaction rate constant

Table 2:   Silver-titanium/alumina test matrix.

w/o Ti	Temperature (°C)	
	1000	1100
1		r,k
3		r,k
5	r,k	r,k

$r = r(t)$ and γ_{SL}^{II}
k = reaction rate constant

B. RESULTS

In this section, spreading kinetics and wetting thermodynamics data for a series of copper-titanium and silver-titanium alloys on alumina are presented. The results of this research are divided into three major sections. The first important result is the generation of experimental curves of spreading radius versus time for the copper-titanium/aluminum and silver-titanium/alumina systems. The other two sets of data generated are input parameters for the theoretical spreading model which were not available in the literature: the solid-liquid interfacial energy (γ_{SL}), and the surface reaction rate constant (k). The equilibrium value of γ_{SL} was determined from the final spreading radius and k was determined from plots of fractional coverage (X) versus time as a function of composition and temperature.

1. Spreading Data

The sessile drop technique is frequently used to study the spreadability and wettability of liquid metals on ceramic and metal substrates [2-16]. In many of these studies, the contact angle is measured as a function of time. The wettability is then evaluated based on the minimum final contact angle or the time required to reach the final contact angle [2-7,9,12-18]. Similarly, the spreadability is evaluated based on the change in the solid/liquid contact area (or contact radius) with time [5,8,12]. In these analyses, the relative spreadability is determined from the maximum contact area (or radius) and the time required to reach the maximum spreading area (or radius). If a liquid drop of known volume assumes a geometry which can be approximated as a spherical cap, then the contact angle can be calculated from the spreading radius and vice versa [19,20,21]. Thus, contact angle and contact radius (or area) measure approximately equivalent phenomena for a spherical cap geometry.

In sessile drop experiments involving a non-reactive liquid on a ceramic substrate (i.e. copper or silver on alumina or silicon carbide [6,22]), the equilibrium contact angle is attained in a few seconds. In this situation, the information obtained from this test consists of either the final contact angle or the final solid-liquid interfacial contact radius. Thus, the rate of sample heating from room temperature to the testing temperature and the total isothermal holding time are not important (except for holding times that are sufficiently long to cause metal volatilization and drop volume loss).

However, when a reactive metal is placed on a ceramic substrate, both the isothermal holding time and the sample heating rate are important. In general, when a reactive metal liquid is placed on a ceramic substrate and held at a constant temperature, the contact angle decreases and the spreading radius increases with time until constant values are reached. Thus, these tests must be conducted for longer times compared to similar non-reactive metals if the equilibrium contact angle or spreading radius are desired. In reactive systems, spreading kinetics data are needed to understand spreading mechanisms. Spreading radius, spreading area, or contact angle versus time are commonly used [2-8,12,13,15,16]. However, if the rate of spreading is related to the rate of reaction product formation, significant non-isothermal spreading can occur during the heating cycle, and the generation of isothermal spreading kinetics data becomes almost impossible without very rapid heating rates.

a. Effect of Initial Sample Configuration

Most sessile drop techniques for metal alloys on ceramics use one of two starting sample configurations. Either the metal is pre-alloyed to a given composition [8,12,13] or two or more pure starting materials are either stacked [3,7,14] or compacted together as powders [9] so that when the metal melts, the desired composition is achieved. For both of these techniques, the liquid metal is in contact with the ceramic substrate before the isothermal testing temperature is reached. Depending upon the heating rate and the difference between the solidus temperature and the testing temperature, some of the isothermal spreading data may be lost.

Ideally, the non-isothermal spreading would be eliminated by placing the liquid metal upon the ceramic substrate at the desired test temperature. Fujii, et al [15] were able to couple liquid aluminum to boron nitride at the desired test temperature by using a back pressure to push an aluminum drop onto a boron nitride substrate. Their primary goal was to break the oxide film on the liquid aluminum surface before testing, but this experimental technique also resulted in complete (starting at time equal to zero seconds) isothermal data of contact angle versus time. Similarly, Ip, et al [18] placed liquid aluminum upon calcia and alumina using a graphite plunger arrangement. Using this method, only the contact angle could be reported since control of the liquid metal volume was difficult.

If an oxide film on the liquid metal is not a factor in spreading, then the metal alloy could be placed in a crucible, heated to the testing temperature and then poured onto the ceramic substrate. However, these techniques have a major experimental complication. Most reactive metals (e.g. titanium) either react with or dissolve the crucible materials. Dissolution results in a depletion of reactive metal from the liquid alloy and a decrease in the drop volume. These changes make test reproducibility extremely difficult. For liquid aluminum alloys, the oxide film barrier hinders reaction between the ceramic substrate and the liquid metal, but it also prevents reaction between the liquid aluminum and its container material. Aluminum alloys of known volume and composition could thus be introduced to a ceramic substrate at the desired testing temperature, while reaction between most other reactive metals and their containers complicate volume and composition control.

In this study, a different experimental approach eliminated the non-isothermal spreading of copper-20 w/o titanium alloys on polycrystalline alumina. In the solid state, the reaction of titanium with alumina is relatively slow. If a piece of titanium is placed on a alumina substrate, heated to 1300°C and then cooled back to room temperature, the titanium slides off the aluminum plate when it is tilted, leaving very little discoloration of the alumina surface where the titanium had rested. Similarly, liquid copper does not react with most ceramics. Therefore, if liquid copper is introduced to a combination of an alumina substrate with solid titanium already placed on the substrate, no significant spreading will occur until the copper dissolves the titanium. By introducing the liquid copper onto the alumina/titanium combination at the desired testing temperature, isothermal spreading kinetics data can be generated if the time for copper-titanium dissolution is small.

The spreading radii for the stacked starting configuration (Figure 6b), plotted as a function of time and temperature, are shown in Figure 7. At all temperatures, the spreading radius increased with time until a temperature-specific constant value was reached. The initial and final radii were smaller at lower temperatures, and the difference between the initial and final radii was larger at lower temperatures. These results are consistent with the data reported in the literature [2,3,5,7-9,12,15,16]. The spreading radius has been reported to increase with time to a constant value at an exponentially decaying spreading rate, while the contact angle exponentially decreases with time to a constant value. It has further been shown that an increase in temperature causes the initial and final spreading radii to increase (or correspondingly, the initial and final contact angles to decrease) while the differences between the initial and final radii (or contact angles) decrease.

Figure 8 shows the spreading radius versus time for the same sample runs, but including the spreading which occurred during the heat-up cycle. The negative times on the plot represent the non-isothermal spreading observed from approximately 1100°C to the testing temperature. All of the samples except the 1000°C sample showed significant spreading during the heat-up cycle from the point where the first liquid contacted the substrate to the isothermal holding temperature. Either the change in the initial spreading radius was only a temperature effect, or significant isothermal spreading kinetics data were lost during the heat-up cycle.

The spreading radii versus time plots for a copper-20 w/o titanium liquid in contact with alumina at 1200°C (for all three different starting configurations) are shown in Figure 9. The spreading which occurred before the sample reached 1200°C is again shown as negative times in Figure 10 (along with the short time spreading data at 1200°C). The pre-alloyed sample formed a liquid and began spreading first, but the stacked sample actually spread more rapidly once a liquid formed. The sample where no liquid contacted the substrate until 1200°C had a much lower spreading radius at t = 0 (upon reaching 1200°C) than the other samples, but all of the samples spread to similar final radii. This observation supports the conclusion that significant isothermal spreading data is lost for the first two sample configurations (Figures 6a and 6b).

b. Copper-Titanium/Alumina Isothermal Data

Next, utilizing the improved initial sample configuration (Figure 6c), a series of spreading radius versus time curves were generated for copper-titanium alloys on alumina as a function of liquid metal composition and temperature (Figures 11 to 13). These curves represent "true" isothermal data and can be used to test the proposed spreading model. All of the curves had a similar shape. The spreading radius increased with time with an exponentially decreasing rate of change. All of the samples had relatively small initial radii (between 2.2 and 4.4 mm) and the radius always increased with time to an equilibrium value. For all of these tests, the equilibrium spreading radius was attained in less than ten ks. For a given composition, the final radii were similar at all three temperatures. Two spreading regimes were observed as a function of composition. At low titanium compositions (at or below 7 w/o titanium), the final spreading radii were only 3.4 to 5.1 mm. Conversely, for higher titanium concentrations, the final radii values ranged from 8.5 to 13.2 mm.

c. Silver-Titanium/Alumina Isothermal Data

Four spreading radius versus time curves were generated for the silver-titanium/alumina system using the sample configuration shown in Figure 6c. The results are shown in Figure 14. For this system, the radius versus time curves are nearly flat. The initial radius for a silver-5 w/o titanium alloy on alumina at 1000°C was 4.70 mm, while the initial radii at 1100°C only varied from 3.1 to 3.2 mm for titanium concentrations from one to five weight percent. The maximum overall radius change was very small (less than 0.5 mm) and for both of the 5 w/o titanium samples, no measurable radius change occurred. The silver-1 w/o and 3 w/o titanium samples exhibited behavior similar to copper-titanium alloys on alumina (increasing radius with time at an exponentially decreasing rate) but the overall change in radius was much less for these samples.

2. Determination of Solid-Liquid Interfacial Energies

The spreading radius for a given drop volume may be modeled by computer simulation. Similarly, for a spherical cap of known volume, a given spreading radius has a unique contact angle. The drop geometry leads to three simultaneous equations that can be solved for the contact angle, θ:

$$V = \frac{1}{2}\pi h\left(r^2 + \frac{h^2}{3}\right) \qquad (1)$$

$$R = \frac{1}{2}\left(\frac{r^2}{h} + h\right) \qquad (2)$$

$$\tan\theta = \frac{r}{R - h} \qquad (3)$$

where V = drop volume
h = drop height
r = spreading radius
R = radius of curvature

The volume of a liquid metal alloy sample weighing 1.2 grams was estimated using the pure metal liquid densities given by Iida [23] and Equation 1. For small contact angles, using the spreading radius to calculate the contact angle has been shown to be more accurate than a direct measurement of the contact angle [19].

Next, if the liquid-vapor and solid-vapor energies are known, the solid-liquid interfacial energy can be calculated at equilibrium by solving the Young Equation for γ_{SL}. For this study, the solid-vapor interfacial energy for alumina/argon from Nikolopoulos [11] was used along with a "rule-of-mixtures" approximation for the liquid-vapor interfacial energy which was calculated using the pure liquid data of Iida [23].

The uncertainties in the volume calculation, the measured radius, and solid-vapor and liquid-vapor interfacial energies lead to uncertainties of 60 to 80 mJ/m^2 for the solid-liquid interfacial energy values and ± 4 degrees for the contact angles.

a. Copper-Titanium/Alumina System

The contact angle versus composition as a function of temperature for copper-titanium alloys on alumina is shown in Figure 15. These value are similar to the contact angles of 14 to 108 degrees reported by Naidich[17] for copper alloys containing 1 to 8 a/o titanium (less than 7 w/o titanium) at 1150°C.

The contact angle is a strong function of composition and a weak function of temperature. A transition from a non-wetting to a wetting contact angle (θ less than ninety degrees) was observed near five weight percent titanium. Four regions of contact angle behavior as a function of titanium concentration were obtained for the 1120°C tests. Pure copper had a non-wetting contact angle of approximately 142 degrees. Small titanium additions (one to five weight percent) resulted in a major improvement in wettability and non-wetting contact angles of 100 to 110 degrees were obtained. Between five and ten weight percent titanium, a transition from a non-wetting to relatively small wetting contact angle occurred. In this region, the contact angle decreased from 110 degrees down to less than fifteen degrees. Above ten weight titanium, the contact angle remained approximately constant (between one and fifteen degrees) up to titanium compositions of thirty weight percent.

The two contact angle plateaus and the wetting transition are similar to results obtained by Kristalis et al [24] for the NiPd-Ti/Al$_2$O$_3$ system. In the NiPd-Ti/Al$_2$O$_3$ system, three wetting plateaus corresponding to the formation of three different titanium oxide reaction products (Ti$_5$O$_9$, Ti$_3$O$_5$, and Ti$_2$O$_3$) were observed as the titanium concentration was varied from one to thirty atomic percent. The plateaus observed in this study for the copper-titanium/alumina system may correspond to two different titanium oxide reaction products. The reaction products were not identified in this study. However, several reaction products with distinctly different colors were observed under the light microscope during the fractional coverage measurements for the samples with compositions corresponding to the transition region of the contact angle curves. An example of the formation of two different reaction products on the same sample is shown in Figure 16. The two phases appeared rusty-orange and dark purple under the light microscope. For the majority of the samples displaying dual reaction products, parallel bands of each reaction product, similar to those shown in Figure 16, were observed.

The resulting equilibrium solid-liquid interfacial energies (γ_{SL}^{II}) are plotted in Figure 17. The γ_{SL}^{II} curves had the same general shape as the contact angle curves. The interfacial ranged from approximately 2320 mJ/m^2 for pure copper at 1120°C to between 35 and 150 mJ/m^2 for titanium concentrations above ten weight percent. For all three temperatures, γ_{SL}^{II} decreased with increasing temperature although the decrease was relatively small for the higher titanium concentrations.

The equilibrium contact angle (θ) and γ_{SL}^{II} values were also calculated for copper-20 w/o titanium alloys in contact with alumina as a function of temperature and sample configuration. The results are listed in Table 3. Unlike what was found in previous studies [4-7,9,14-16], the contact angle does not decrease significantly with temperature. Instead, these values are approximately constant from 1100 to 1300°C.

Table 3: The equilibrium contact angle (θ) and the solid-liquid interfacial energy (γ_{SL}^{II}) for copper-20 w/o titanium on polycrystalline alumina.

T (°C)	Starting Configuration	q (°)	γ_{SL} (mJ/m^2)
1000	stacked	33.5	355
1100	stacked	16.2	117
1200	pre-alloyed	24.1	128
	pre-alloyed	12.9	42
	stacked	10.6	30
	stacked	6.6	15
	Cu(l) on Ti(s)	8.1	20
	Cu(l) on Ti(s)	12.9	41
1300	stacked	12.1	-17

Although conventional sessile drop tests cannot be used to analyze the isothermal spreading kinetics, these tests can be used to evaluate the equilibrium contact angle and the solid-liquid interfacial energy at long times. (Assuming oxidation of the drop does not hinder spreading, and depletion of the liquid metal by reaction and volatilization does not significantly decrease the drop volume). The similar contact angle and solid-liquid interfacial values obtained at 1200°C for the three starting configurations support this conclusion.

With the exception of the 1000°C and 1300°C tests, the γ_{SL} values are very small (less than 150 mJ/m^2) and positive. (The small negative value for 1300°C reflects the uncertainties in the calculation.) These results support the hypothesis that the formation of a TiO$_x$ reaction product at the interface is the major factor in the improved wetting and spreading of these alloys on alumina. Very low values of solid-liquid interfacial energy cannot be obtained between a metallic liquid and an ionocovalent solid because of the differences in bond character. However, oxides of titanium are known to take on a metallic character [25]. A liquid metal in contact with a metallic solid could have very similar bonding and a much lower interfacial energy.

The initial contact angles for copper-20 w/o titanium alloys on alumina (upon dissolution at 1200°C) were 105 and 97 degrees for the Cu(l) on Ti(s)/alumina sample configuration. These values suggest that only small improvements in the wettability of alumina are obtained by adding titanium to copper until a reaction occurs. Once reaction occurs, the contact angle is actually the contact angle of copper-titanium on a titanium oxide with alumina surrounding the drop and not the contact angle of copper-titanium on alumina. The contact angle of high purity copper on polycrystalline alumina at 1200°C and an oxygen partial pressure of approximately 10^{-5} Pa is 116 degrees [20] which is only 11 to 19 degrees higher than the initial contact angle for copper-20 w/o titanium on alumina at this temperature and oxygen partial pressure. The contact angle after the titanium-alumina reaction is complete is 90 to 100 degrees lower than either of these angles.

The $\gamma_{SL}{}^1$ values were estimated assuming initial contact angles of pure copper on alumina. For this study, a contact angle of 116 ± 4 degrees was used for pure copper on polycrystalline alumina because the data was obtained using the same furnace set-up and atmosphere control as the current study [20].

The high initial contact angles obtained for copper-20 w/o titanium alloys on alumina support this assumption. The same γ_{SV} and γ_{LV} values used in the γ_{SL}^{II} calculation were used in this calculation. The resulting γ_{SL}^{I} value were approximately 2317 mJ/m^2 at 1120°C, 2347 mJ/m^2 at 1160°C and 2377 mJ/m^2 at 1200°C.

b. Silver-Titanium/Alumina Data

The equilibrium contact angle for the four silver-titanium/alumina samples is shown in Figure 18. All of the 1100°C samples had a non-wetting contact angle between 93 and 120 degrees while the 1000°C sample with a five weight percent titanium concentration had an 85 degree contact angle. No wetting transitions or plateaus were observed but this was expected since the titanium concentrations were below the critical titanium concentrations where a transition occurred in the copper-titanium/alumina system. Higher titanium concentrations could not be tested in this temperature range because higher concentrations result in a solid plus liquid two phase mixture (see Figure 3) The γ_{SL}^{II} values were calculated based on the contact angle are plotted in Figure 19. In this composition range, γ_{SL}^{II} of between 1460 and 1935 mJ/m^2 were obtained. These values are similar to the interfacial energy values obtained for copper-titanium alloys on alumina at 1120°C with corresponding titanium concentrations.

The γ_{SL}^{I} values were estimated assuming initial contact angles of pure silver on alumina were obtained. Limited data is available for the contact angle of pure silver as a function of temperature [26,21-27]. The data that is available shows a strong dependence on oxygen partial pressure, surface roughness, and test method. As a first approximation, the contact angle of 144 degrees at 1000°C reported by MacDonald and Eberhardt [22] and the experimentally determined contact angle of 119 degrees at 1100°C were used in the estimation of γ_{SL}^{I}.

The previous contact angles correspond to $_{SL}^{I}$ values of 2337 mJ/m^2 at 1000°C and 1935 mJ/m^2 at 1100°C.

3. Determination of Reaction Rate Constant, k

A necessary parameter for the evaluation of the proposed spreading models is the reaction rate constant, k. It should be stressed that this is the rate constant for the formation of a monolayer of reaction product on the surface of the ceramic substrate which is not the same as the rate constant for parabolic thickening. In general, the monolayer formation rate constant should be greater than the parabolic rate constant since the surface layer must form before the reaction product can significantly thicken. The reaction rate constant values would only be similar if the surface nucleation rate is very slow and the bulk diffusion rates of the reacting species are very high. In this case, a few islands of reaction product will nucleate and grow into the substrate. The completion of the surface monolayer will then occur by the growing together of the large product "islands" and the growth rate would be approximately the same in all directions.

While the values for the parabolic thickening rate have been reported for several systems the rate of surface coverage has not been quantitatively studied. This is because the commercially successful alloys have very rapid coverage rates and thus one hundred percent coverage is achieved in only a few seconds. Therefore, these k values must be experimentally determined. In this study, an attempt was made to measure k as a function of temperature and composition for the same systems for which spreading data were generated.

First, utilizing the improved sessile drop technique, an experiment was performed for verification of product nucleation and growth at the ceramic-metal interface. Molten copper was dropped onto solid titanium stacked on alumina at approximately 1200°C. The liquid metal was allowed to spread for approximately 120 seconds. After cooling, the metal was dissolved using nitric and hydrochloric acids and the alumina surface was examined using optical microscopy.

The microstructure at several metal/ceramic contact radii is shown in Figure 20. The initially covered surface is completely covered with reaction product. Outside of the initial radius, islands of reaction product are observed. The fraction of the surface covered with reaction product decreases as the relative coverage time decreases (as the radial distance increases). Finally, no reaction product was observed outside of the liquid metal/substrate contact area. The presence of reaction product islands beneath the drop supports reaction product nucleation and growth. The absence of reaction product ahead of the drop refutes a metal volatilization and condensation mechanism.

The fractional coverage as a function of time for an Avrami-type surface nucleation and growth mechanism (2-D) is:

$$X = 1 - \exp(-kt)^2 \qquad (4)$$

This equation can be rearranged to solve for k:

$$k = \exp\left\{ \frac{\ln\left[\ln\left(\frac{1}{1-X}\right) - 2\ln t \right]}{2} \right\} \qquad (5)$$

If the fractional coverage is known for a given time, the reaction rate constant can be calculated. In this study, pairs of fractional coverage time data for copper-titanium and silver-titanium alloys on alumina were used to calculate k. At each composition and temperature, the mean value of k and the standard deviation of the mean were calculated.

Several factors were considered in the determination of k. First, the spreading rate is extremely sensitive to atmospheric conditions. Therefore, X(t) was measured using the same furnace and atmospheric conditions as the spreading radius measurements. Second, the same driving force as the spreading drop should be achieved. This precludes the use of an immersion method (dipping the substrate into molten metal) since the immersion free energy change differs from the spreading free energy change by an $(A \gamma_{Lv})$ term. Therefore, the same sessile drop geometry used for the spreading measurements was used for the X(t) measurements.

a. Copper-Titanium/Alumina System

The fractional coverage of an alumina substrate by reaction product versus time for copper-titanium alloys is shown in Figures 21-31. The error bars represent the standard deviation of the mean for ten measurements performed for each sample condition. The three different symbols on each plot represent results from three different sample runs. There is considerable scatter in the data but the increase in fractional coverage with time is still apparent. Several of the plots show a clear increase in fractional coverage with time. For example, 10 w/o titanium at 1160°C (Figure 27), 20 w/o titanium at 1160°C (Figure 28), and 10 w/o titanium at 1200°C (Figure 30) all exhibit a continuous increase in coverage with increasing time. Other test conditions displayed an increase in coverage with time for a given sample but with a poor correlation between samples. Samples exhibiting this type of behavior are 5 w/o titanium at 1120°C (Figure 22), 7 w/o titanium at 1120°C (Figure 23), 10 w/o titanium at 1120°C (Figure 24), 20 w/o titanium at 1120°C (Figure 25), 5 w/o titanium at 1160°C (Figure 26), 5 w/o titanium at 1200°C (Figure 29) and 20 w/o titanium at 1200°C (Figure 31). Only a 3 w/o titanium samples at 1120°C (Figure 21) showed a significant decrease in coverage with time and this was only a single data point.

Representative fractional coverage micrographs are shown in Figure 32. The reaction product initially forms small islands (Figure 32a). The islands grow together (Figure 32b) until eventually islands of non-reacted alumina are surrounded by reaction product (Figure 32c). Finally, the entire surface is covered by reaction product (Figure 32d). The large standard deviation in some of the fractional coverage measurements is partially due to the non-uniform growth of the islands. The reaction product islands or alumina islands shown in figures 32b and 32c can be very large and are randomly dispersed in a given coverage region.

Two other coverage anomalies are shown in Figure 33. First, for some samples, the fractional coverage is greater at the outer radius of a given coverage region (Figure 33a). It appears that the drop sometimes takes a relatively large step forward, and the nucleation and growth process starts at the outer edge of the drop and moves inward. In several cases, multiple rings of this type were observed beneath the drop. Also, surface irregularities appear to influence the uniformity of the surface coverage. On several samples, a nearly circular region was observed where the fractional coverage was lower inside the circle and greater at the edge of the circle (Figure 33b and c). The circle is hypothesized to be an artifact from the processing or inspection of the alumina plates. It is further hypothesized that edge of the circular region acts as a non-heterogeneous nucleation site while the circular region has fewer nucleation sites than the base alumina substrate. This type of behavior supports the proposed nucleation and growth mechanism. If evaporation/condensation ahead of the drop controlled spreading, then a uniform coverage would be expected beneath the drop.

Based on the above X(t) data, the reaction rate constant was calculated from Equation 107. The average k-values and the standard deviation of the mean for the copper-titanium/alumina system are listed in Table 4. The k values ranged from 0.012 to 0.036 s^{-1}. The standard deviation of the mean was very large. In one case (5 w/o Ti at 1200°C) it was larger than k itself.

Table 4:   The average surface reaction rate constant (k) for copper-titanium alloys on polycrystalline alumina. The uncertainties are the standard deviation of the mean.

T(°C)	Ti (w/o)	k (s^{-1})
1120	3	0.012 ± 0.007
1120	5	0.014 ± 0.006
1120	7	0.029 ± 0.016
1120	10	0.017 ± 0.016
1120	20	0.023 ± 0.017
1160	5	0.022 ± 0.016
1160	10	0.036 ± 0.004
1160	20	0.020 ± 0.013
1200	5	0.022 ± 0.027
1200	10	0.016 ± 0.005
1200	20	0.043 ± 0.015

b. Silver-Titanium/Alumina System

The fractional coverage versus time for silver-titanium alloys on alumina is shown in Figure 34. The error bars represent the standard deviation of the mean for fifteen measurements at each test condition. All four test conditions exhibited similar coverage behavior. The fractional coverage was very small (less than five percent) for the first 300 seconds. The coverage then increased with increasing time. The behavior was similar to the copper-titanium/alumina fractional coverage data except the coverage rate was much slower for the silver-titanium/alumina system. While greater than 99 percent coverage was frequently obtained in less than 100 seconds for copper-titanium alloys on alumina, complete reaction product coverage was not obtained for silver-titanium alloys on alumina after 600 seconds.

Representative fractional coverage micrographs are shown in Figure 35. The island formation is similar to the islands formed for copper-titanium alloys on alumina except that a more needle-like product growth is observed for the silver-titanium alloys on alumina.

For this system, the k values ranged from 0.0012 to 0.0027 s^{-1} which are an order of magnitude smaller then the copper-titanium/alumina k values. The average k values and the standard deviation of the mean are given in Table 5.

Table 5:   The average surface reaction rate constant (k) for silver-titanium alloys on polycrystalline alumina. The uncertainties are the standard deviation of the mean.

T($^\circ$C)	Ti (w/o)	k (s^{-1})
1000	5	0.0016 ± 0.0010
1100	1	0.0021 ± 0.0010
1100	3	0.0012 ± 0.0011
1100	5	0.0027 ± 0.0015

c. Variable Reaction Rate Exponent (n) Values

In all of the previous k calculations, the reaction rate exponent (n) in Equation 107 was assumed to be equal to 2. A more general solution of this equation for k is:

$$k = \exp\left\{ \frac{\ln\left[\ln\left(\frac{1}{1-X}\right) - n\ln t\right]}{n} \right\} \tag{6}$$

The validity of the n equal to 2 assumption can be tested by varying n and solving for k. If n=2 is the best fit for k, then it should have the lowest standard deviation of the mean. The effect of variations in n on k were studied for copper-20 w/o titanium alloy samples on alumina at 1120°C. This test condition was chosen because a large number of fractional coverage versus time data pairs were available (11 total). The reaction rate exponent was varied form 1.00 to 3.00 by increments of 0.25 along with an n-value of 4.00. For each n-value, the mean value of k and its standard deviation were determined. A summary of the values is given in Table 6. For n between 1.50 and 3.00, the standard deviation is approximately a constant value (0.017 to 0.018). The smallest standard deviation for k is obtained at n equal to 2.00 (± 0.0017) but this is extremely close to the standard deviations obtained for n equal to 1.50, 1.75, 2.25, 2.75 and 3.00 (± 0.018). For n-values outside of this region, the standard deviation increases slightly. Based on these results, n equal to 2 is a reasonable assumption but values of n from 1.50 to 3.00 could used.

Table 6:   Values for the surface reaction rate constant (k) obtained by varying n at 1120°C for a copper-20 w/o titanium alloy on alumina. The uncertainty of k is the standard deviation of the mean.

n	$k\ (s^{-1})$
1.00	0.025 ± 0.026
1.25	0.023 ± 0.020
1.50	0.023 ± 0.018
1.75	0.023 ± 0.018
2.00	0.023 ± 0.017
2.25	0.023 ± 0.018
2.50	0.023 ± 0.018
2.75	0.023 ± 0.018
3.00	0.024 ± 0.018
4.00	0.024 ± 0.019

C. DISCUSSION

In this section, the resulting radius versus time curves predicted by the model are presented. The theoretical curves generated using the computer model are compared to the experimentally generated isothermal data. The significance of the results and the shortcomings of the comparison are discussed.

In addition, the utilization of the in situ videotaping technique revealed major differences in the spreading behavior of a reactive liquid metal drop on a ceramic substrate depending on the initial sample configuration. An qualitative analysis of the melting and spreading behavior of the three different sample configurations is presented. The analysis gives further insight into the mechanisms governing spreading.

1. Theoretical Model Versus Experimental Data

The theoretical model was evaluated for both the copper-titanium/alumina and silver-titanium/alumina systems using the computer model previously developed. For each test condition, the computer model was evaluated three times. The average γ_{SV}, γ_{LV}, γ_{SL}^{I} and γ_{SL}^{II} were used in all three runs while three different k values were evaluated. The three different k values evaluated were the mean value of k, an upper bound value of k, and a lower bound value of k. The upper bound of k was the mean value of k plus the standard deviation of the mean and the lower bound was the mean value minus the standard deviation. For the case where the standard deviation was greater than the mean, the smallest k-value obtained for that test condition was used for a lower bound. The errors in V, γ_{SV}, γ_{LV}, γ_{SL}^{I} and γ_{SL}^{II} were not incorporated into the evaluation of the model, since they only change the initial and starting radii and they do not affect the general shape of the curves.

The computer model and the experimental spreading data were compared for eleven conditions in the copper-titanium/alumina system and for two conditions in the silver-titanium/alumina system.

a. Copper-Titanium/Alumina System

While the model was tested for both the copper-titanium/alumina and silver-titanium/alumina systems, the best comparisons were made for the copper-titanium alloys with high titanium concentrations on alumina because of the large changes in spreading diameter. The predicted spreading radius versus time curves (with upper and lower limits) for the copper-titanium/alumina system along with the corresponding experimental data are plotted in Figures 36 to 49. The error bars represent the uncertainty in the measurement of the spreading radius. For curves displaying large changes in radii, the error bars are not visible on the graphs because they are smaller than the data points. In general, the experimental data match the shape of the predicted curves but the experimental data do not always fall within the bounds of the theoretical model predictions. An excellent correlation is obtained at 1160°C for copper-5 w/o titanium on alumina (Figure 44). Relatively good fits are obtained at 1120°C for 3 w/o titanium and 20 w/o titanium (Figures 36 and 43); at 1160°C for 20 w/o titanium (Figure 46); and at 1200°C for 20 w/o titanium (Figure 48). At 1120°C for 10 w/o titanium and at 1200°C for 5 w/o titanium (Figures 42 and 47, respectively), the experimental data points are between the theoretical results for the average k and the lower bound k but these two sample conditions exhibited very large uncertainties in k. The experimental data points for the remaining four conditions (Figures 37, 38, 45 and 49) were usually below the theoretical result for the lower bound k. A condition where all of the data points are below the theoretical lower bound suggests that either the model is incorrect or that the surface reaction rate is lower than the calculated value.

b. Silver-Titanium/Alumina System

The predicted spreading radius versus time curves for the silver-titanium alloys on alumina are shown in Figures 50 and 51. The corresponding experimental spreading data are also shown on these curves. The error bars again represent the uncertainty in the measurement of the spreading radius. Experimental data for the silver-1 w/o titanium system at 1100°C compared rather poorly with the theoretical model while data for the silver-3 w/o titanium system displayed a close fit between the experimental data and the theoretical curves. However, the small changes in the spreading radii with time hinder the ability to compare the experimental and theoretical results. Although experimental spreading radius data were available for four test conditions, only two were analyzed. The silver-5 w/o titanium tests at 1000 and 1100°C were not evaluated because the radius changes were less than the uncertainty in the radius measurements, and because squares of pure titanium were found inside the metal drops during dissolution of the drops for fractional coverage measurements. The presence of pure titanium suggests that the metal drops were within the liquid plus solid two-phase field, which invalidates the density, geometry and titanium concentration assumptions used in evaluating the spreading model.

c. Validity of Model

Based on the theoretical spreading curves and the spreading data presented for copper-titanium and silver titanium alloys on alumina, the proposed spreading model can not be conclusively verified, but the results support the mechanism proposed in the model. The theoretical curves exhibit general shapes similar to the experimental curves, and the experimental and theoretical spreading times are of the same order of magnitude. Interfacial reaction product island development and growth micrographs obtained during the determination of the fractional coverage as a function of time further support the proposed spreading model.

Several factors hinder the conclusive verification of the model. First, the model is only a first approximation, and to simplify the model, homogeneous reaction product nucleation and growth was assumed. However, the alumina surfaces were heterogeneous, containing both grain boundaries and surface defects. The preferred growth regions and preferred growth directions shown in Figures 16, 32, 33 and 35 illustrate the non-homogeneous nature of the reaction product formation. For non-homogeneous nucleation and growth, slower reaction rates would be expected at longer times (after the preferred nucleation sites are exhausted), and the fractional coverage with time would deviate from a simple two-dimensional Avrami-type equation.

Conclusive verification of the model was also hindered by experimental reproducibility problems. The theoretically predicted spreading rates are exponentially related to the reaction rate constant (k) and large deviations in k were obtained. In addition, multiple spreading tests at the same test condition also displayed large experimental variability (Figures 43 and 49).

Another difficulty encountered in the evaluation of the model was the uncertainty in the interfacial energy value inputs used to evaluate the computer model. The reaction product- liquid interfacial energy value (γ_{SL}^{II}) was determined based on the final spreading radius, which results in a forced match between the final theoretical and experimental radii. However, the uncertainty of $\pm$ 60 to 80 mJ/m^2 in the other interfacial energy values could significantly shift the initial radius, and thus the overall curve. The model could be more closely scrutinized if more reliable interfacial energy data (and especially totally independent γ_{SL}^{II} values) were available, but current experimental and theoretical capabilities preclude the generation of more reliable interfacial energy data for reactive metal/ceramic systems.

d. Importance of Initial Sample Configuration

The initial spreading kinetics can be qualitatively analyzed based upon the time at which the first liquid forms, and upon the titanium concentration of the liquid during the thermal cycle. Both factors are strongly influenced by the starting sample configuration. In this section, the proposed melting and dissolution paths for the three initial sample configurations are discussed for the copper-titanium/alumina system. Frames from the in situ videotapes are utilized to support the proposed melting and dissolution mechanisms.

e. Pre-Alloyed Copper-Titanium Alloy

According to the binary copper-titanium phase diagram (Figure 2) [28], copper-20 w/o titanium should be completely molten at 885°C. However, in this study, melting was not observed at temperatures less than approximately 1000°C. This may have been due to the oxide layer formed on the outside of the metal sample. When a liquid forms, it is not outwardly visible until the thermal expansion of the liquid drop breaks the oxide layer. Once this occurs, the liquid has the same composition as the initial solid (weight fraction, X_{Ti} = 0.20) which corresponds to a mole fraction of approximately 0.26. For this configuration, non-isothermal spreading was observed for between 2.2 and 3.4 ks, while the sample was heated from approximately 1000°C to 1200°C.

The sessile drop morphology for a typical spreading cycle is shown in Figure 52 for a pre-alloyed copper-titanium alloy on alumina at 1200°C. The camera was tilted slightly along the tube furnace axis to obtain a clearer contrast between the metal and the substrate. The alloy melted before 1200°C and formed a non-wetting contact angle (Figure 52b). Before 1200°C was attained, the drop had already achieved a wetting contact angle (Figure 52c-d) and considerable spreading had already occurred before 1200°C was attained (Figure 52e). Finally, the drop stopped spreading and eventually formed a pancake-like shape (Figure 52f).

f. Titanium Stacked on Copper

The melting of solid titanium in contact with solid copper is more complicated, but it can still be qualitatively explained by the binary phase diagram (Figure 2). In this system, the first liquid formed has a composition between two intermetallics, Ti_2Cu + TiCu. This liquid forms at the copper-titanium interface. Then as the copper and titanium mutually dissolve into the liquid, the copper concentration increases until only CuTi remains and the liquid resolidifies. In reality, the resolidification of the drop was not observed because of the slower dissolution kinetics of the solid copper by the liquid alloy relative to the dissolution of the solid titanium by the same liquid. The experimentally observed dissolution and melting path is shown schematically in Figure 53. First, a liquid formed between the copper and titanium at approximately 982°C. The actual temperature at which this was observed was higher than that predicted from the phase diagram because the layer must be nearly 0.2 millimeters thick before it can be easily resolved. Next, the liquid front moved into both the copper and the titanium. Surprisingly, the titanium dissolved more quickly than the copper and eventually there was only a copper-titanium liquid on top of a solid piece of copper. As more solid copper dissolved, the liquid volume increased and the liquid flowed off the solid copper. At this point, a titanium-rich liquid contacted the alumina substrate and began spreading (at approximately 1120°C). Finally, at 1140°C, the entire drop was liquid.

While the difference between the temperature at which liquid first contacted the alumina substrate and the isothermal holding temperature was somewhat less than that for the pre-alloyed samples, the extent of spreading was somewhat greater. This result was unexpected because the time required to reach the isothermal holding temperature after the first liquid was formed was similar for both the stacked and pre-alloyed sample configuration. (The stacked samples took 2.0 and 3.0 ks to equilibrate, while the pre-alloyed samples required 2.2 and 3.4 ks.) However, the stacked samples initially formed a titanium-rich liquid with a titanium mole fraction as high as 0.56. Assuming the activity is to be approximately equal to the mole fraction, one can conclude that the stacked sample configuration resulted in an initial titanium activity as high as 0.56. That activity decreased with time to 0.26 as the drop became completely molten. In contrast, the pre-alloyed samples constantly maintained a titanium activity of 0.26 in the liquid. The higher initial titanium activity of the stacked sample configuration could explain the more rapid spreading observed during the heat-up cycle.

Representative photographs from the spreading cycle for a copper-20 w/o titanium alloy at 1200°C for a titanium stacked on copper initial test configuration are shown in before the copper was completely dissolved and most of the spreading was finished before 1200°C was attained.

g. Liquid Copper on Solid Titanium

When the liquid copper was introduced to the solid titanium at 1200°C, the liquid was always at the desired temperature. (As an aside, the copper always melted at a temperature between 1085 and 1105°C and crept down between the plates and fell at 1200°C. This observation supports the conclusion that it was the formation of an oxide layer and not temperature measurement inaccuracies which led to the higher melting temperatures for the copper-titanium alloy.) The liquid was not at the desired composition until the solid titanium was completely dissolved, but complete dissolution only required 10 to 25 seconds. In addition, the spreading was expected to be slower during dissolution, because the liquid was copper-rich. This resulted in a lower titanium concentration and activity. Thus the isothermal data lost was even less than the 10-to-25-s dissolution time because the titanium-lean liquid should manifest slower spreading kinetics.

The dissolution of titanium by liquid copper and the formation of a copper-20 w/o titanium alloy on alumina at 1200°C is shown in Figure 55. Utilizing this starting configuration, the copper dissolves the titanium at the desired test temperature in only a few seconds. An initially non-wetting contact angle is achieved. Several steps in the resulting isothermal spreading cycle are shown in Figure 56. At long times, the drop flattens but it is spherical throughout the majority of the spreading cycle.

Overall, the best isothermal kinetics data were obtained by introducing the liquid copper to the solid titanium at the desired test temperature. This resulted in the loss of less than twenty seconds of spreading data while the solid titanium dissolved into the liquid copper. Significant data (on the order of kiloseconds) was lost during the heat-up cycles for both the pre-alloyed and initially stacked metal sample configurations. The worst experimental condition occurred for the stacked metal samples where a high activity, titanium-rich liquid alloy was formed during heating.

D. CONCLUSIONS

A non-empirical theoretical spreading model for reactive metals on ceramics was derived based on the progress of the nucleation and growth of islands of reaction product at the ceramic-metal interface. The spreading radius as a function of time was predicted as a function of time utilizing a computer program to numerically solve for the fraction of surface covered by reaction product with time, based on Avrami-type surface nucleation and growth kinetics. The only inputs to the model were the drop volume, four interfacial energies (substrate/liquid, reaction product/liquid, substrate vapor and liquid/vapor), and the reaction rate constant.

The theoretical curves and the experimental data exhibited a relatively good correlation. A conclusive verification of the model was hindered by large uncertainties in the experimentally determined k values and spreading rate data, and the large uncertainties in the interfacial energy values published in the literature.

Reaction rate constants of between 1×10^{-2} and 5×10^{-2} s^{-1} were obtained for copper-titanium alloys on alumina. Within the uncertainty in the measurements, the k values were not a function of composition and temperature (between 1120° and 1200°C, and 3 to 20 w/o titanium). The k values for silver-titanium alloys were determined to be between 1×10^{-3} and 3×10^{-3} s^{-1}, values which are an order of magnitude lower than the k values obtained for copper-titanium alloys on alumina.

Micrographs of the formation and growth of reaction product islands at the ceramic-metal interface support the proposed spreading mechanism.

The use of the conventional sessile drop configurations was shown to result in the loss of significant spreading kinetics data during the heat-up cycle for reactive metal liquids on ceramic substrates. The non-isothermal spreading of copper-titanium alloys on alumina was avoided by introducing the liquid copper to the solid titanium/alumina at the test temperature. Because relatively rapid spreading began shortly after the metal contacted the substrate, any sample configuration which allowed liquid metal/ceramic contact below the test temperature could not be used to generate true "isothermal" data.

Without these data, the modeling of reactive metal spreading is complicated since time and temperature are both variables.

The determination of the equilibrium contact angle and the solid-liquid interfacial energy was not significantly affected by the starting sample configuration. Regardless of the starting sample configuration, similar contact angles (7 to 24 degrees) were obtained for copper-20 w/o titanium on polycrystalline alumina at 1200°C. The slightly lower energy values for the pre-alloyed copper-titanium samples may have been caused by sample oxidation during heating. However, this is not conclusive because of the relatively large experimental errors in these measurements.

By introducing the liquid copper to the solid titanium at the testing temperature, the contact angle for copper-20 w/o titanium on polycrystalline alumina was found to be 97 to 105 degrees before the titanium reacted with the alumina. This supports the hypothesis that additions of titanium to copper or silver melts do not significantly increase the wettability of copper on alumina, but that the liquid alloy wets a titanium oxide reaction product.

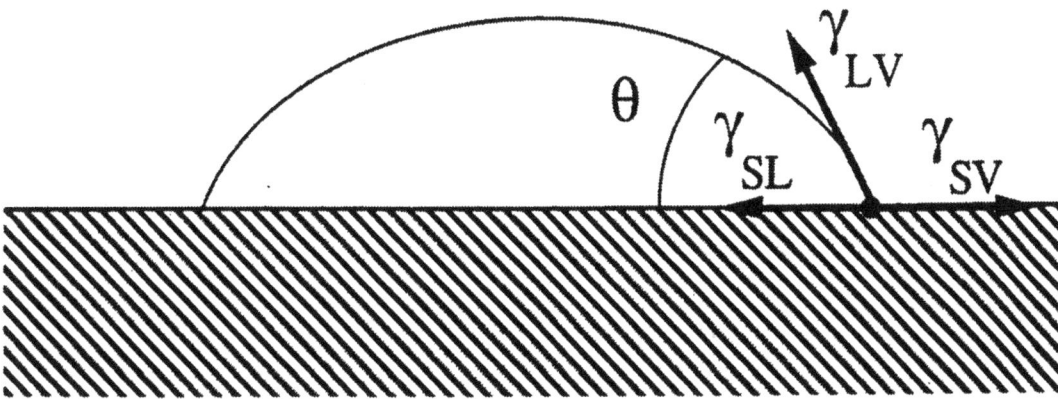

Figure 1: Sessile drop schematic diagram illustrating contact angle and relevant interfacial energies.

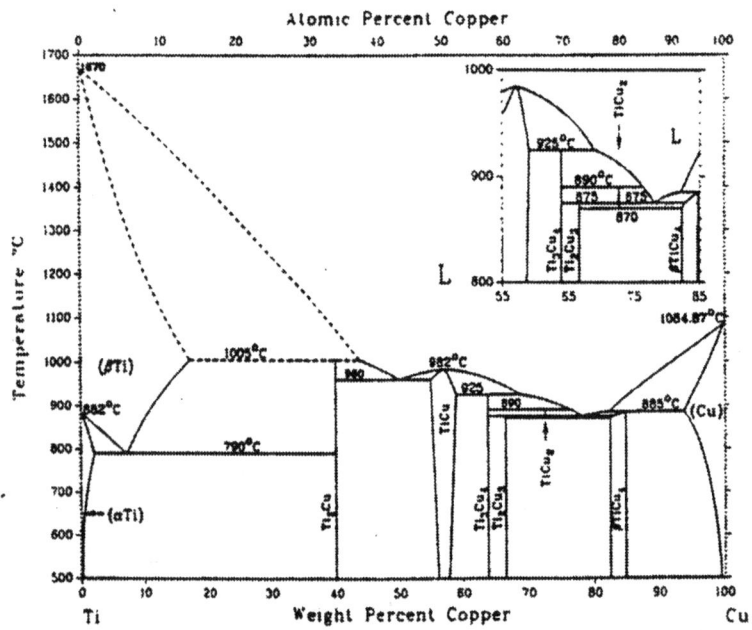

Figure 2: Copper titanium phase diagram (Massalski)

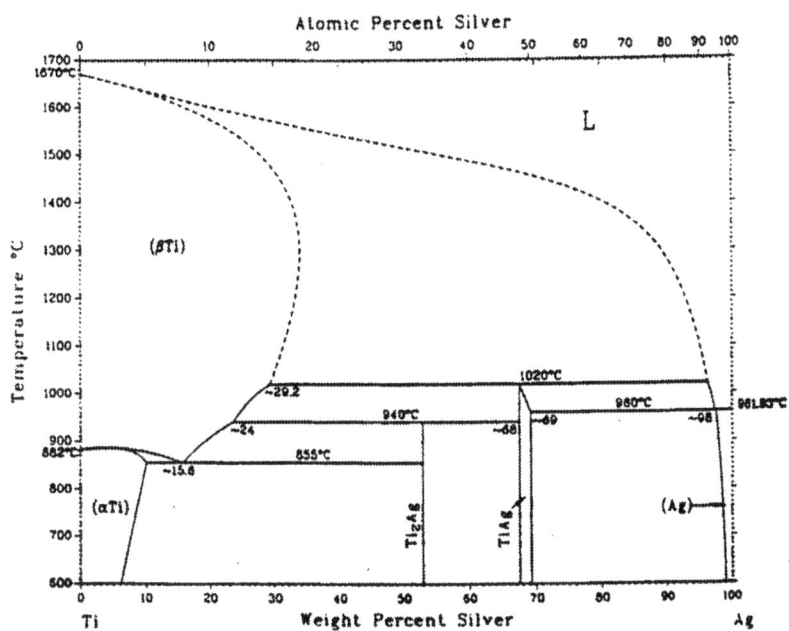

Figure 3: Silver titanium phase diagram (Massalski)

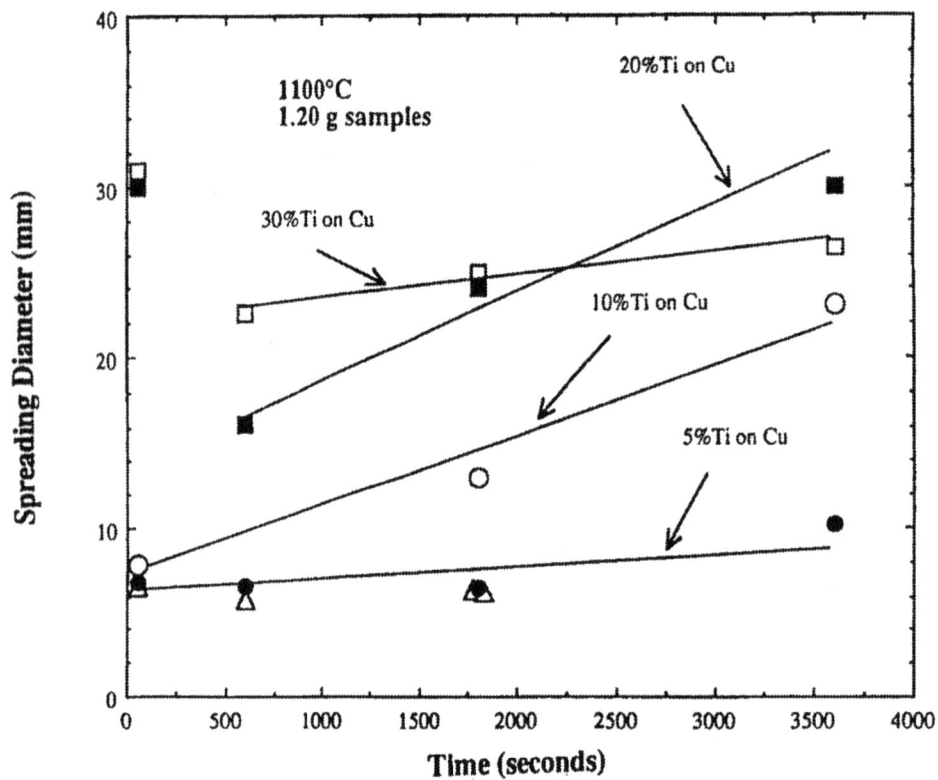

Figure 4: Initial vacuum furnace spreading data for copper-titanium alloys on alumina.

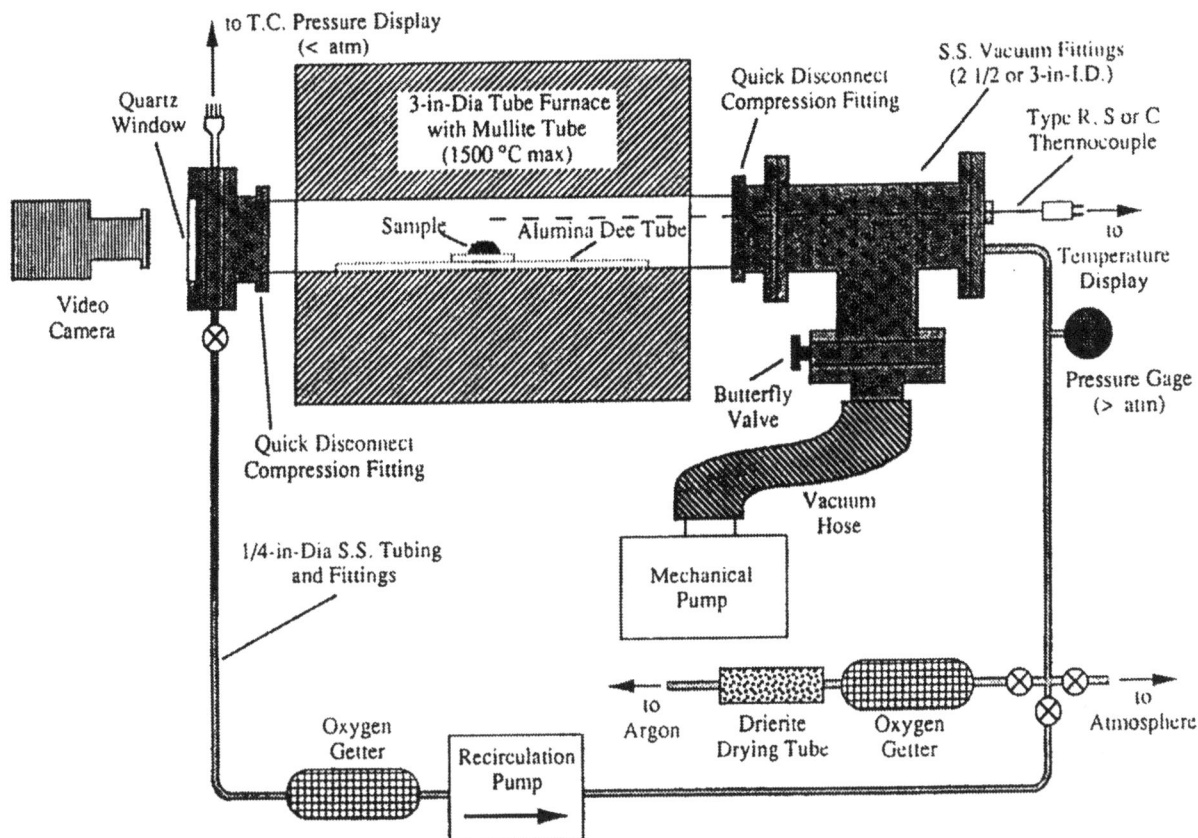

Figure 5: Schematic diagram of experimental set-up used for spreading kinetics data measurements.

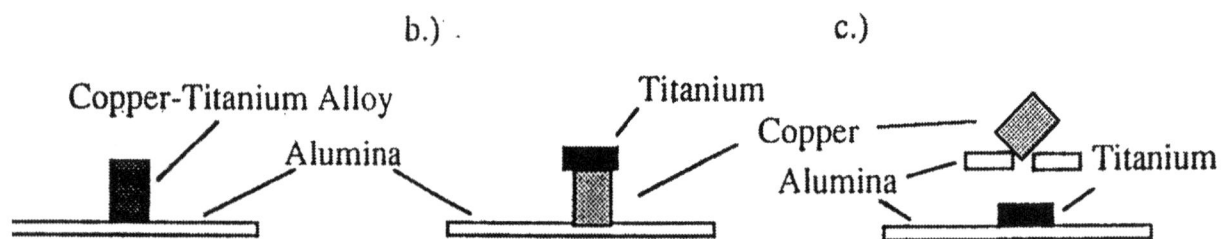

Figure 6: Initial room temperature sample configurations for sessile drop experiments(copper-titanium alloys on alumina), a.) pre-alloyed copper-titanium on alumina, b.) stacked titanium on copper on alumina, and c.) titanium on alumina with copper suspended above titanium.

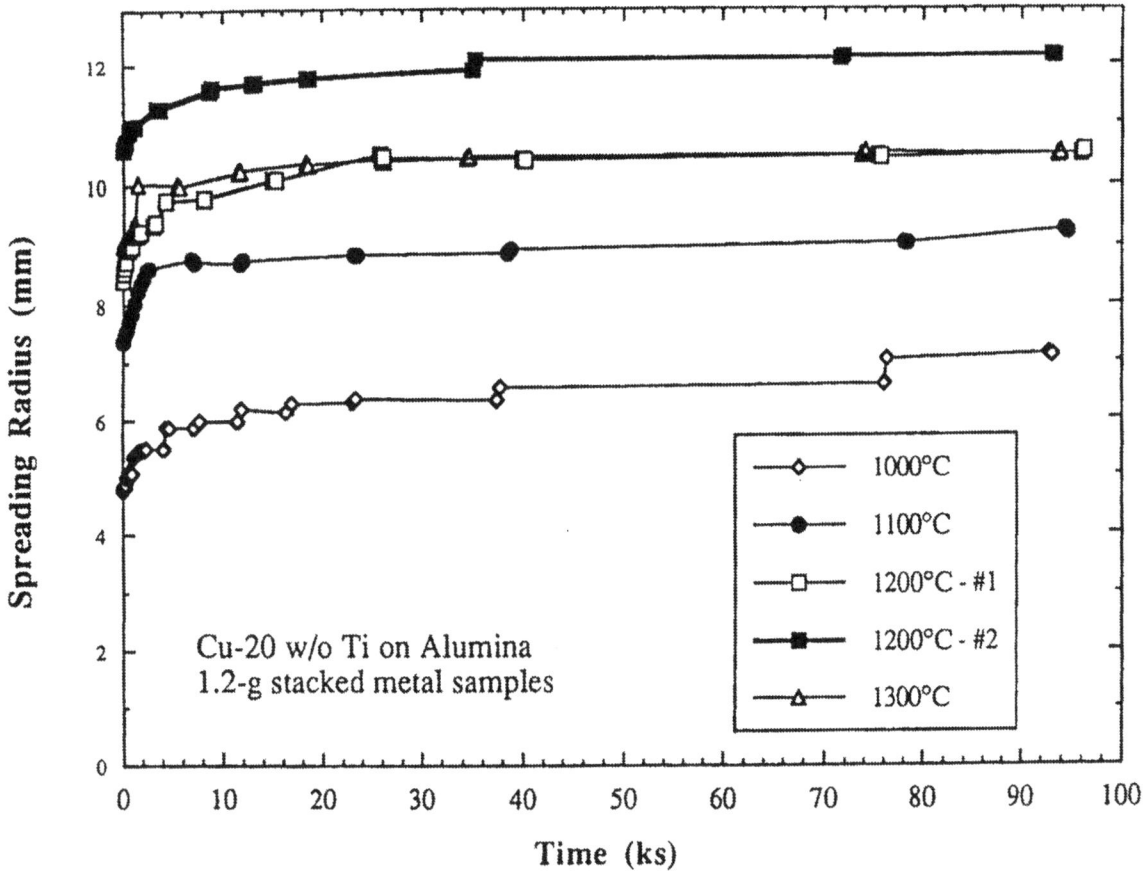

Figure 7: Spreading radius versus time for Cu-20 w/o Ti alloys on polycrystalline
alumina at 1000, 1100, 1200, and 1300°C. The initial sample configuration
was titanium stacked on copper stacked on alumina.

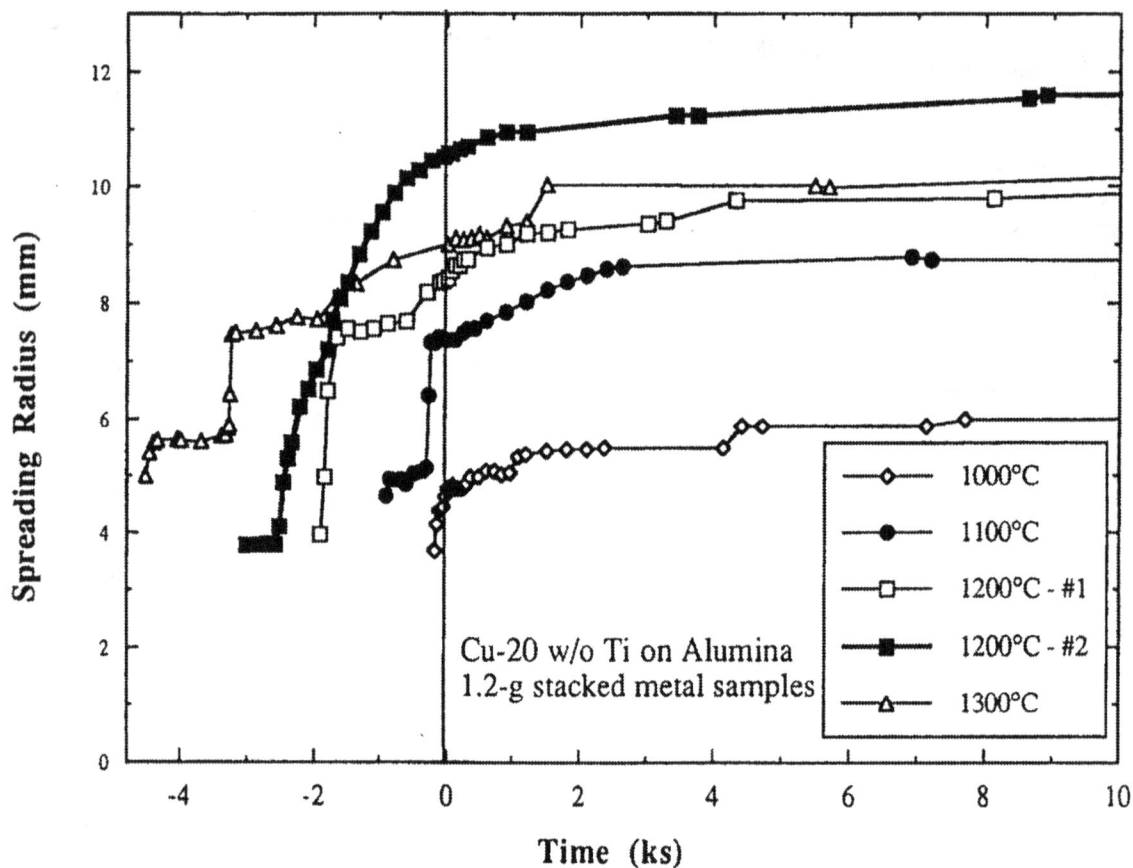

Figure 8: Spreading radius versus time for Cu-20 w/o Ti alloys on polycrystalline
alumina at 1000, 1100, 1200, and 1300°C. The negative times represent the
time required to heat the samples from the temperature at which the first liquid
forms to the isothermal testing temperature. The initial sample   configuration
was titanium stacked on copper stacked on alumina.

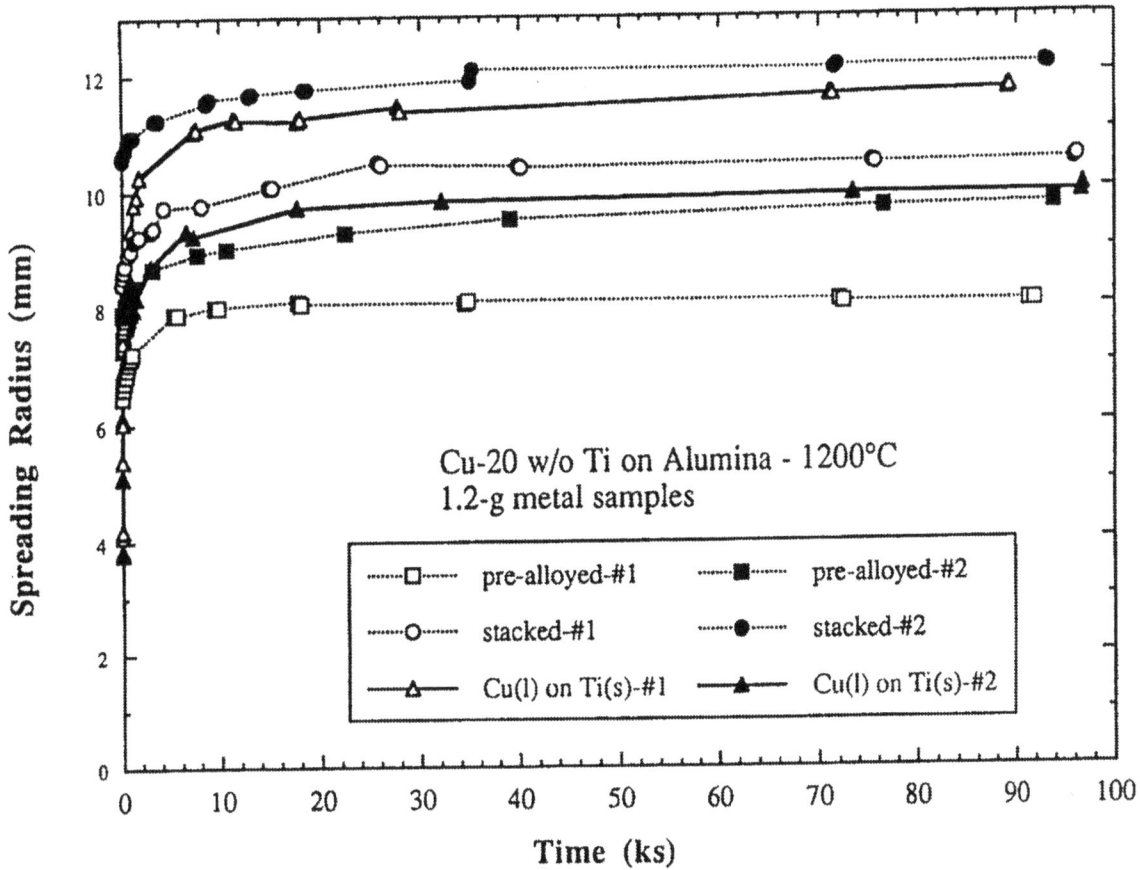

Figure 9: Spreading radius versus time for Cu-wo w/o Ti alloys on polycrystalline
alumina at 1200°C. Three starting configurations are plotted: 1) prealloyed
Cu-Ti (see Figure 6a), 2) Ti stacked on Cu (Figure 6b), and 3)   Cu(l) on Ti(s)
(Figure 6c).

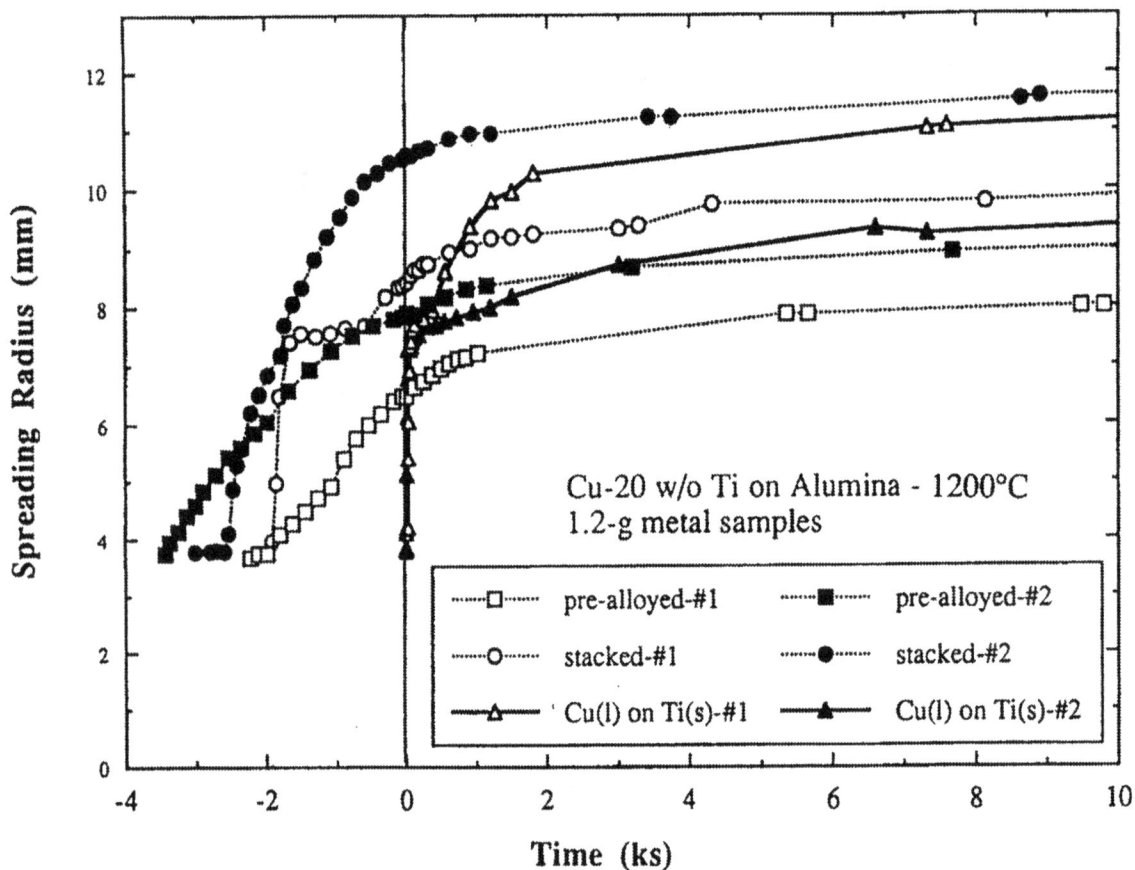

Figure 10: Spreading radius versus time for Cu-20 w/o Ti alloys on polycrystalline alumina at 1200°C. The negative times represent the time required to heat the samples from the temperature at which the first liquid forms to the isothermal testing temperature. Three starting configurations are plotted: 1) pre-alloyed Cu-Ti (see Figure 6a), 2) Ti stacked on Cu (Figure 6b), and 3) Cu(l) on Ti(s) (Figure 6c).

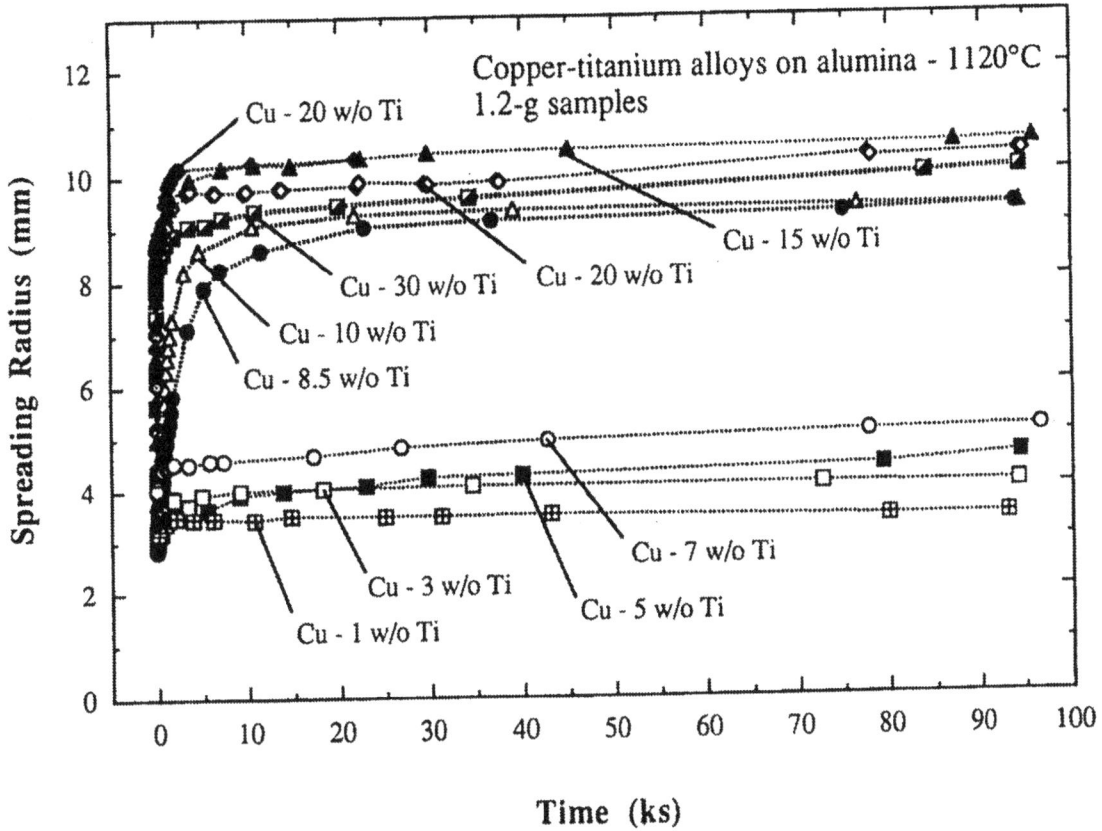

Figure 11: The spreading radius versus time as a function of composition for copper-titanium alloys on alumina at 1120°C. The starting configuration shown in Figure 6c was used.

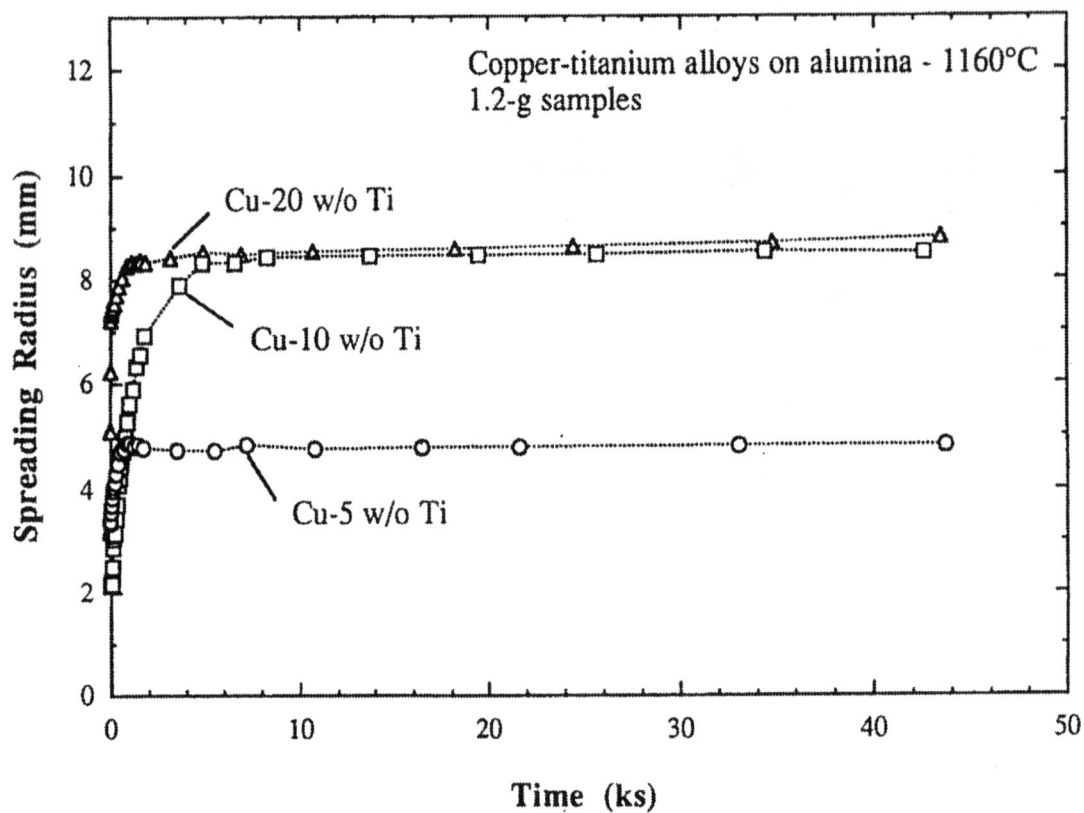

Figure 12: The spreading radius versus time as a function of composition for copper-titanium alloys on alumina at 1160°C. The starting configuration shown in Figure 6c was used.

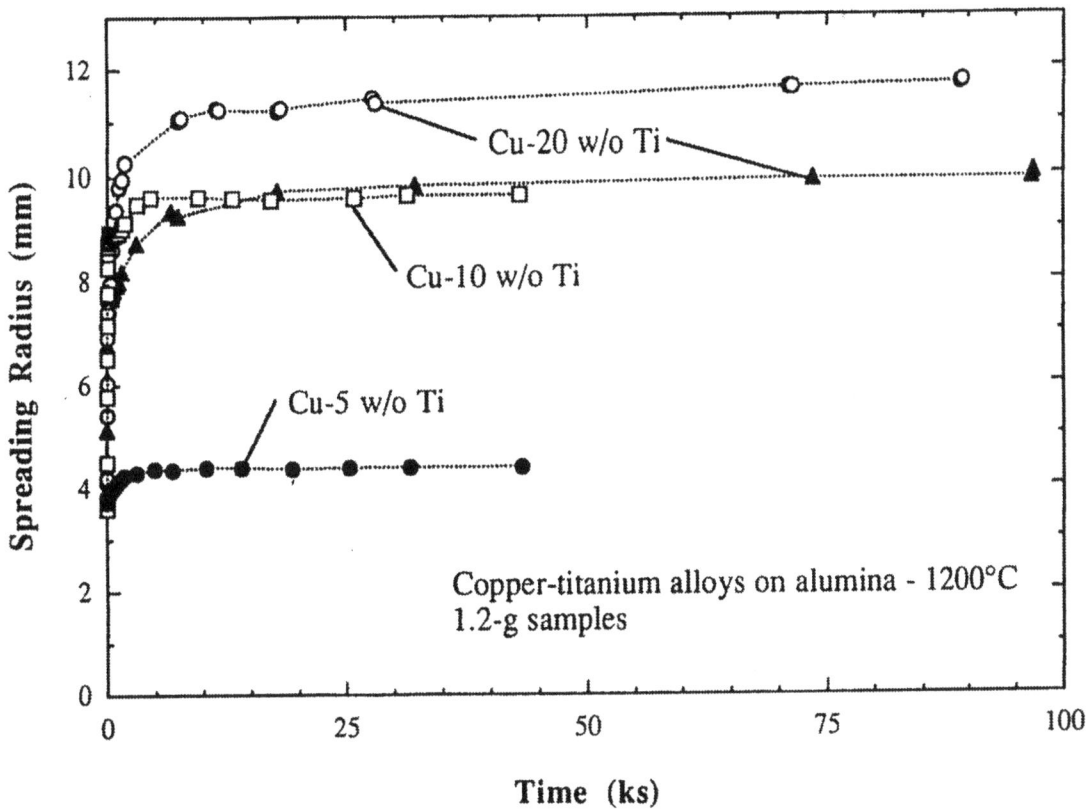

Figure 13: The spreading radius versus time as a function of composition for copper-titanium alloys on alumina at 1200°C. The starting configuration shown in Figure 6c was used.

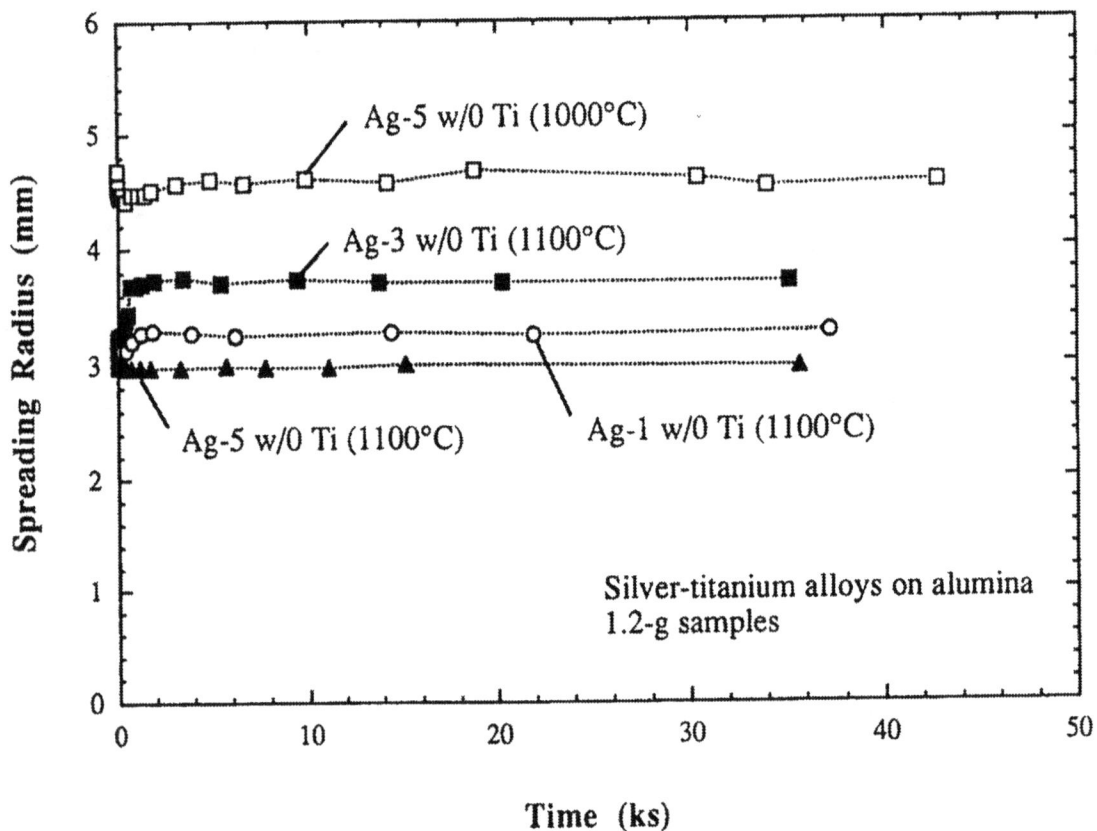

Figure 14: The spreading radius versus time as a function of composition for silver-titanium alloys on alumina at 1000 and 1100°C. The starting configuration shown in Figure 6c was used.

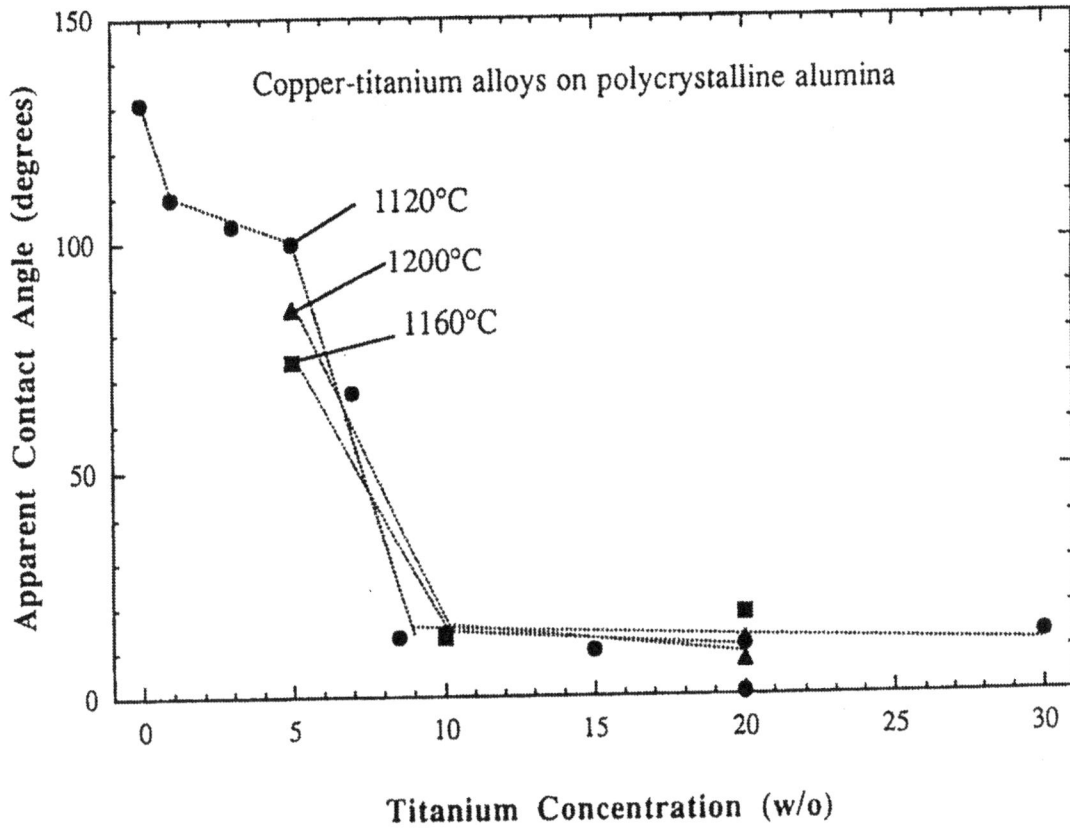

Figure 15: The apparent equilibrium contact angle (q) versus titanium concentration for copper-titanium alloys on alumina.

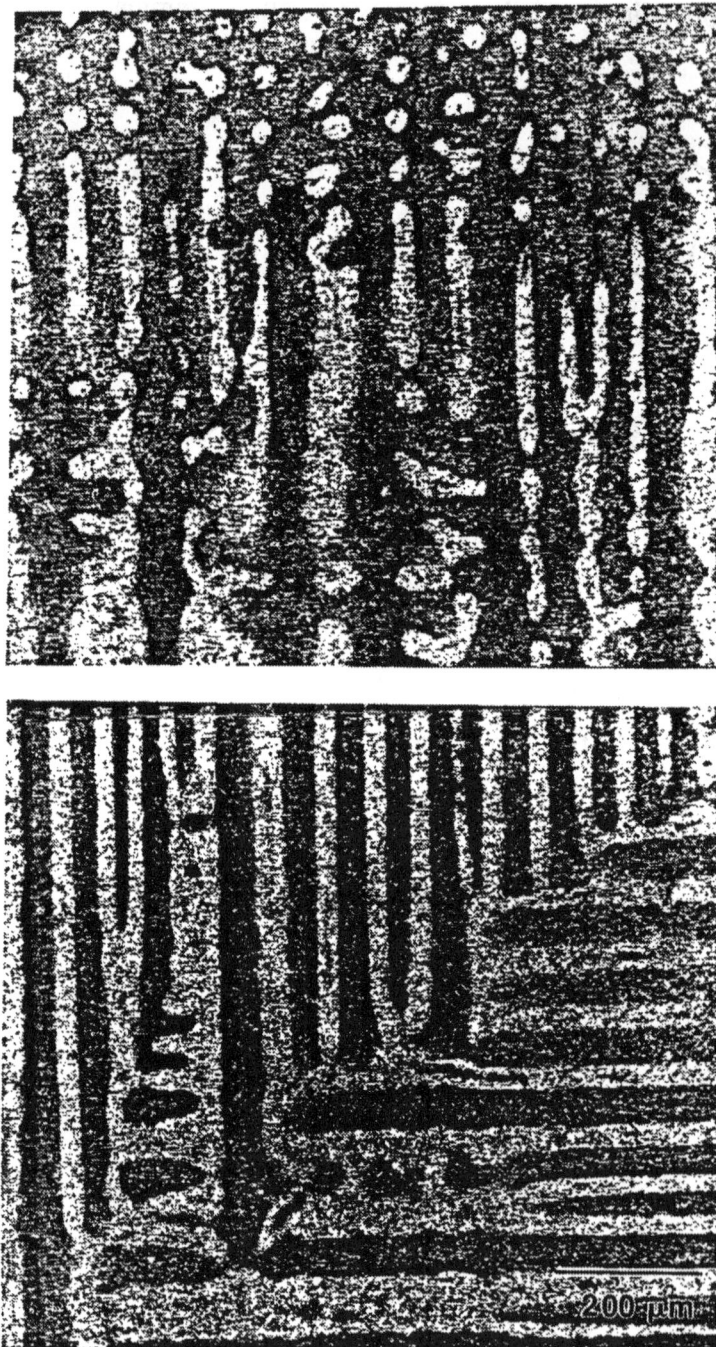

Figure 16: Micrograph of alumina surfaces after reaction showing dual product
formation and preferred growth directions (copper-3w/o titanium on alumina)
(100x).

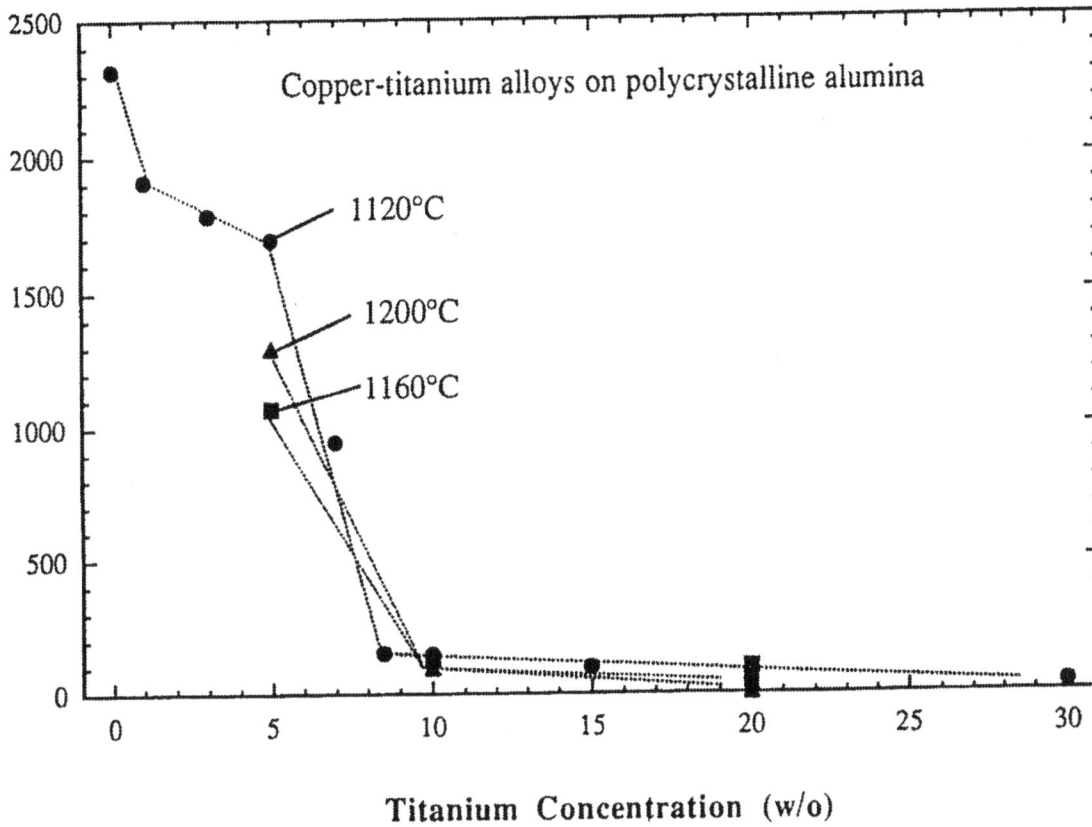

Figure 17: The equilibrium solid-liquid interfacial energy (γ_{SL}^{II}) versus composition or copper-titanium alloys on alumina.

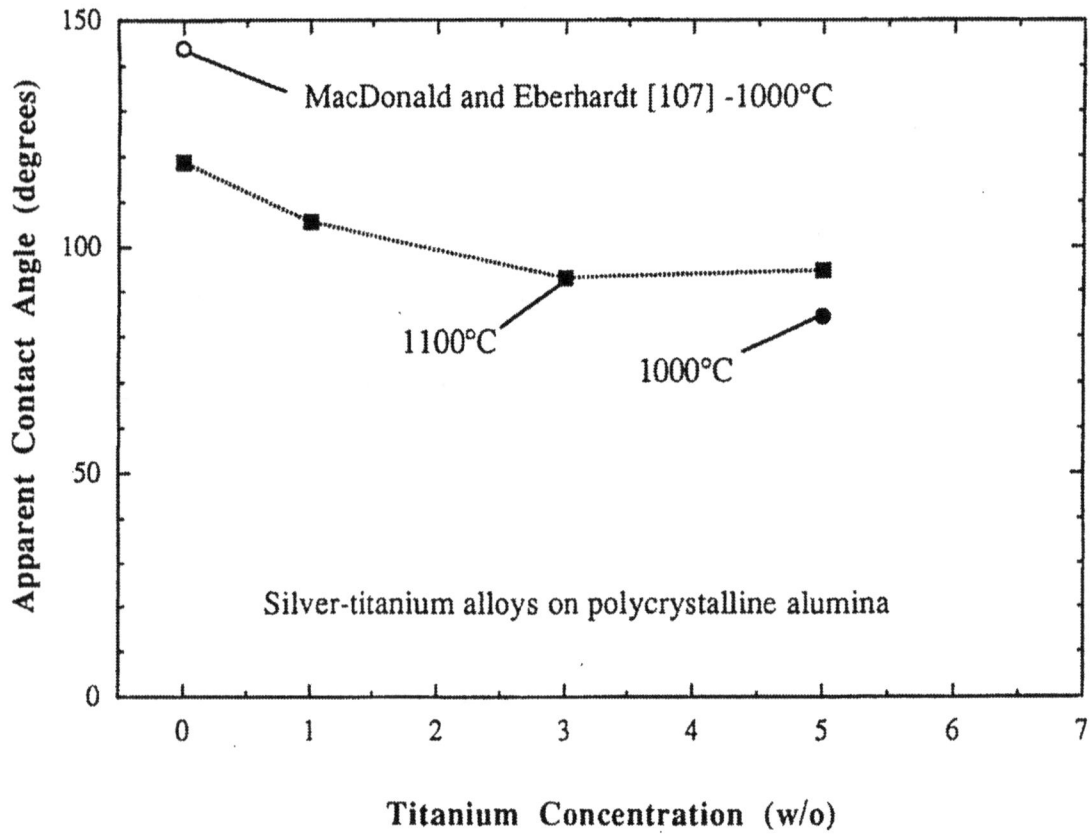

Figure 18: The apparent equilibrium contact angle (θ) versus titanium concentration for silver-titanium alloys on alumina.

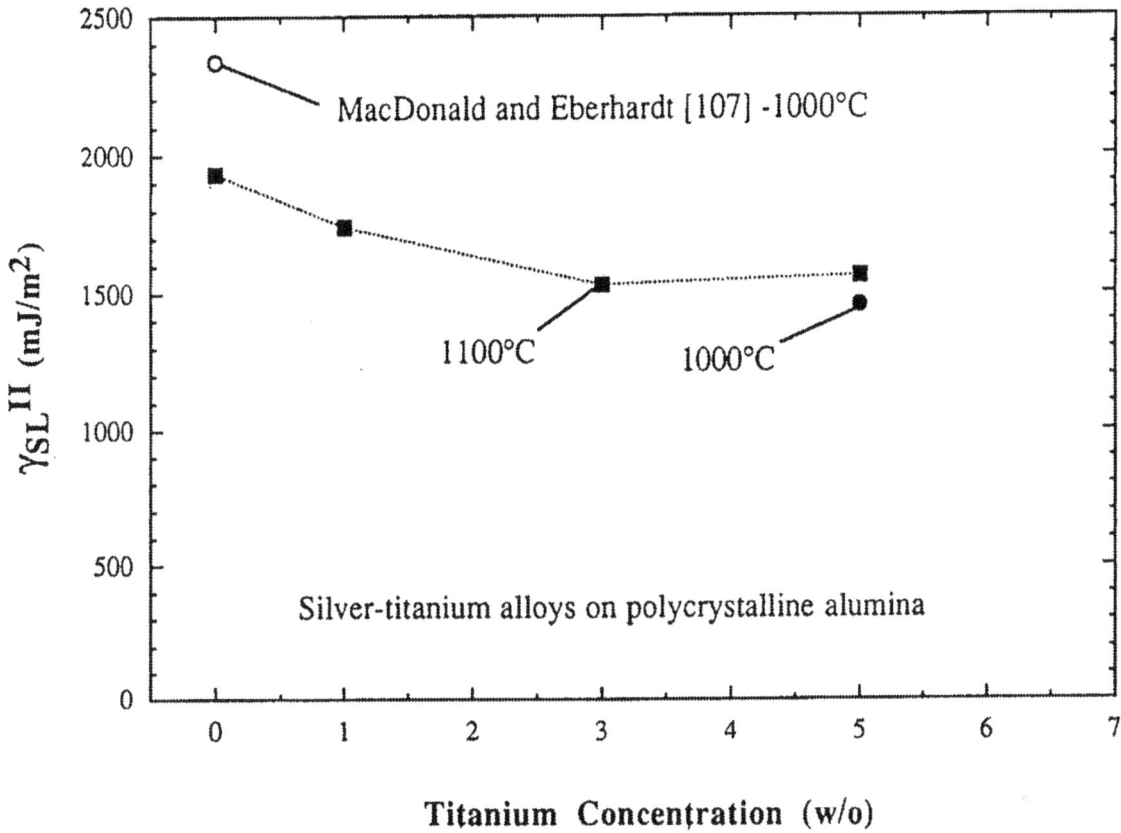

Figure 19: The equilibrium solid-liquid interfacial enerty (γ_{SL}) versus composition   for
silver-titanium alloys on alumina

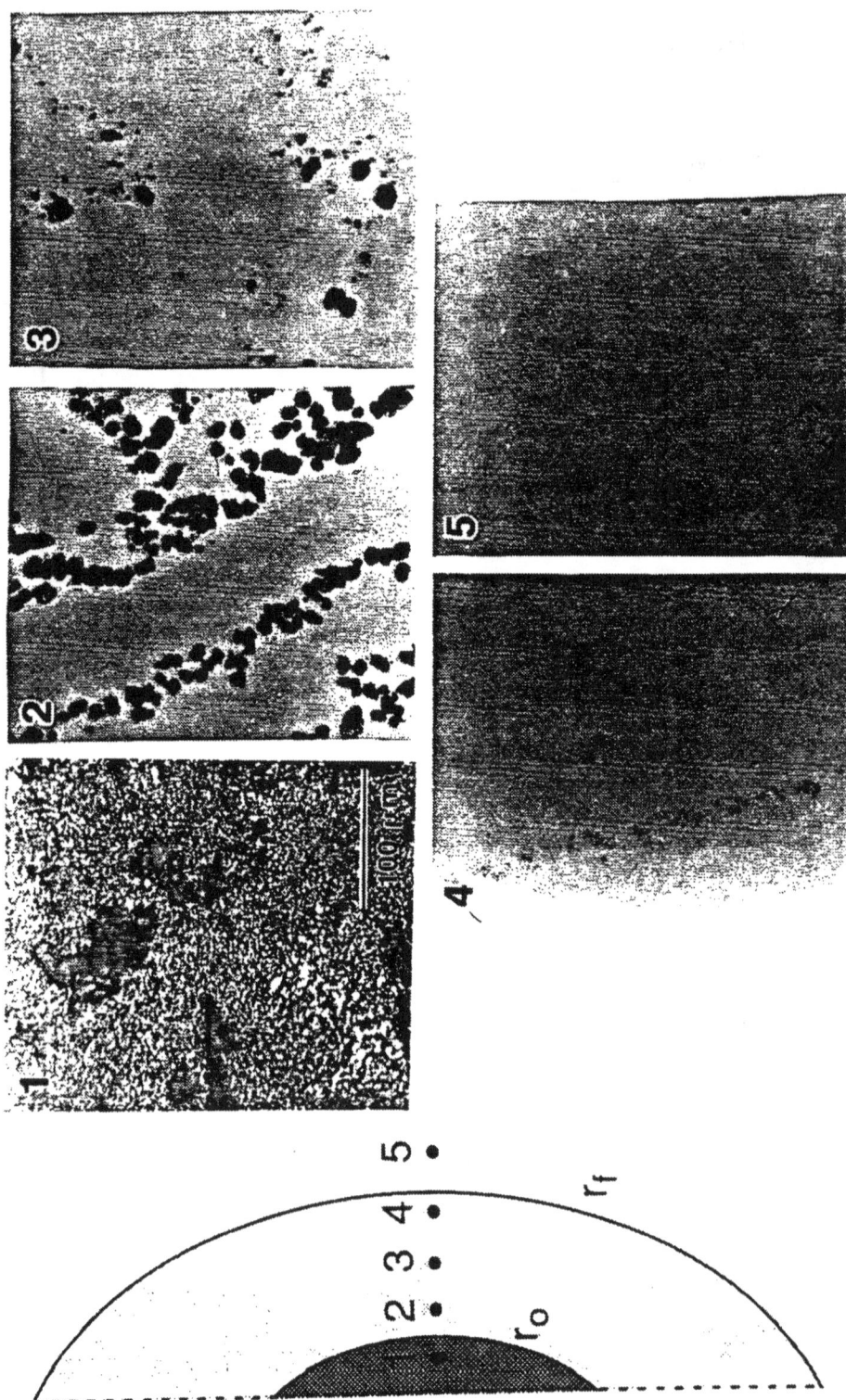

Figure 20: Microstructure of alumina surface after dissolution of copper-titanium drop (200X). The initially covered area, $r' > r_o$, is completely covered with reaction product (1). Reaction product islands are observed in the spreading region, $r_o < r < r_f$ (2-4). No reaction product is observed outside of the spreading area, $r > r_f$ (5).

163

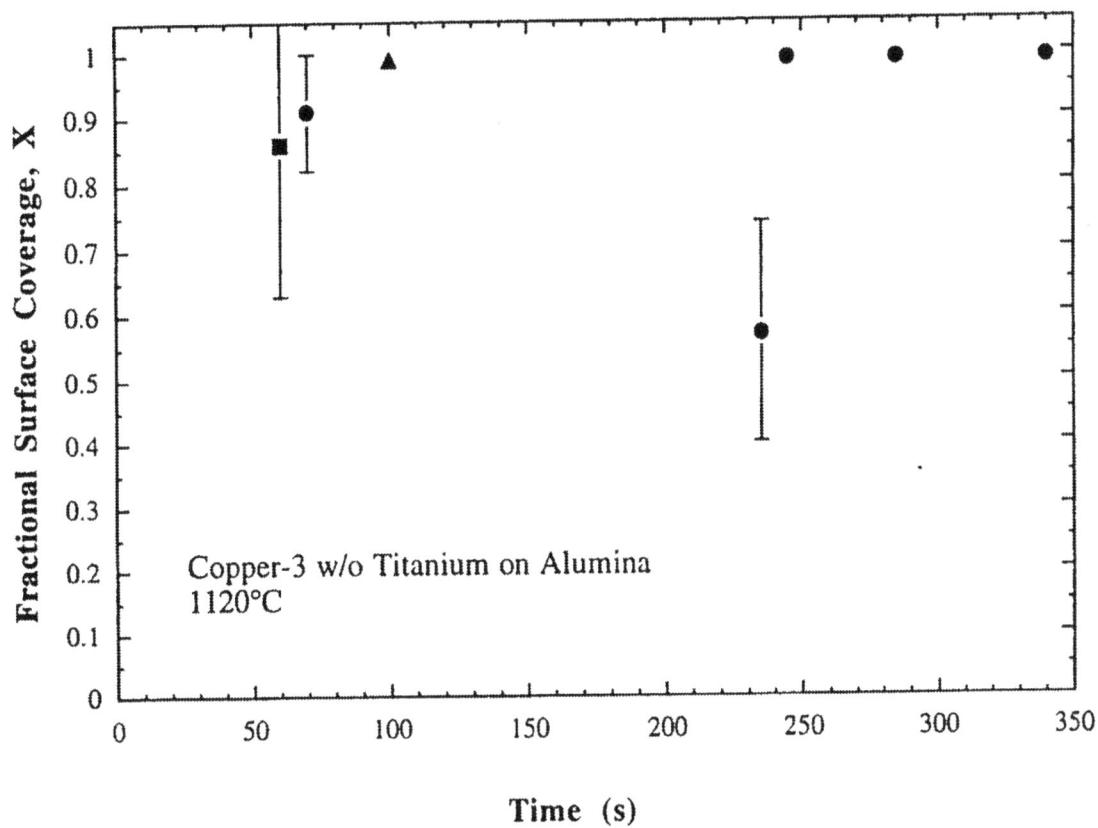

Figure 21: Fractional coverage of reaction product on alumina surface as a function of time for a copper-3 w/o titanium alloy at 1120°C. The 3 different symbols represent 3 independent sample runs. The error bars are the standard deviation of the mean for ten measurements.

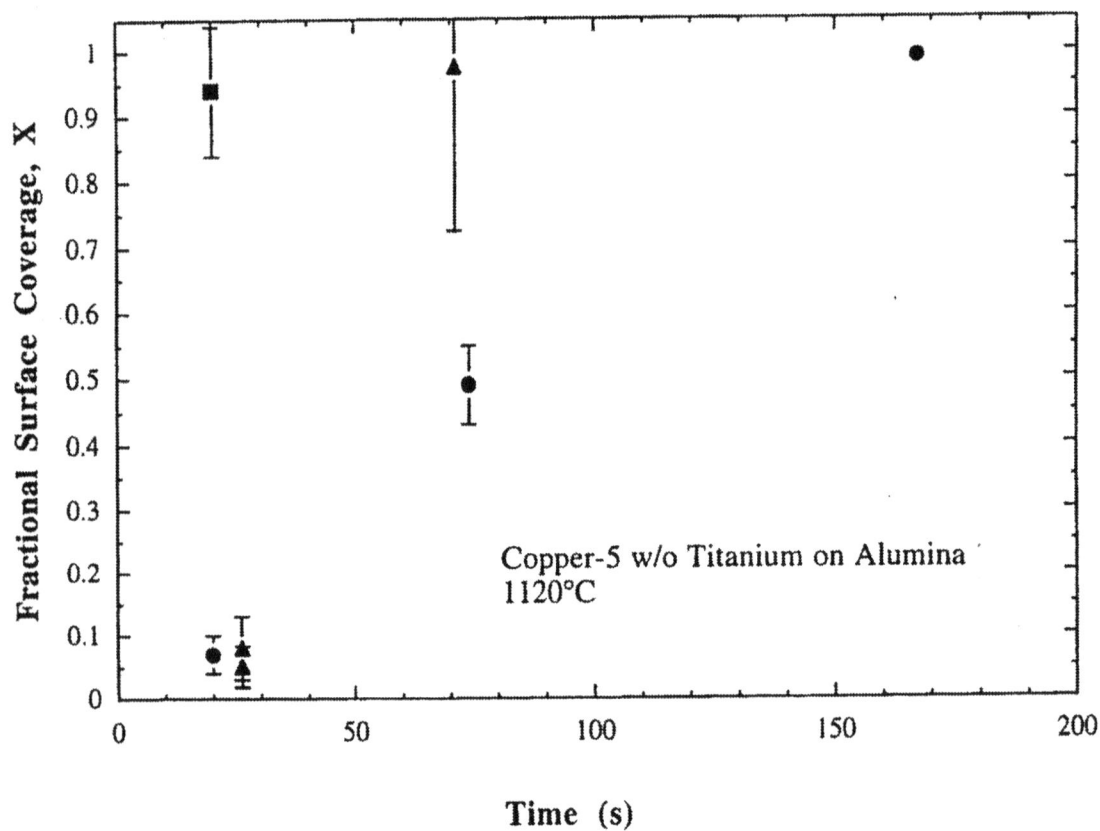

Figure 22: Fractional coverage of reaction product on alumina surface as a function of time for a copper-5 w/o titanium alloy at 1120°C. The 3 different symbols represent 3 independent sample runs. The error bars are the standard deviation of the mean for ten measurements.

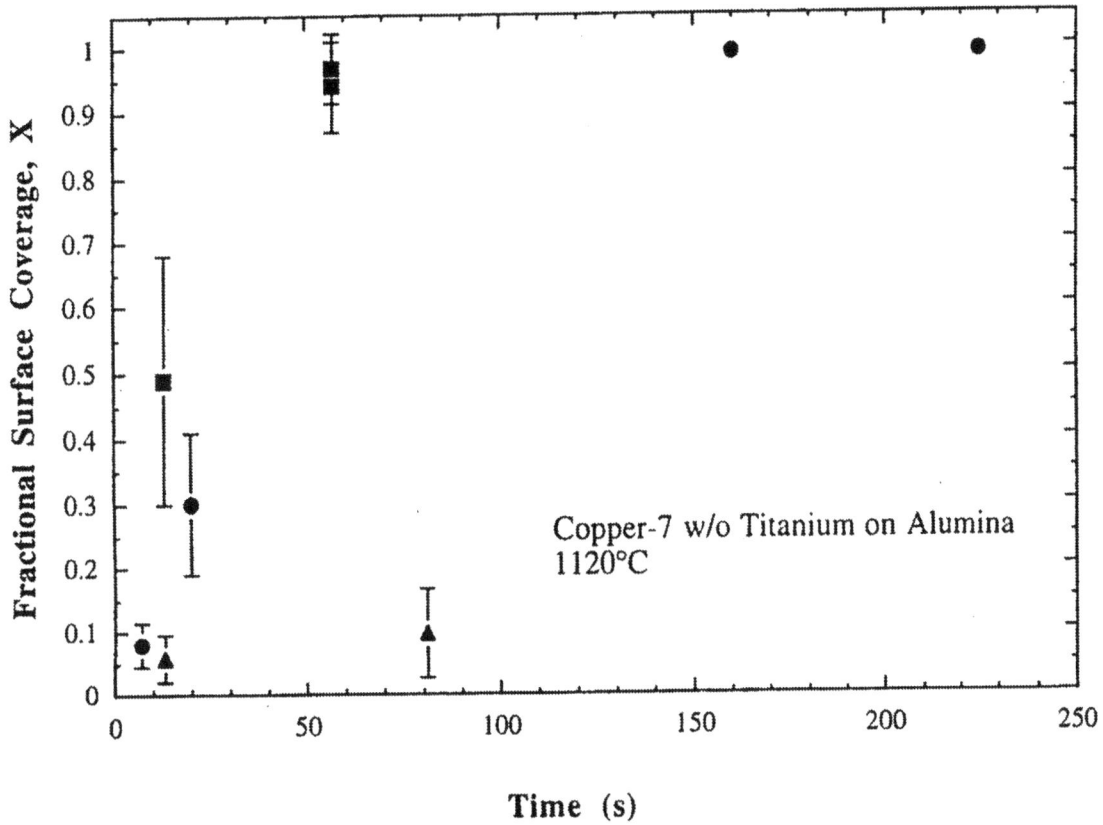

Figure 23: Fractional coverage of reaction product on alumina surface as a function of time for a copper-7 w/o titanium alloy at 1120°C. The 3 different symbols represent 3 independent sample runs. The error bars are the standard deviation of the mean for ten measurements.

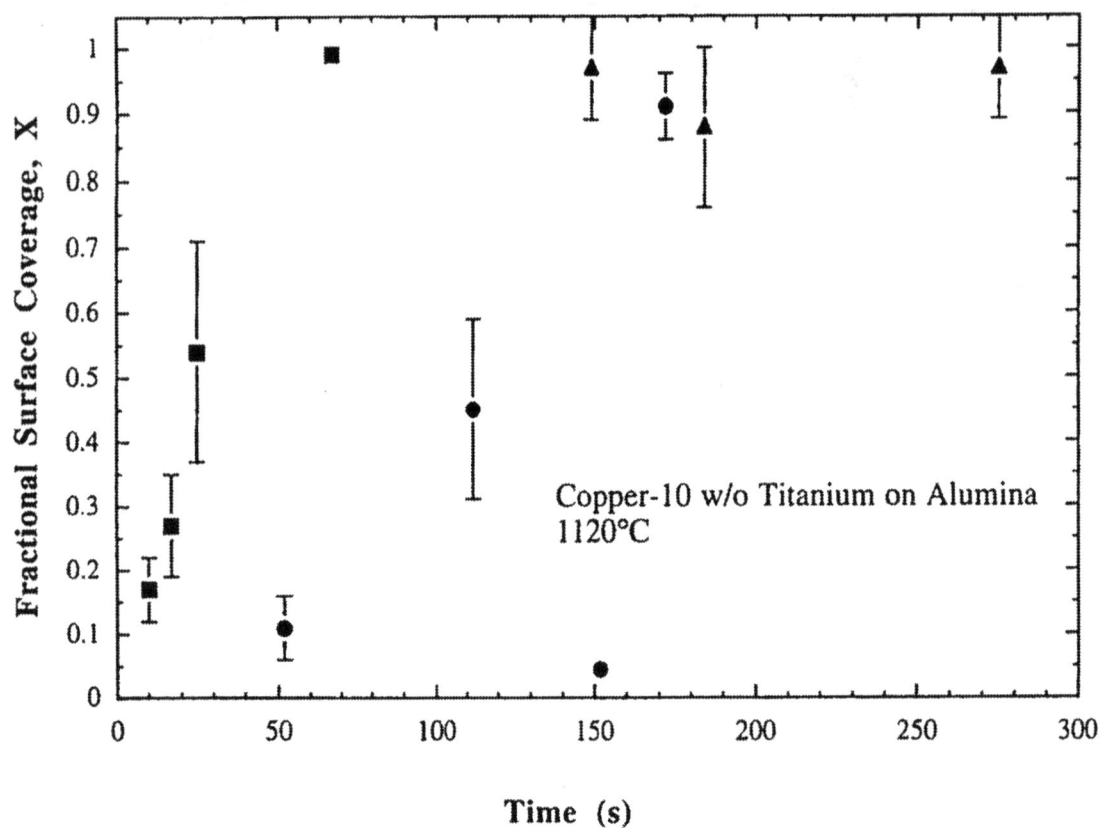

Figure 24: Fractional coverage of reaction product on alumina surface as a function of time for a copper-10 w/o titanium alloy at 1120°C. The 3 different symbols represent 3 independent sample runs. The error bars are the standard deviation of the mean for ten measurements.

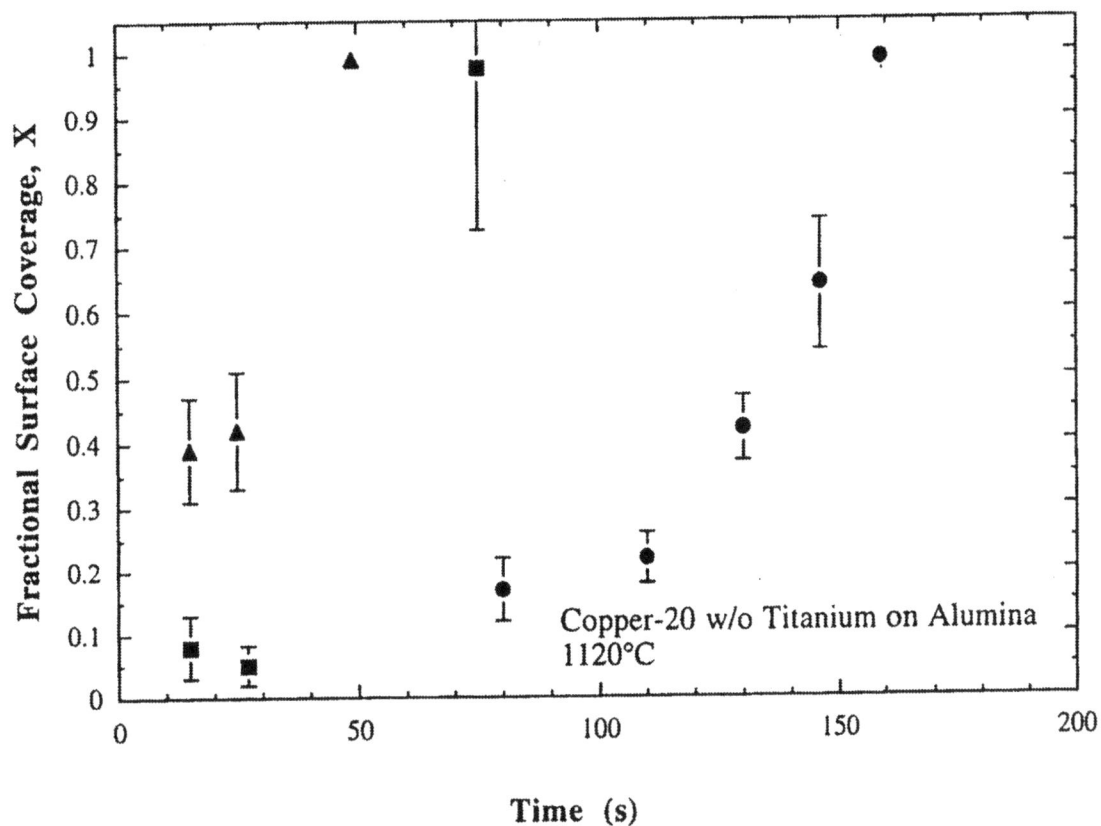

Figure 25: Fractional coverage of reaction product on alumina surface as a function of time for a copper-20 w/o titanium alloy at 1120°C. The 3 different symbols represent 3 independent sample runs. The error bars are the standard deviation of the mean for ten measurements.

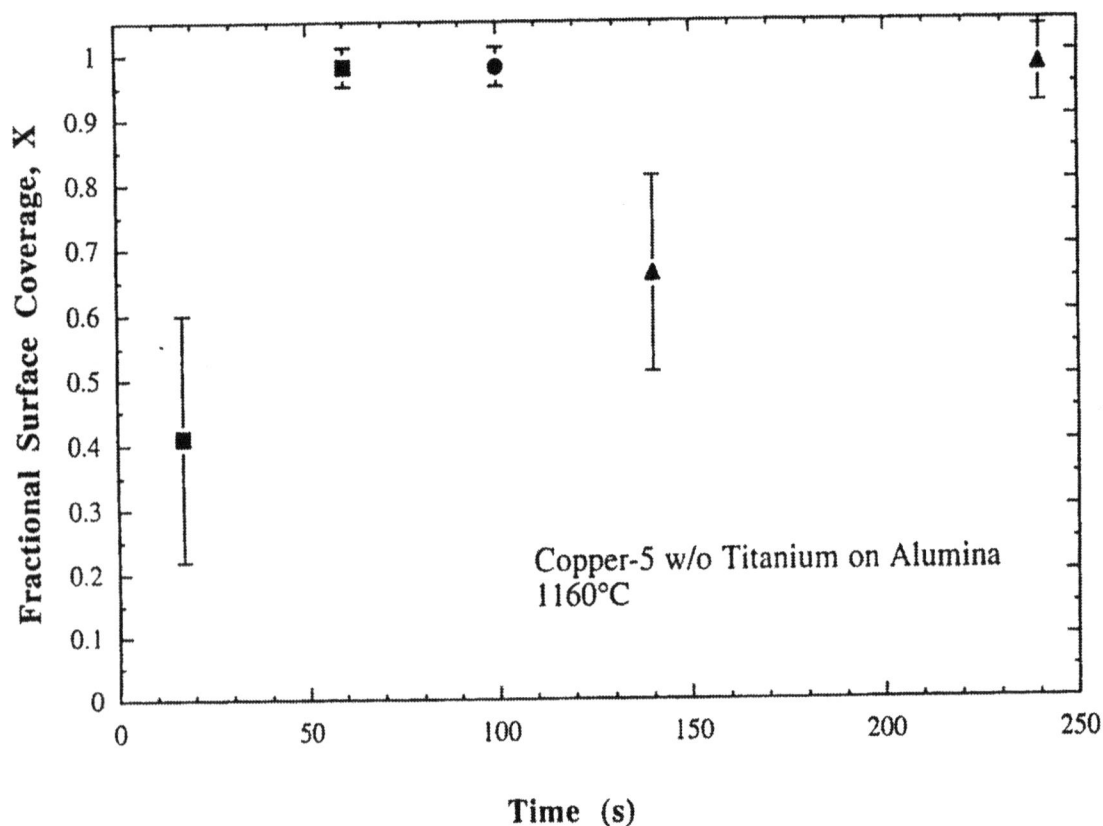

Figure 26: Fractional coverage of reaction product on alumina surface as a function of time for a copper-5 w/o titanium alloy at 1160°C. The 3 different symbols represent 3 independent sample runs. The error bars are the standard deviation of the mean for ten measurements.

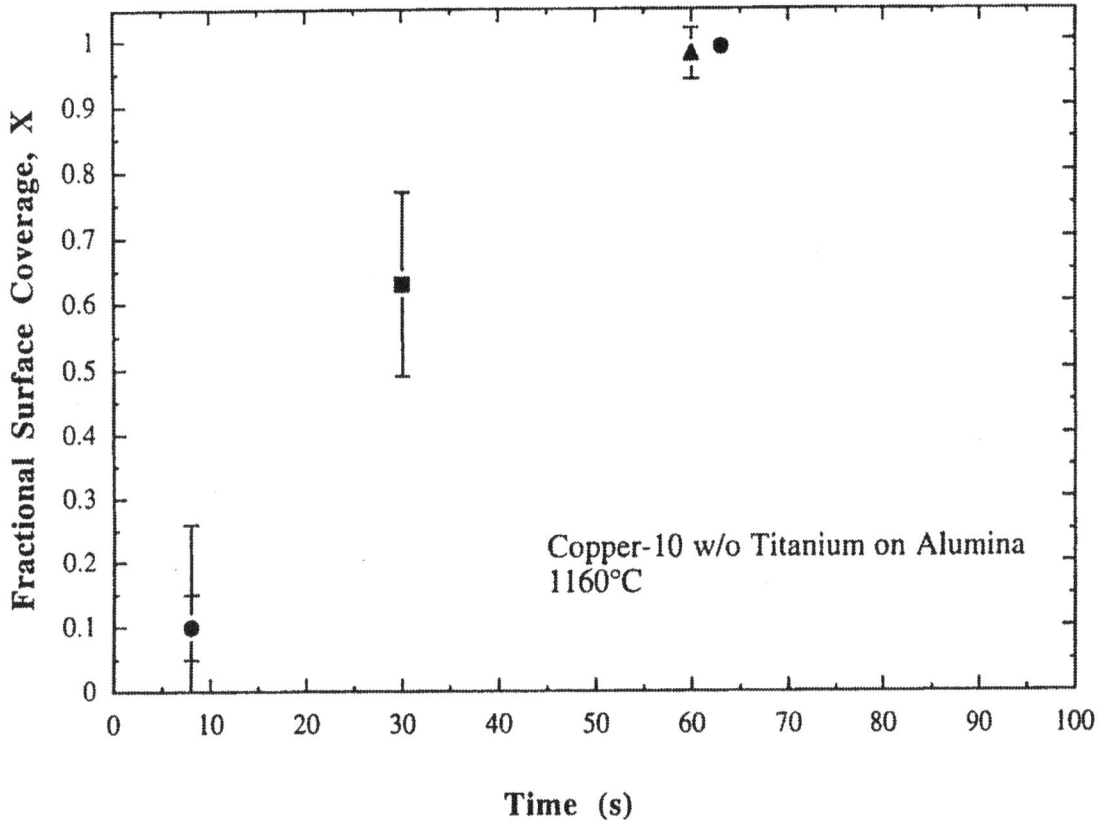

Figure 27: Fractional coverage of reaction product on alumina surface as a function of time for a copper-10 w/o titanium alloy at 1160°C. The 3 different symbols represent 3 independent sample runs. The error bars are the standard deviation of the mean for ten measurements.

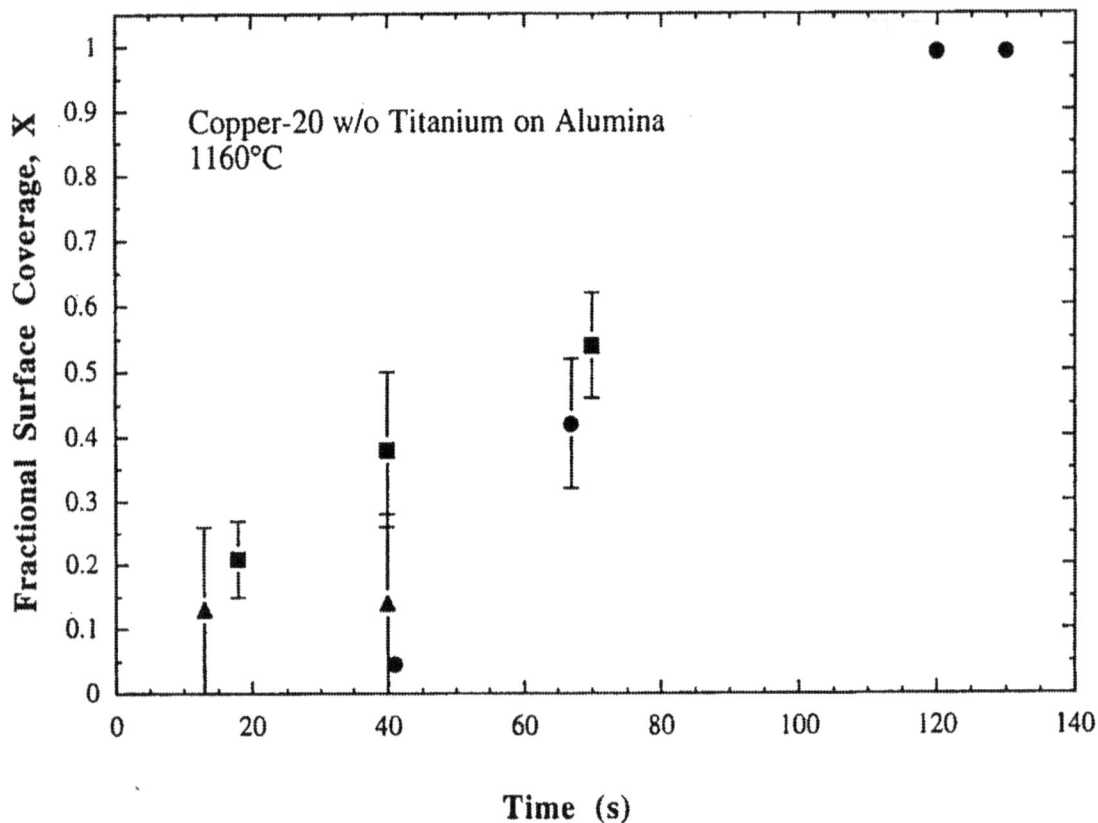

Figure 28: Fractional coverage of reaction product on alumina surface as a function of time for a copper-20 w/o titanium alloy at 1160°C. The 3 different symbols represent 3 independent sample runs. The error bars are the standard deviation of the mean for ten measurements.

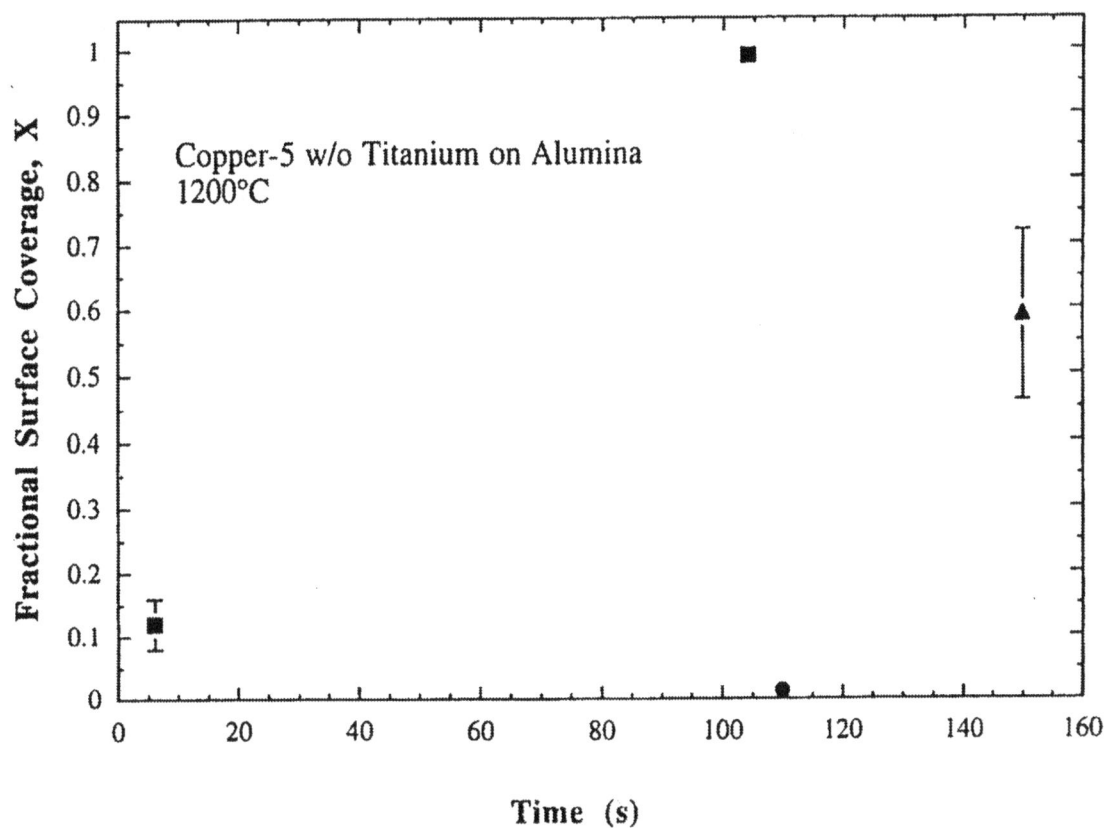

Figure 29: Fractional coverage of reaction product on alumina surface as a function of time for a copper-5 w/o titanium alloy at 1200°C. The 3 different symbols represent 3 independent sample runs. The error bars are the standard deviation of the mean for ten measurements.

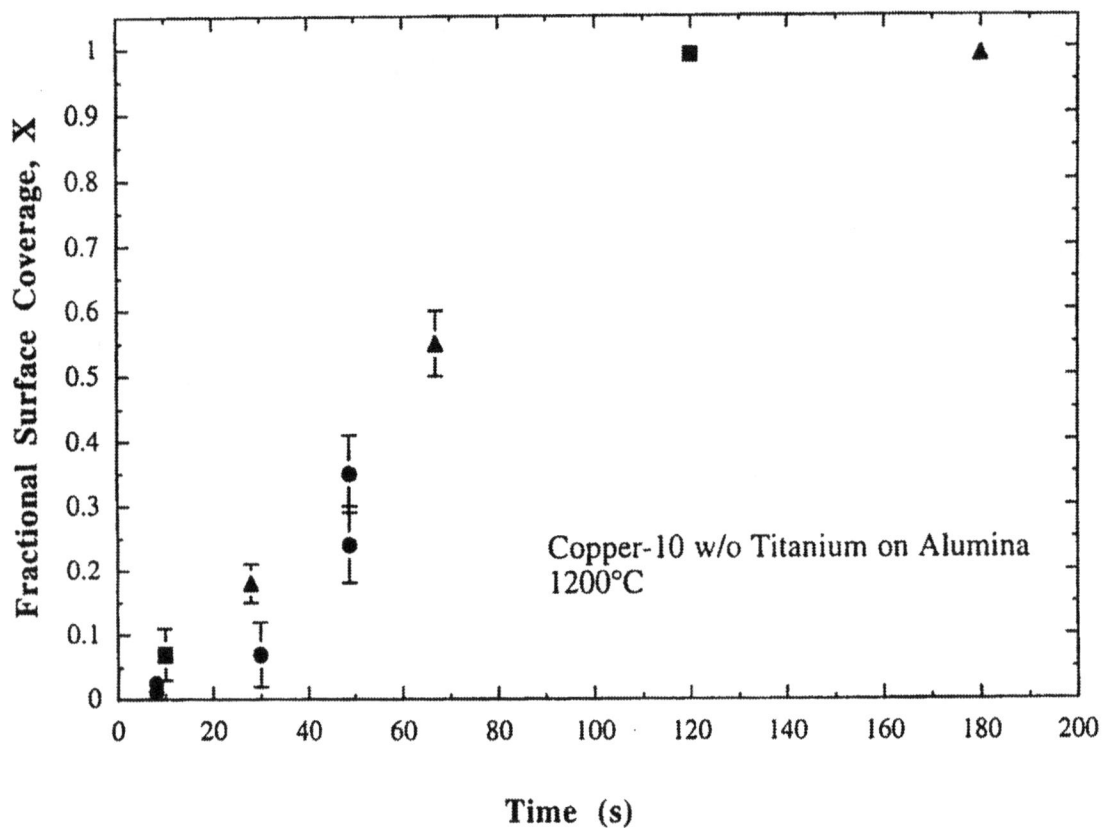

Figure 30: Fractional coverage of reaction product on alumina surface as a function of time for a copper-10 w/o titanium alloy at 1200°C. The 3 different symbols represent 3 independent sample runs. The error bars are the standard deviation of the mean for ten measurements.

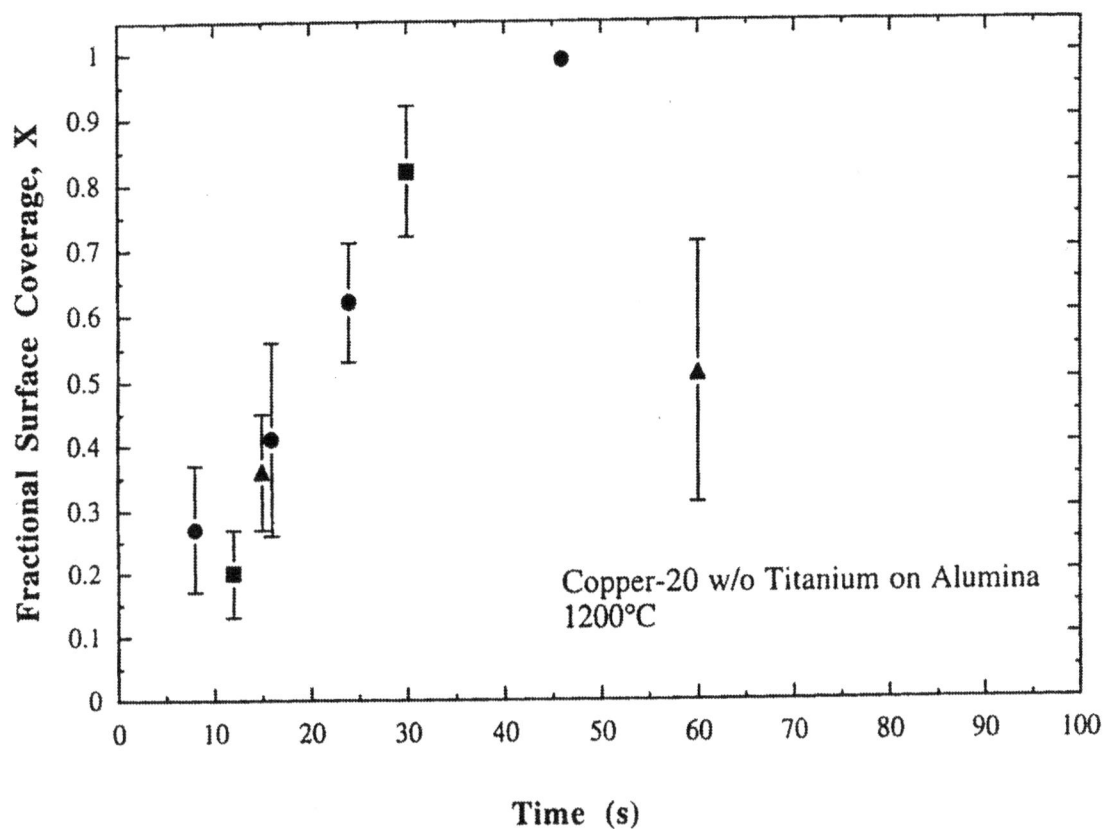

Figure 31: Fractional coverage of reaction product on alumina surface as a function of time for a copper-20 w/o titanium alloy at 1200°C. The 3 different symbols represent 3 independent sample runs. The error bars are the standard deviation of the mean for ten measurements.

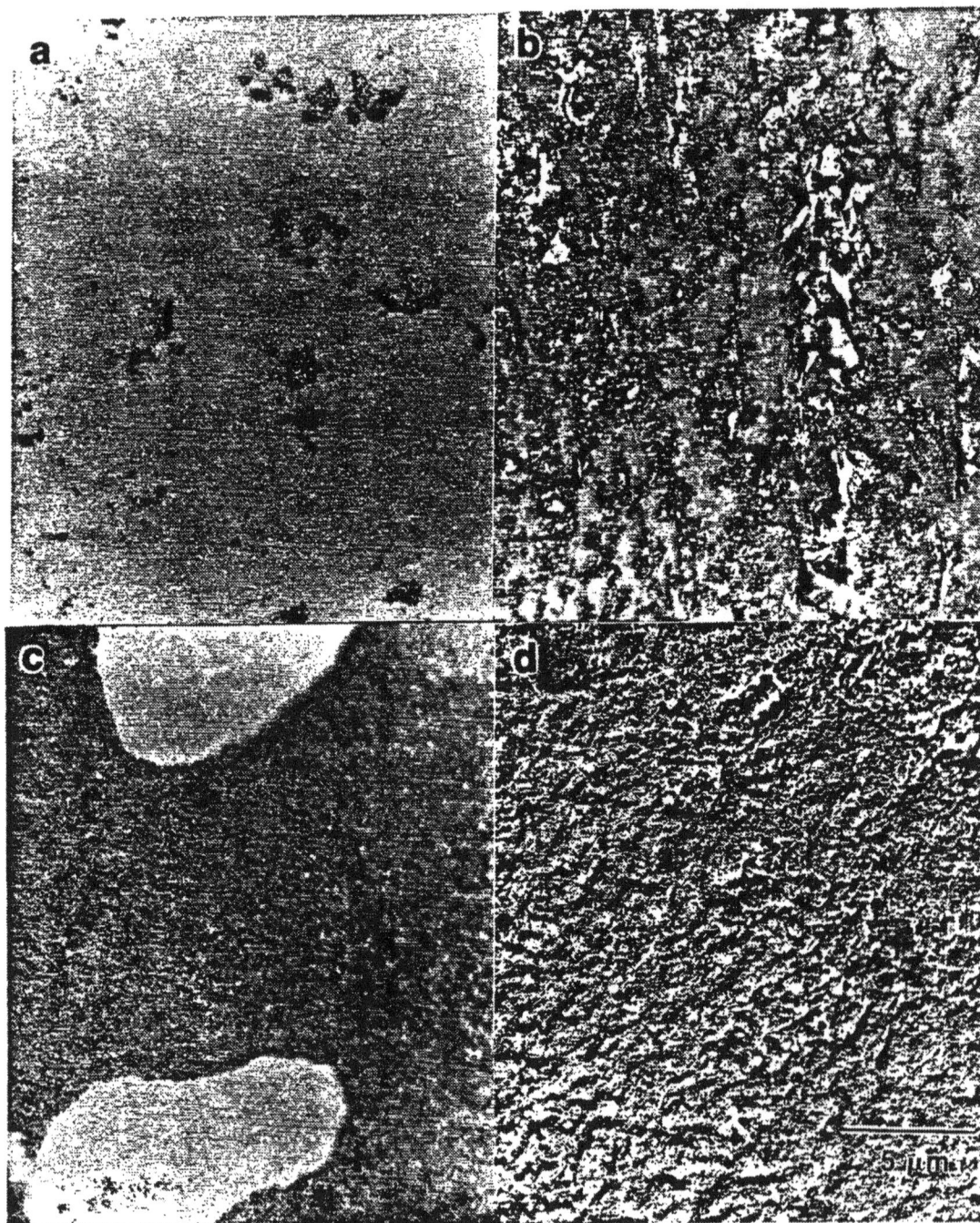

Figure 32: Micrographs of fractional coverage of reaction product on alumina surfaces for copper-titanium alloys (400x), a.) initial islands form, b.) islands grow together, c.) alumina islands are left, and d.) complete coverage is obtained.

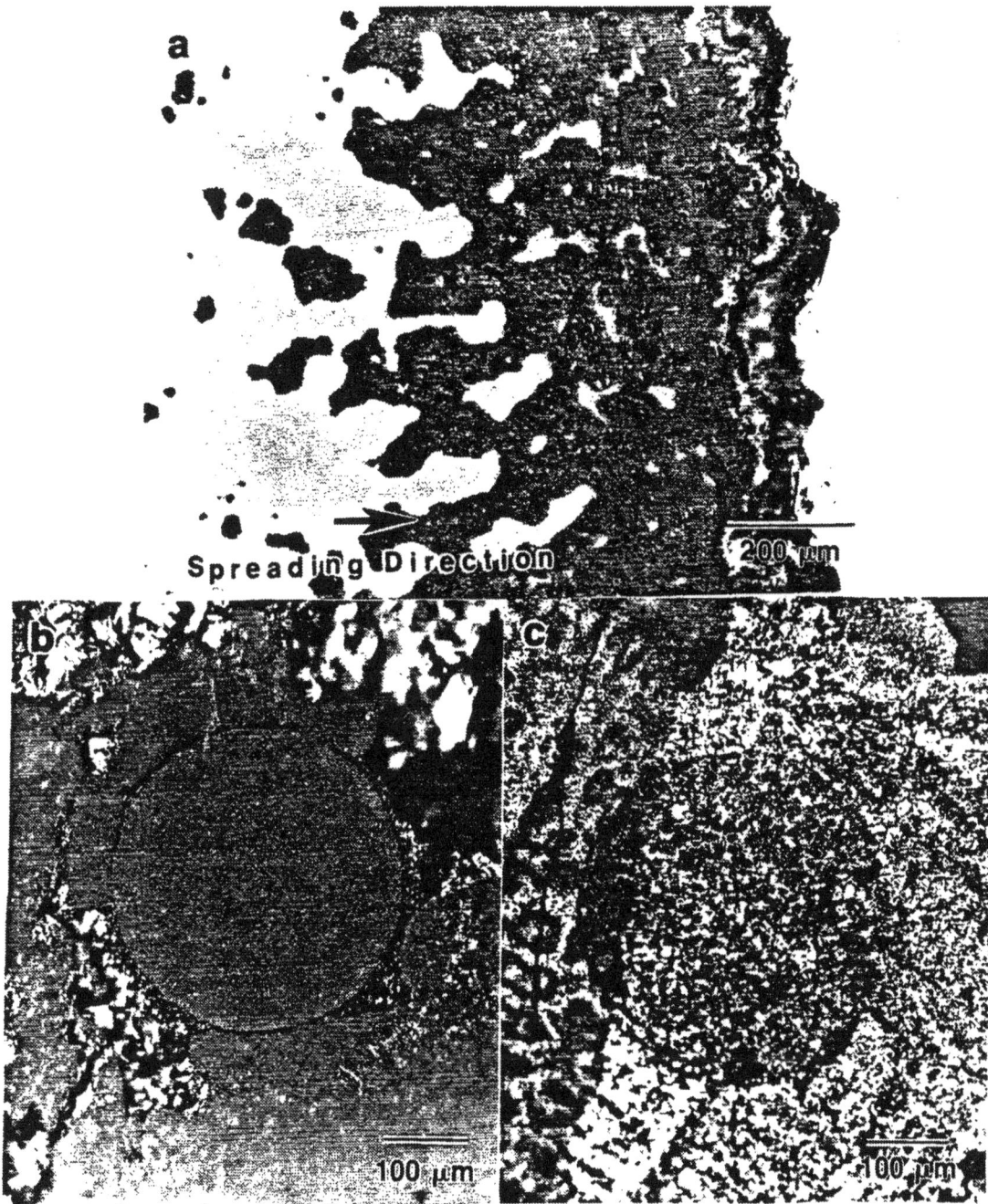

Figure 33: Anomolous spreading behavior for copper-titanium alloys on alumina, a.)
enhanced reaction at the edge of the drop (100x), and b.), c.) circular region
with non-uniform product formation (125x).

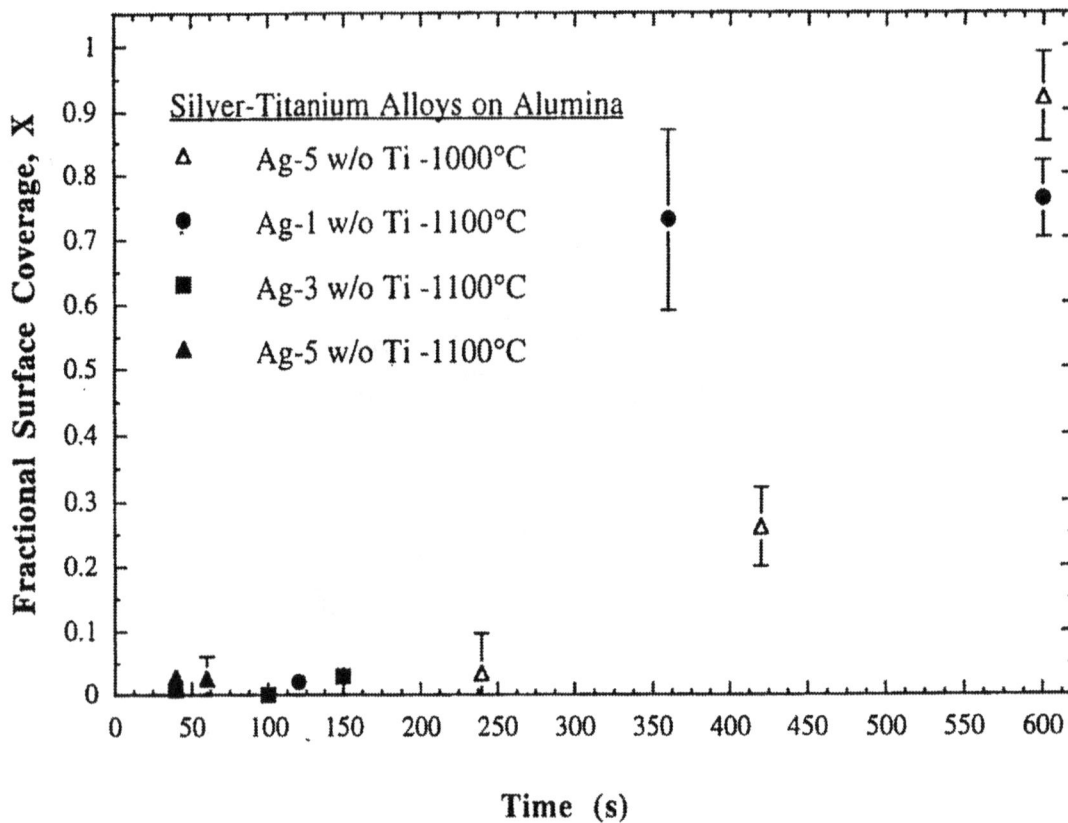

Figure 34: Fractional coverage of reaction product on alumina surface as a function of time for silver-titanium alloys at 1000 and 1100°C. The error bars are the standard deviation of the mean for fifteen measurements.

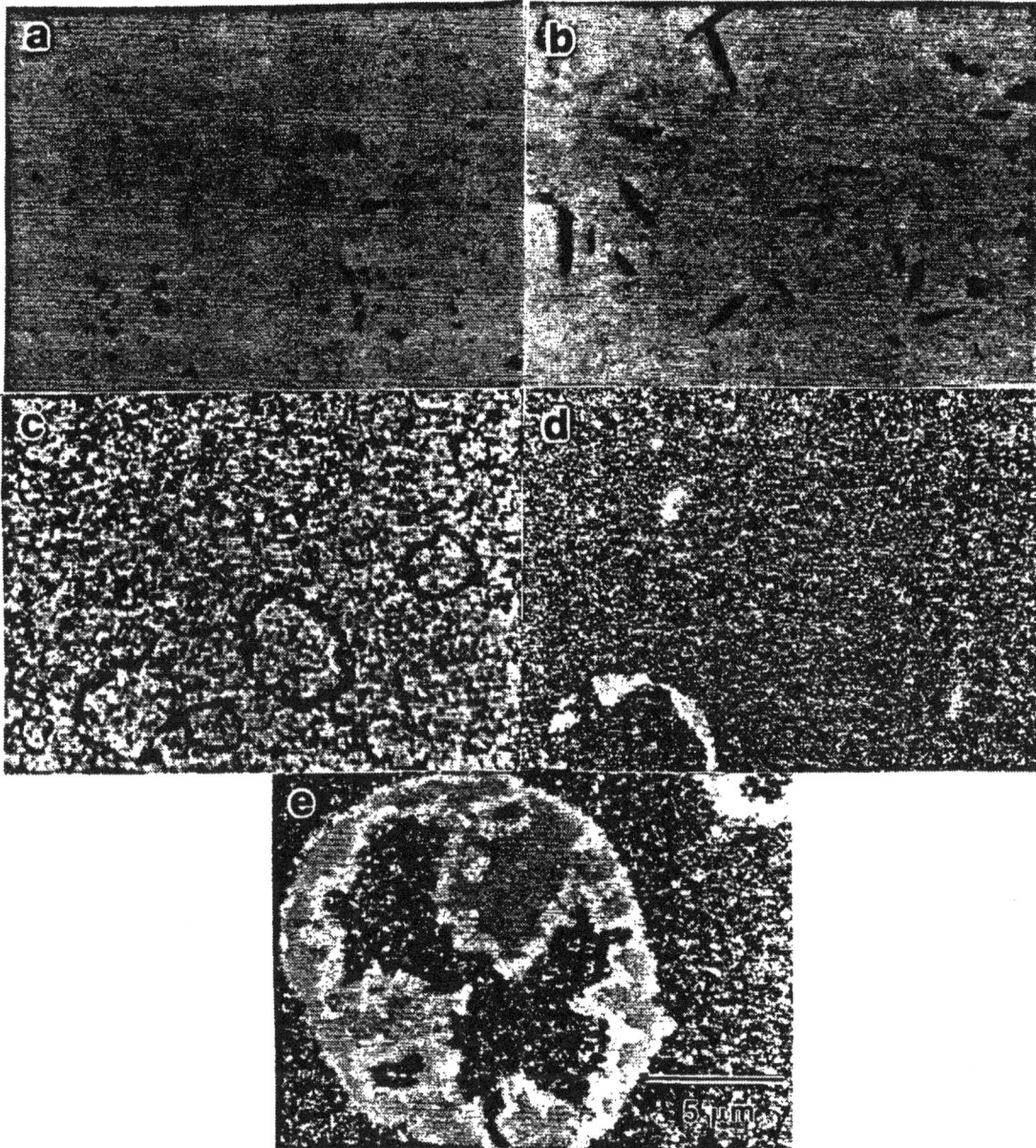

Figure 35: Micrographs of fractional coverage of reaction product on alumina surface for silver-titanium alloys (400x): a.) small islands of reaction product form, b.) in some cases, needle-like product islands grow with time, c.) the islands grow together until d.) the surface is nearly covered with reaction product, e.) large unreacted islands are sometimes observed in nearly completely reacted regions.

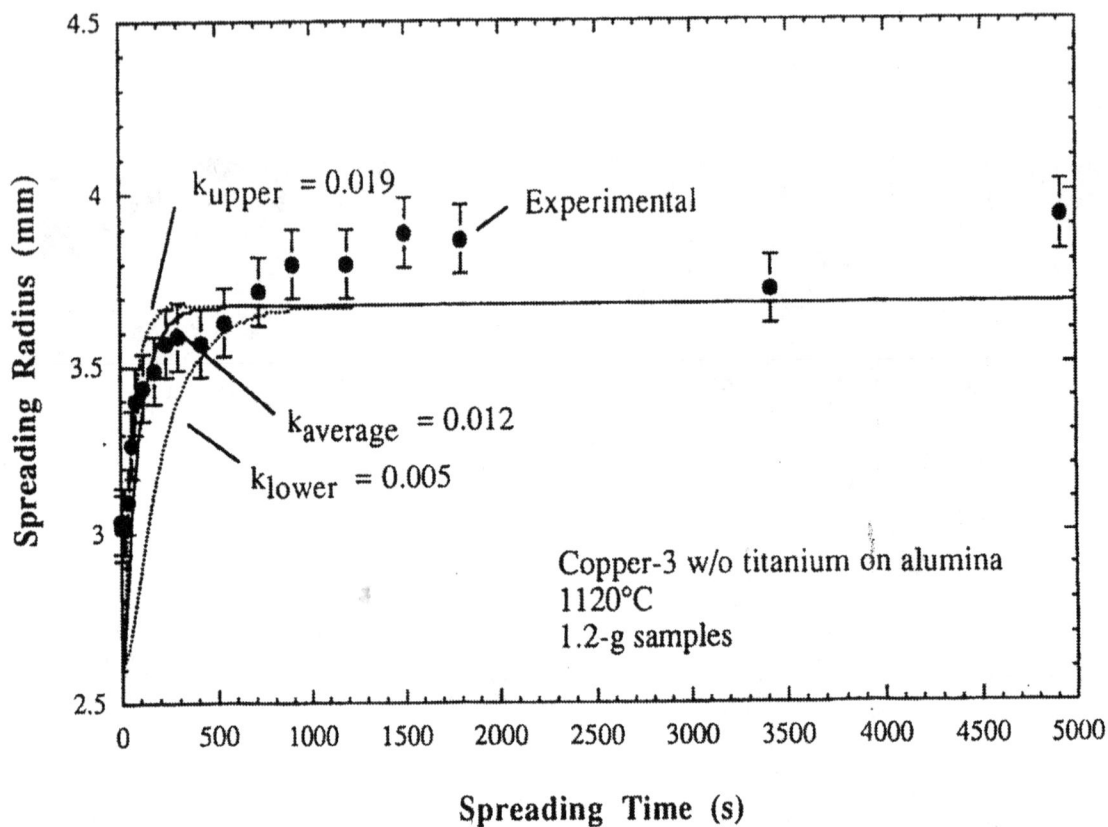

Figure 36: Theoretical spreading radius curves and experimental data for copper-3 w/o titanium on alumina at 1120°C. The error bars on the experimental data represent the uncertainty in the radius measurements.

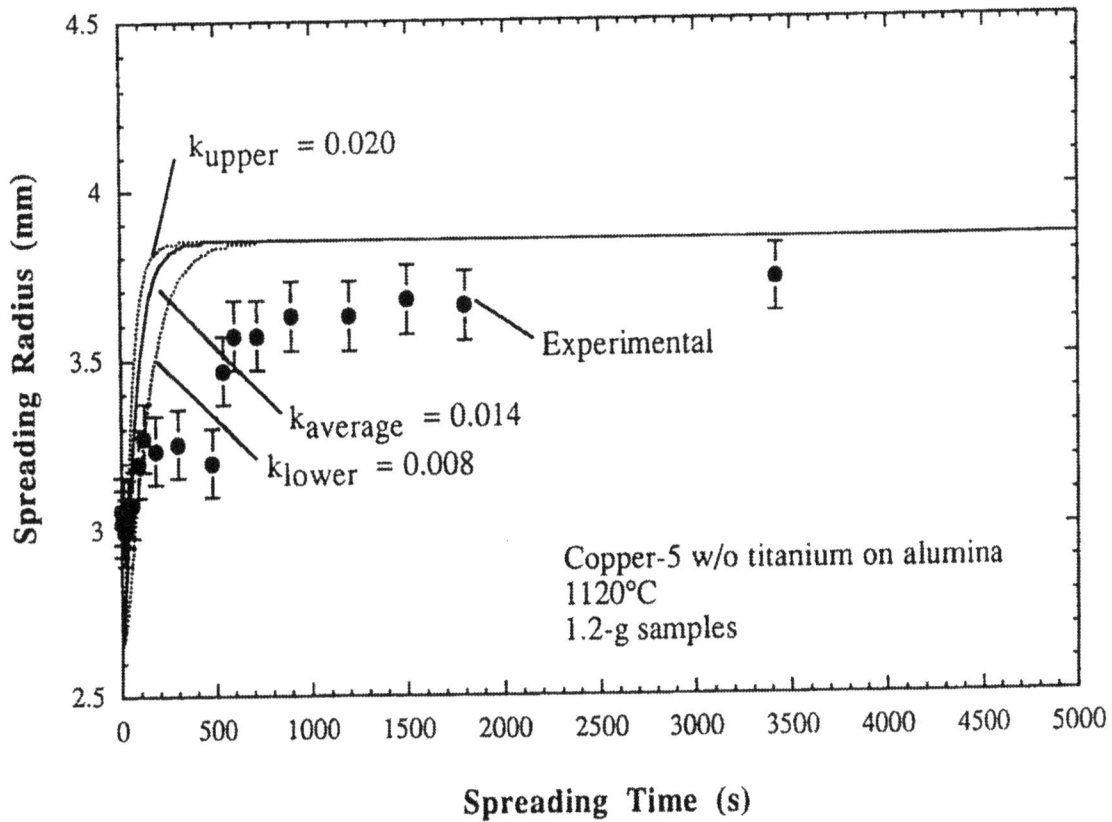

Figure 37: Theoretical spreading radius curves and experimental data for copper-5 w/o titanium on alumina at 1120°C. The error bars on the experimental data represent the uncertainty in the radius measurements.

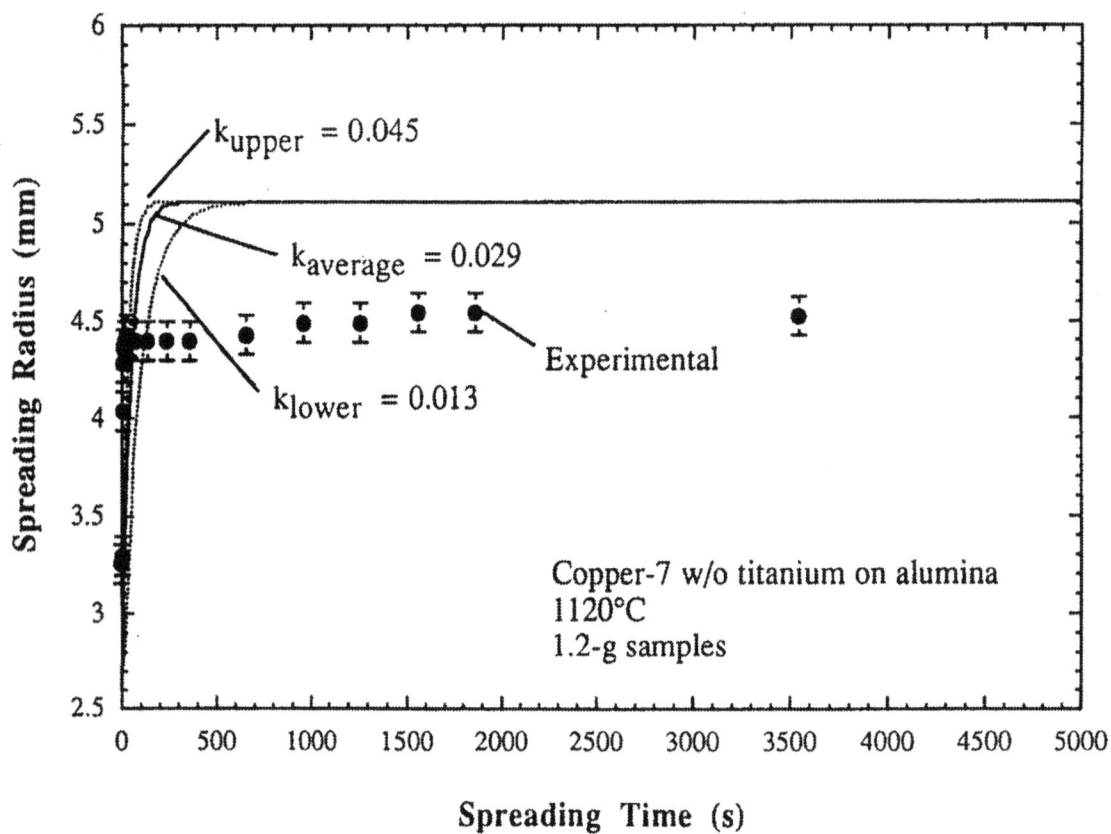

Figure 38: Theoretical spreading radius curves and experimental data for copper-7 w/o titanium on alumina at 1120°C. The error bars on the experimental data represent the uncertainty in the radius measurements.

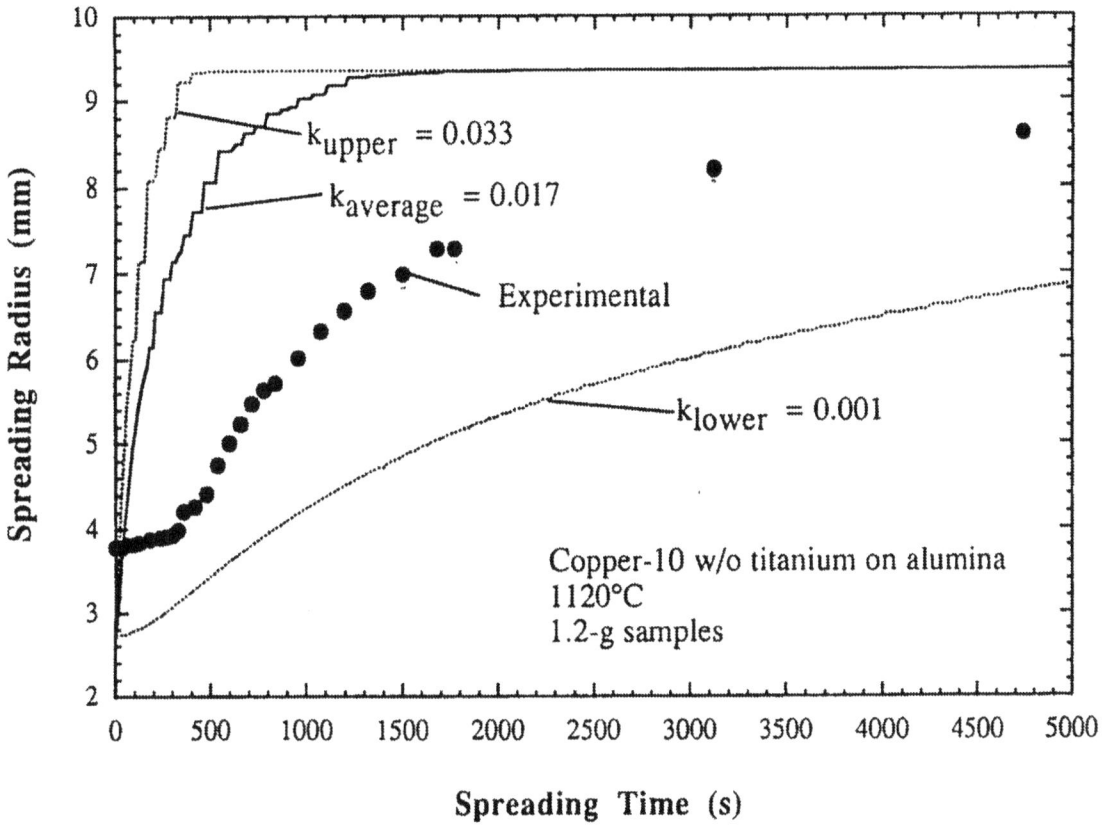

Figure 39: Theoretical spreading radius curves and experimental data for copper-10
w/o titanium on alumina at 1120°C. The error in the radius measurement
is smaller than the data point height.

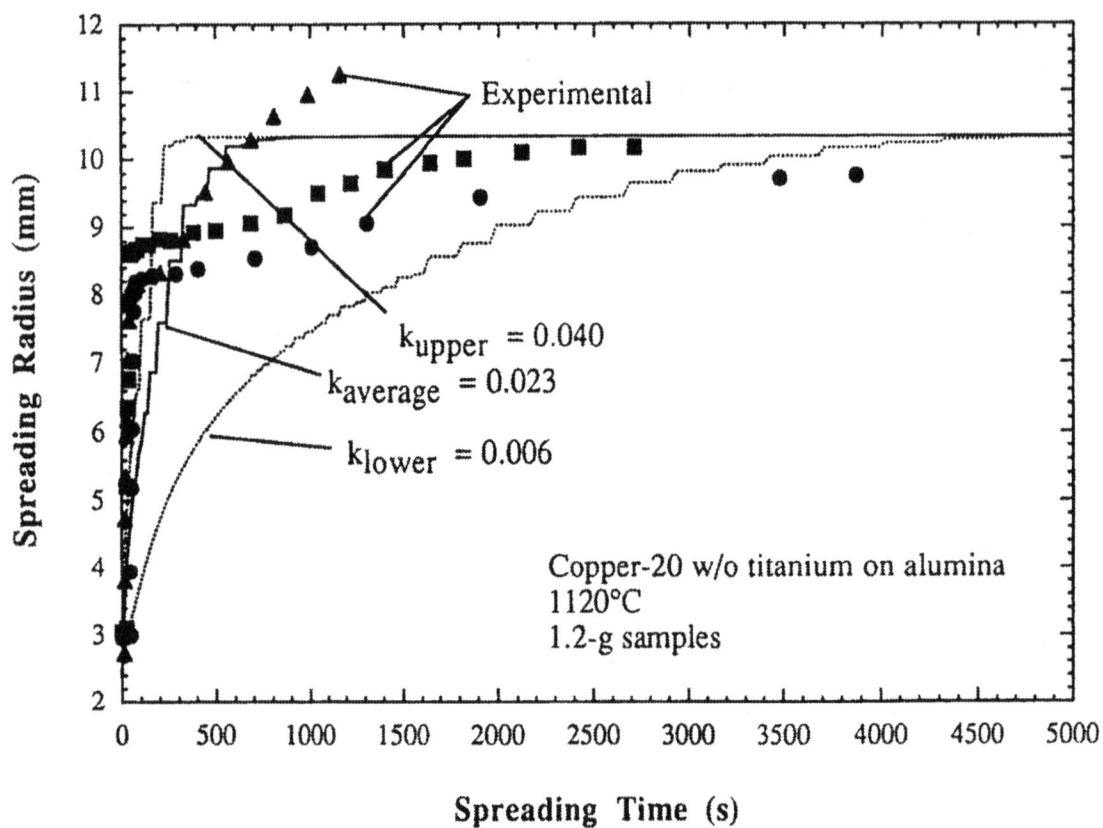

Figure 40: Theoretical spreading radius curves and experimental data for copper-20 w/o titanium on alumina at 1120°C. The error in the radius measurement is smaller than the data point height.

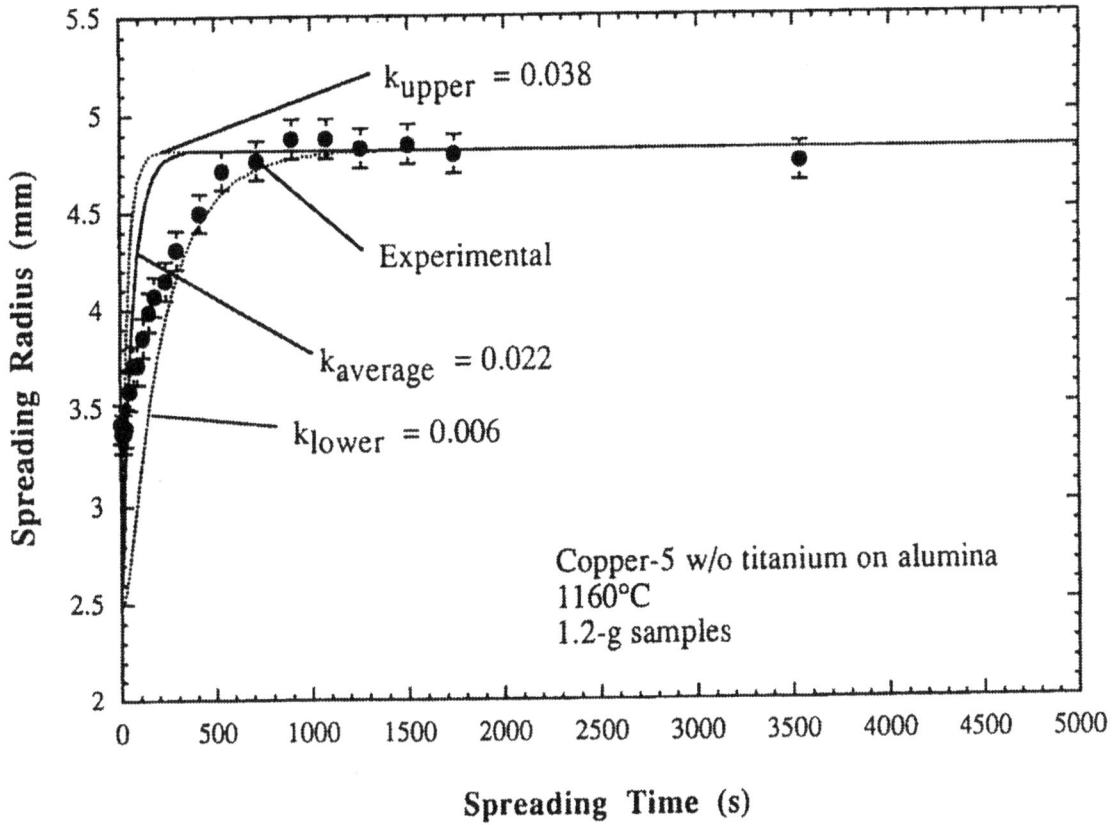

Figure 41: Theoretical spreading radius curves and experimental data for copper-5 w/o titanium on alumina at 1160°C. The error bars on the experimental data represent the uncertainty in the radius measurements.

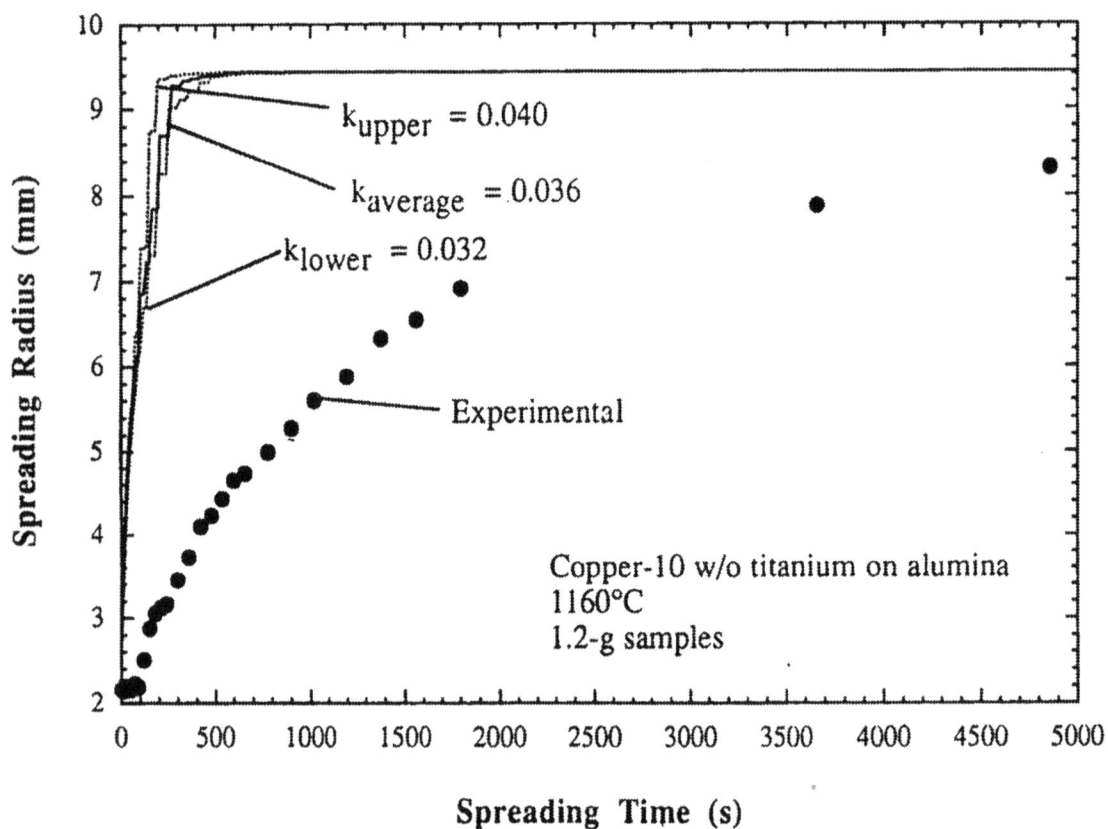

Figure 42: Theoretical spreading radius curves and experimental data for copper-10 w/o titanium on alumina at 1160°C. The error in the radius measurement is smaller than the data point height.

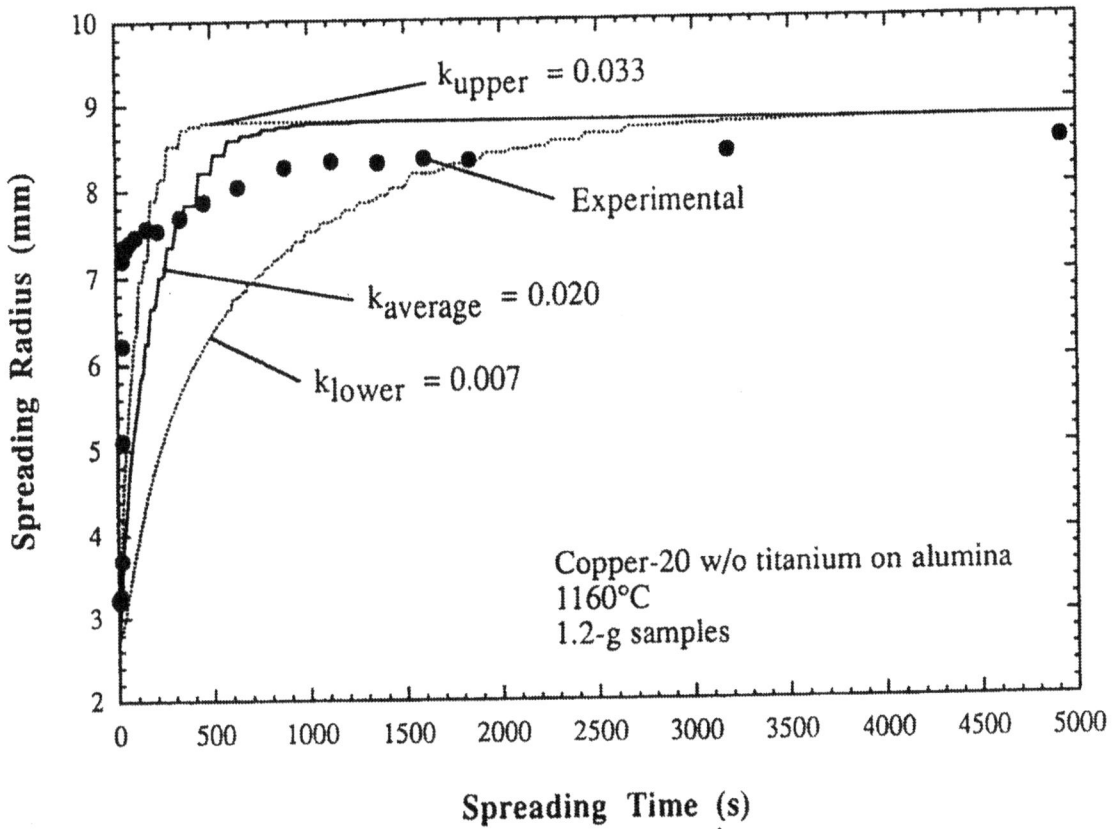

Figure 43: Theoretical spreading radius curves and experimental data for copper-20 w/o titanium on alumina at 1160°C. The error in the radius measurement is smaller than the data point height.

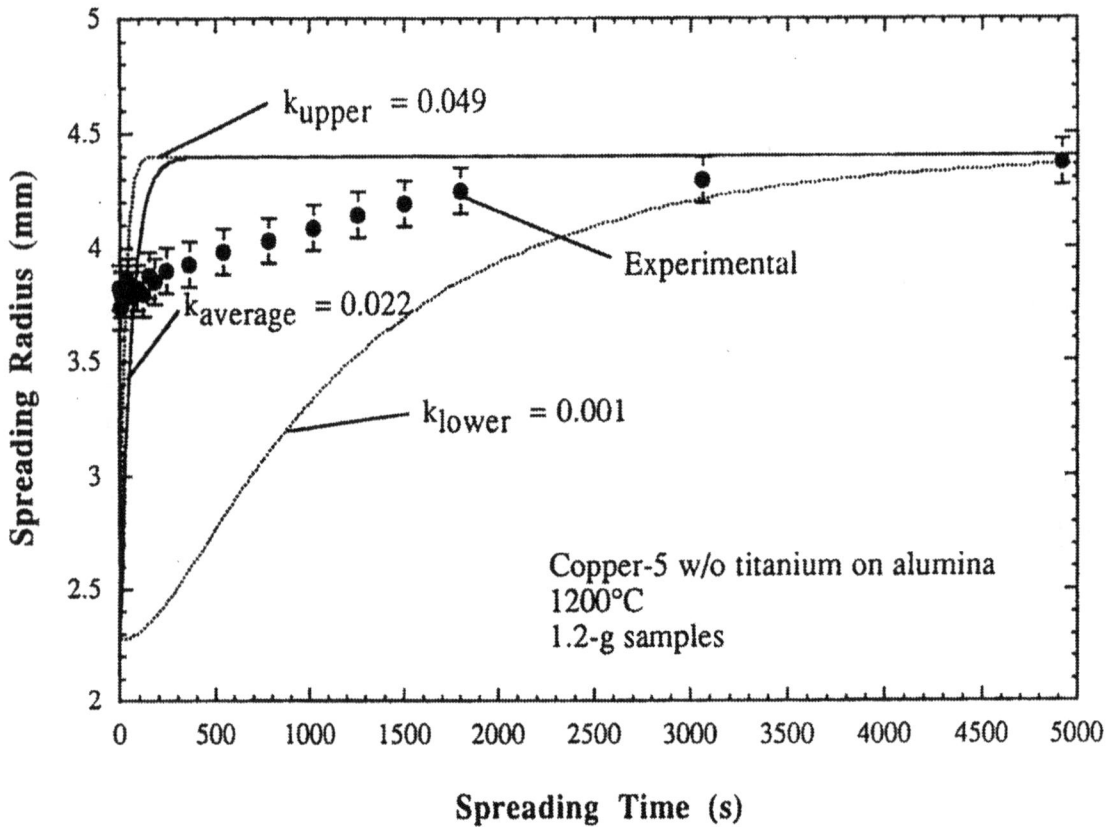

Figure 44: Theoretical spreading radius curves and experimental data for copper-5 w/o titanium on alumina at 1200°C. The error bars on the experimental data represent the uncertainty in the radius measurements.

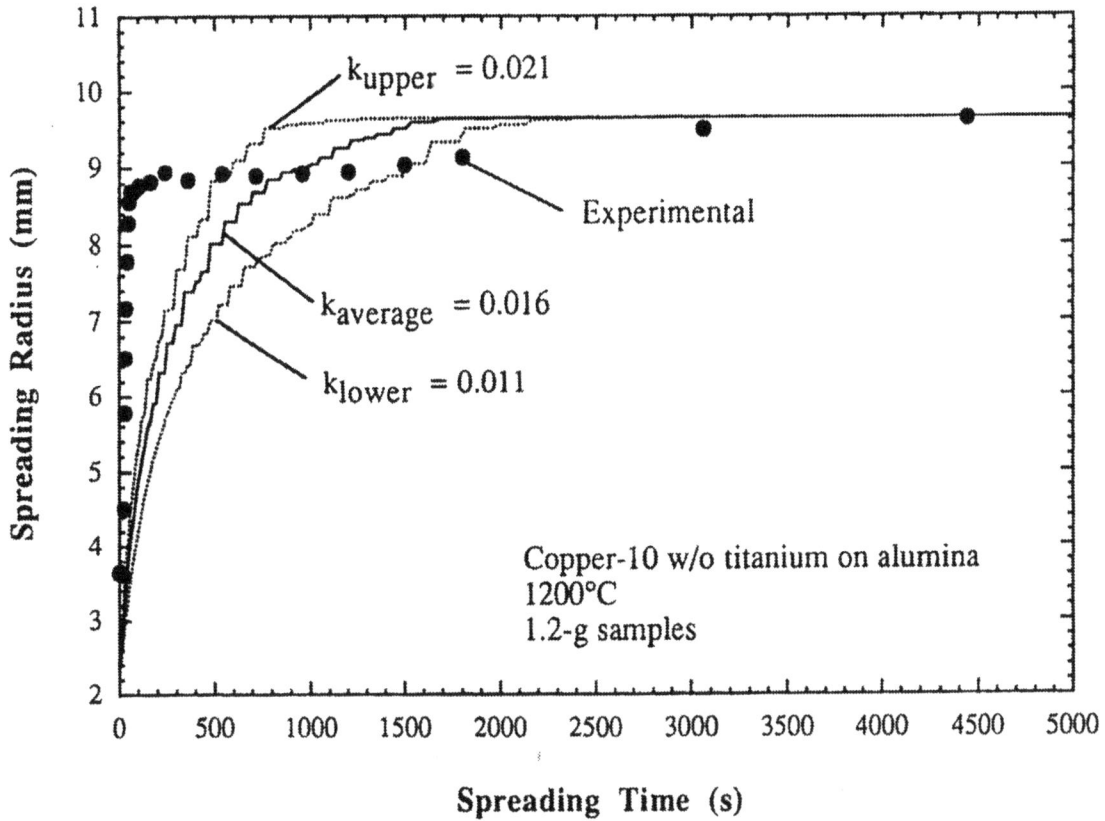

Figure 45: Theoretical spreading radius curves and experimental data for copper-10 w/o titanium on alumina at 1200°C. The error in the radius measurement is smaller than the data point height.

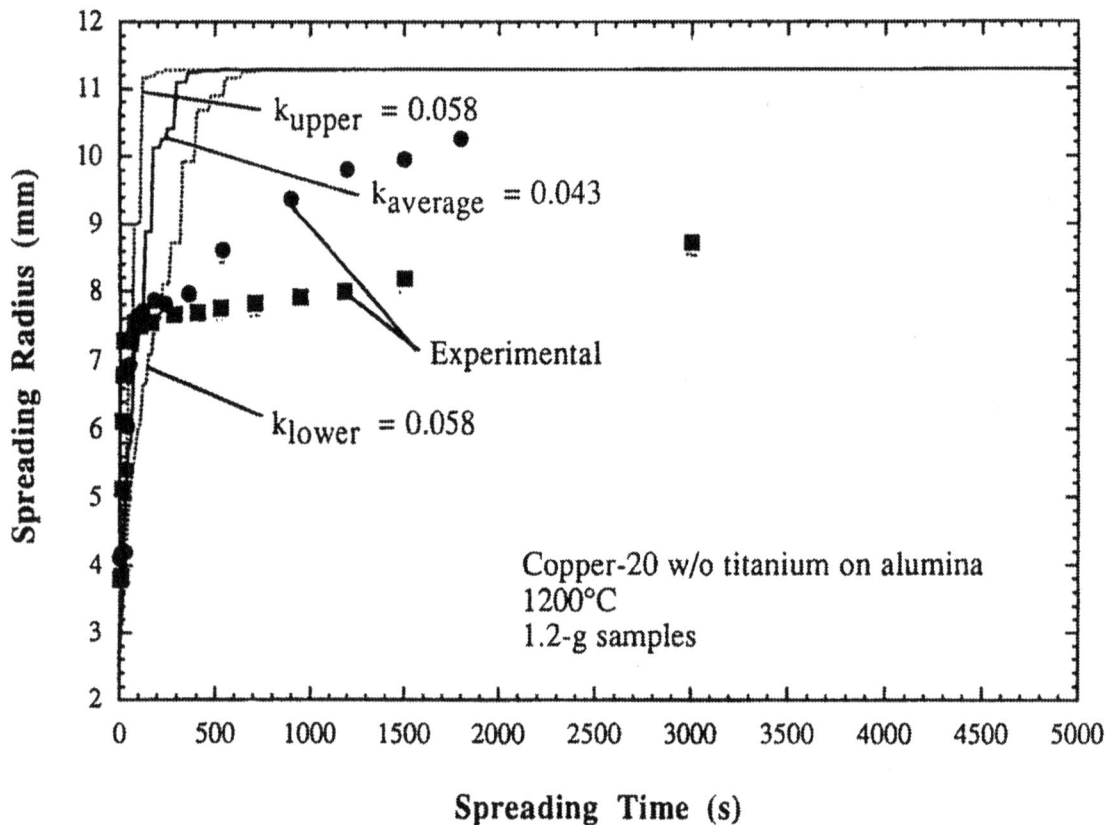

Figure 46: Theoretical spreading radius curves and experimental data for copper-20
w/o titanium on alumina at 1200°C. The error in the radius measurement
is smaller than the data point height.

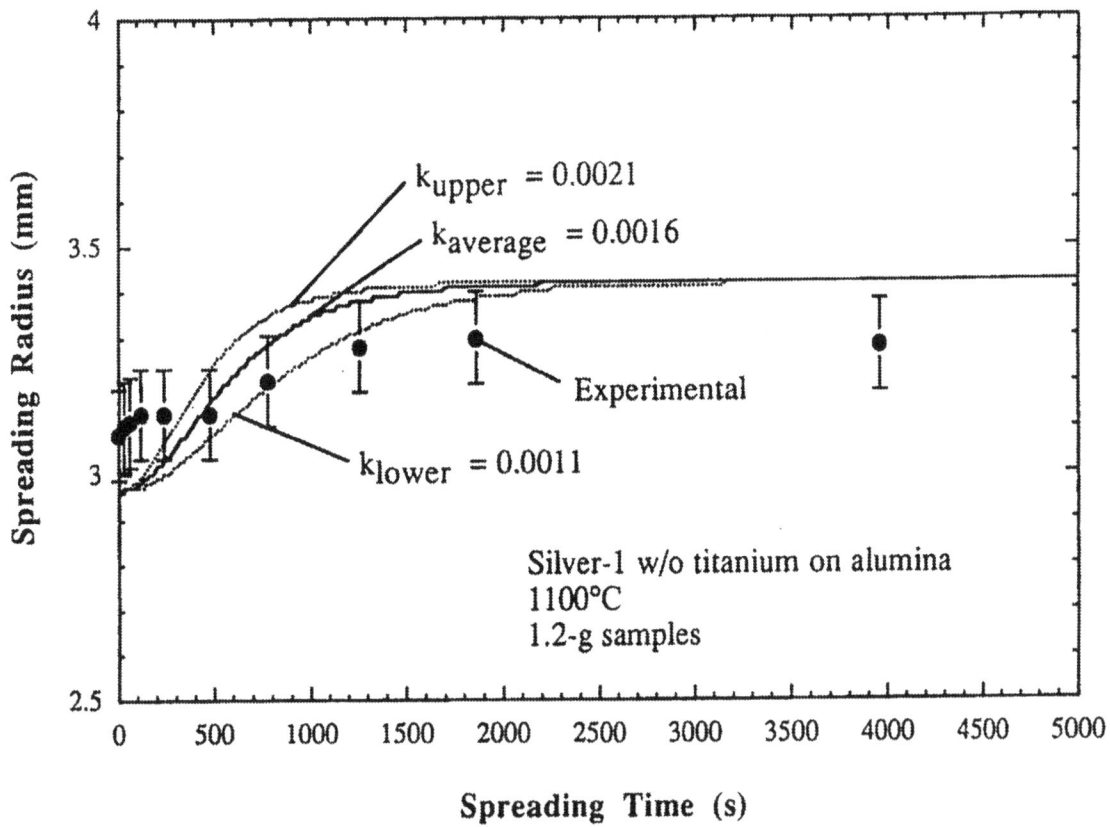

Figure 47: Theoretical spreading radius curves and experimental data for silver-1 w/o titanium on alumina at 1100°C. The error bars on the experimental data represent the uncertainty in the radius measurements.

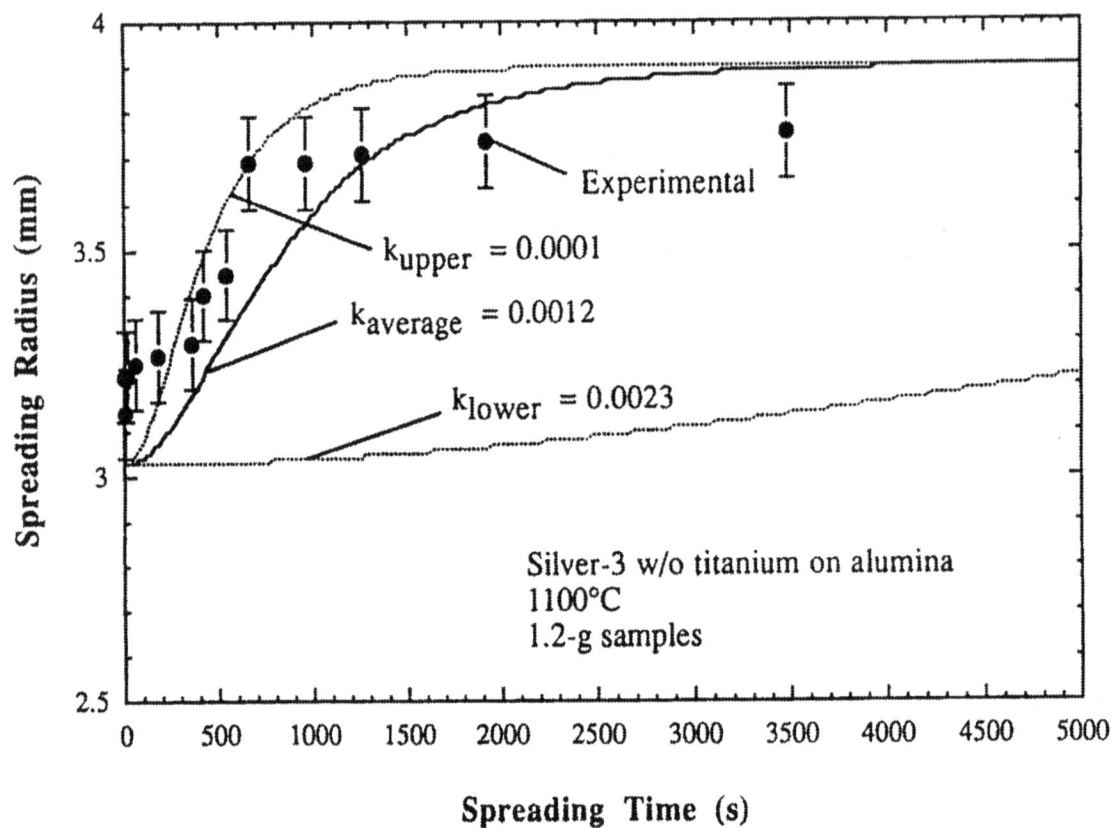

Figure 48: Theoretical spreading radius curves and experimental data for silver-3 w/o titanium on alumina at 1100°C. The error bars on the experimental data represent the uncertainty in the radius measurements.

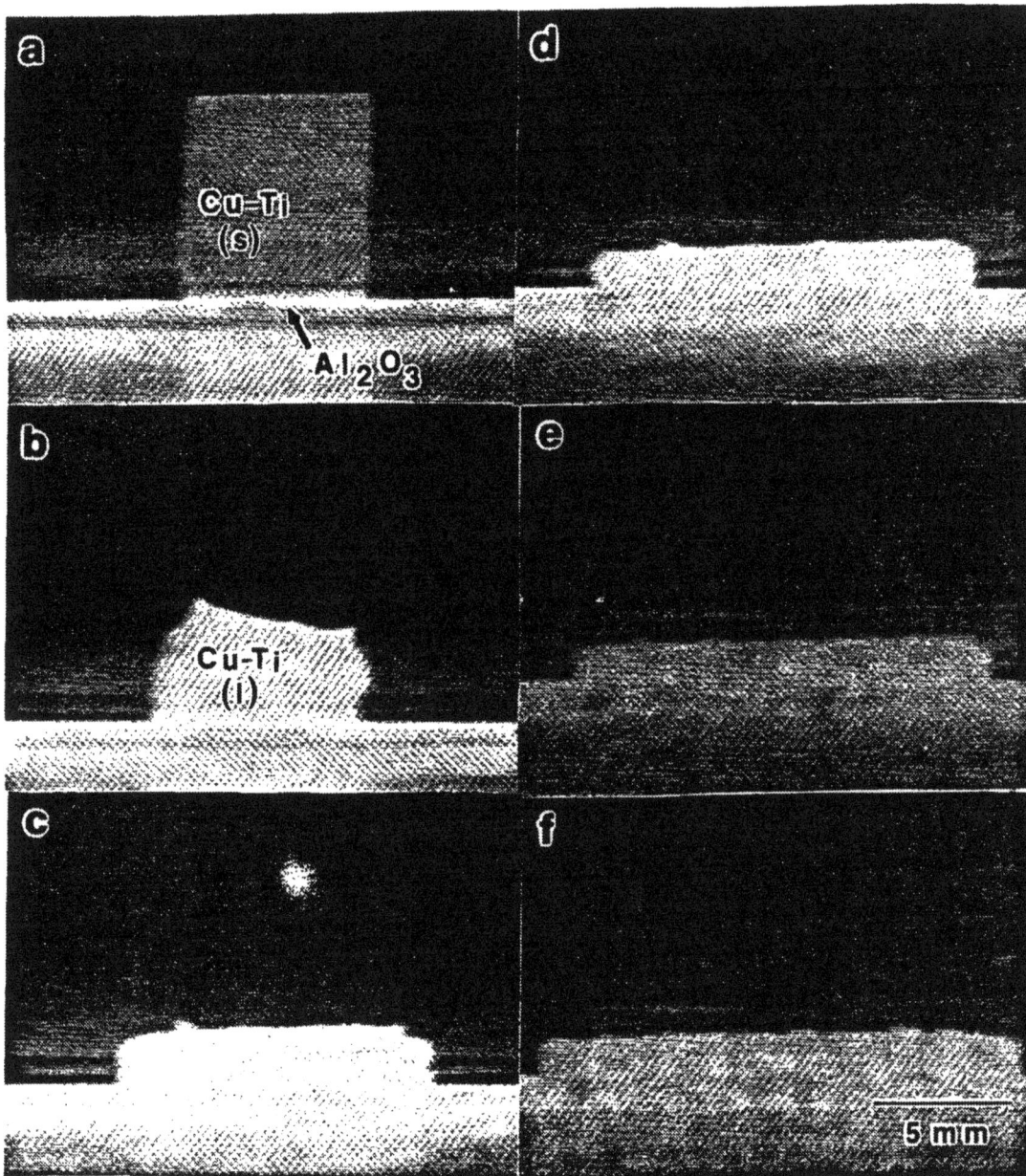

Figure 49: Spreading of a pre-alloyed copper-20 w/o titanium alloy on alumina at 1200°C: a.) initial configuration, b.) alloy melting (≈1010°C), c.) significant spreading during the heat-up cycle (1100°C), d.) 1150°C, e.) upon reaching 1200°C, and f.) after holding at 1200°C for 7 ks, very little spreading has occurred.

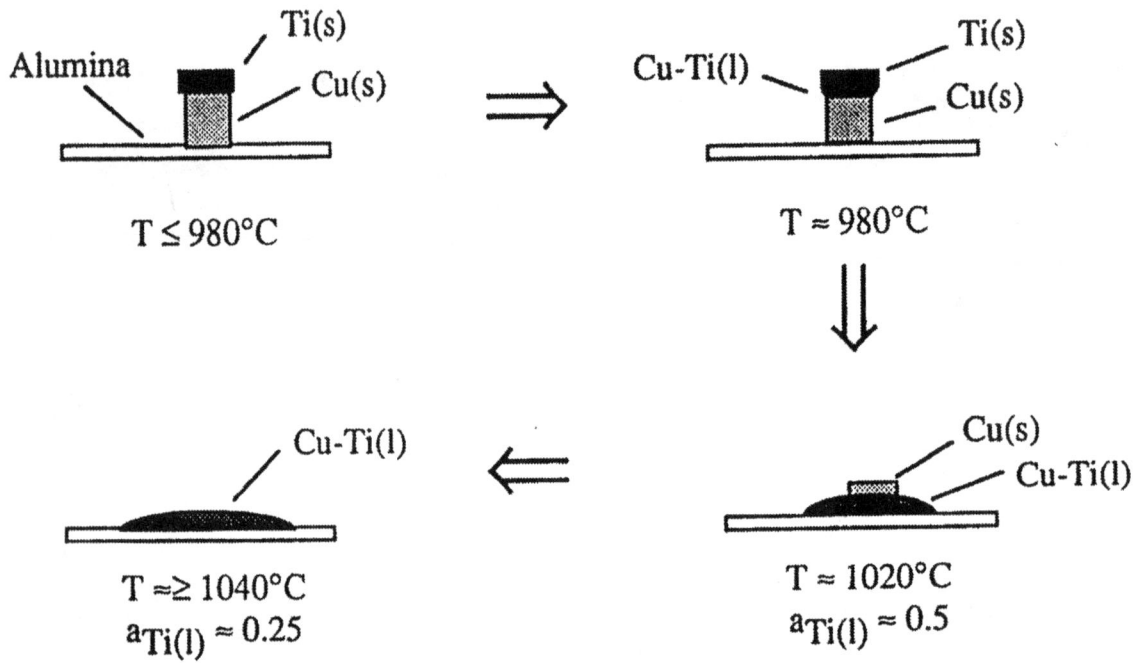

Figure 50: Schematic diagram of the changes which occur when a titanium/copper/alumina stack is heated from room temperature to 1200°C.

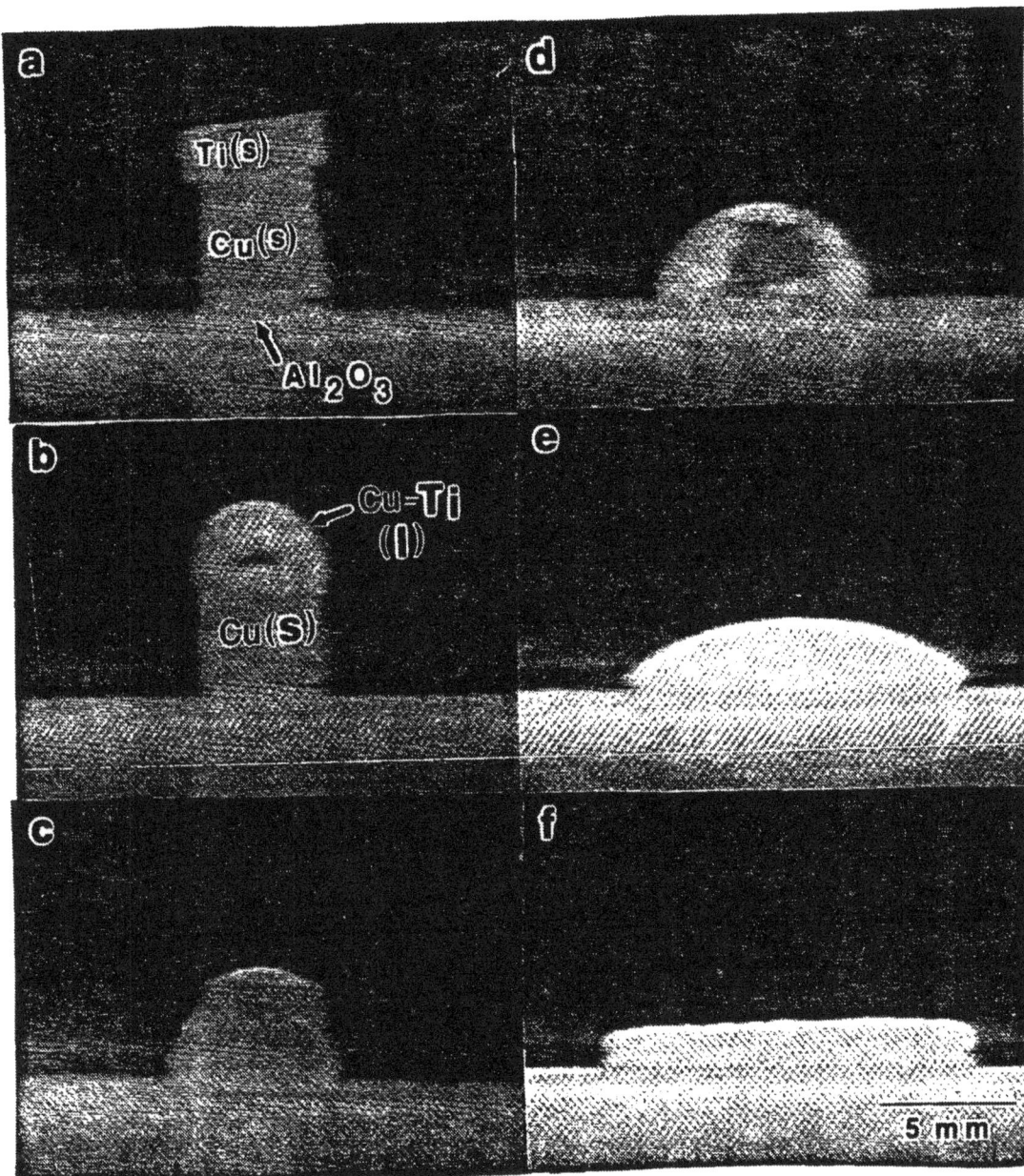

Figure 51: Spreading of a copper-20 w/o titanium alloy on alumina at 1200°C for the titanium stacked on copper starting configuration: a.) initial configuration, b.) high Ti concentration liquid forms (≈975°C), c.) the high-Ti liquid falls onto the slumina substrare (≈990°C), d.) the liquid spreads before the Cu is completely dissolved (≈1005), e.) spreading continues during the heat-up cycle (1100°C), and f.) spreading is nearly complete when 1200°C is attained.

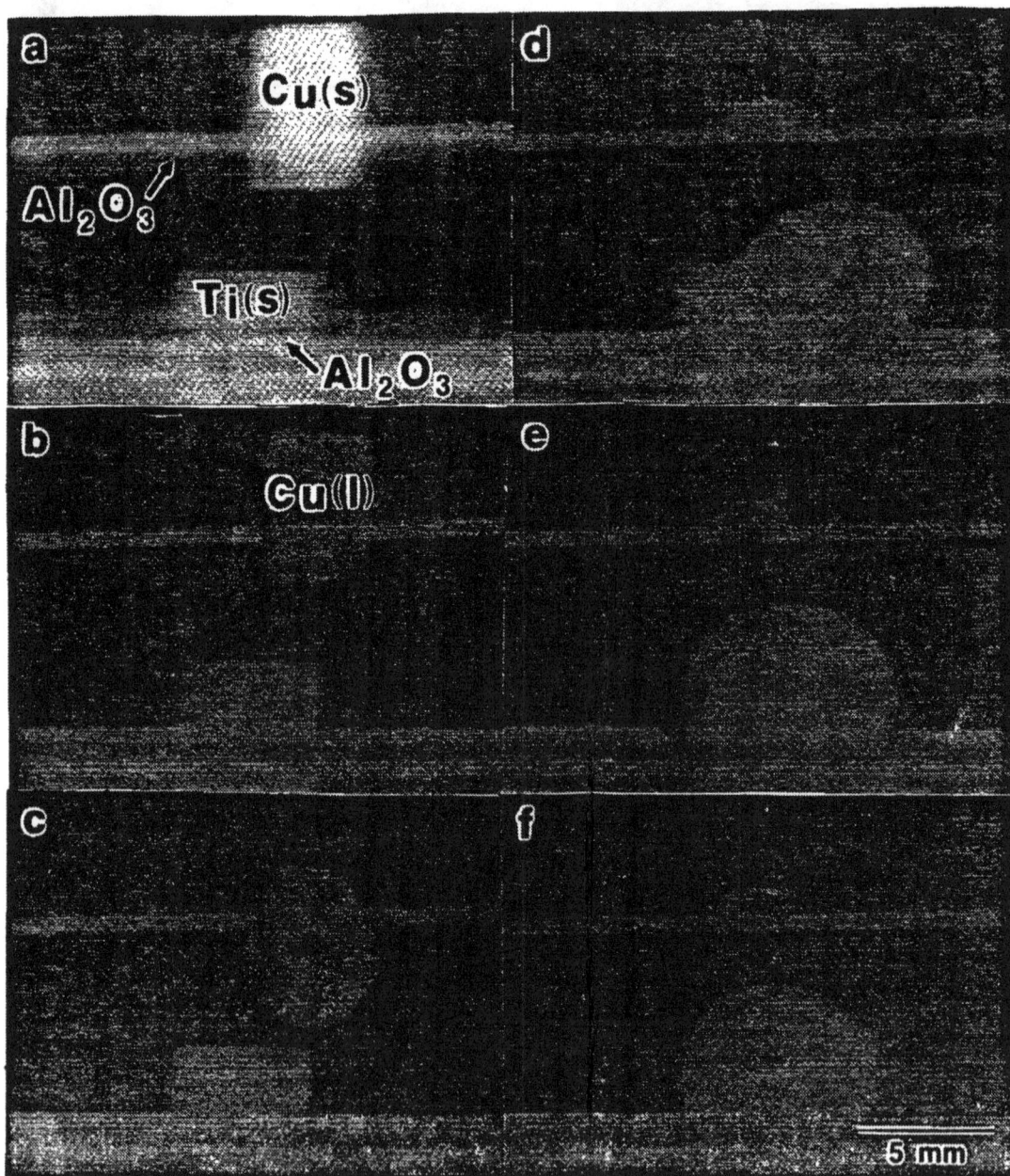

Figure 52: Formation of copper-20 w/o titanium alloy using the improved sessile drop
configuration: a.) initial configuration, b.) copper melts (≈1100°C), c.) liquid
copper creeps through gap between plates, d.) copper falls onto Ti (1200°C),
e.) Cu dissolves Ti (~8s after contact), and f.) dissolution is complete (~24s
after contact).

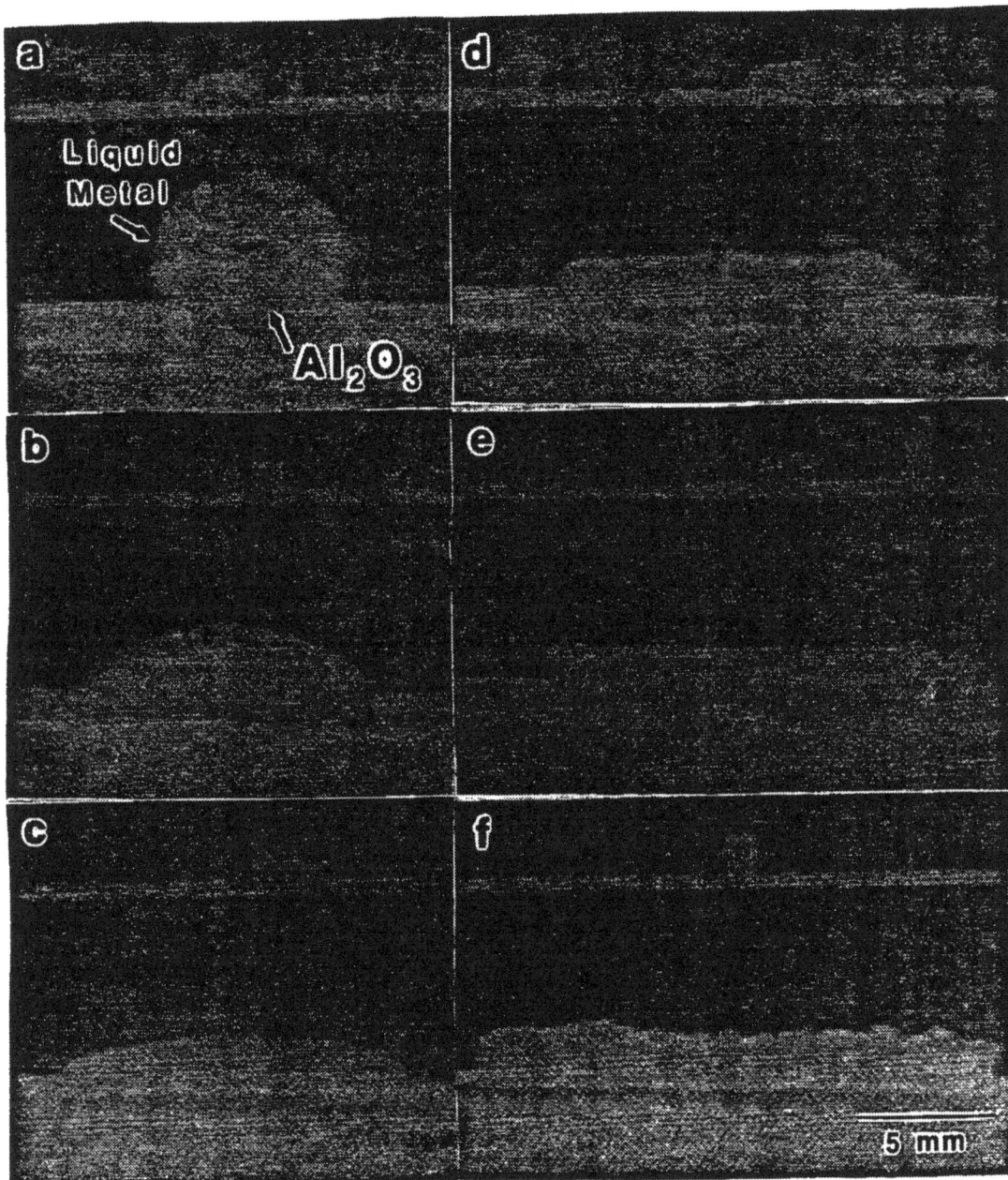

Figure 53: Spreading of copper-20 w/o titanium alloys on alumina at 1200°C for the mproved sessile drop configuration: a.) at 1200°C (t = 0 s), b.) t = 40 s, c.) t = 300 s, e.) t = 1200 s, and f.) t = 10 ks.

D.   REFERENCES

1   "Thin Film Substrate Technical Specifications 10-2-0692", Coors Ceramics
        Company-Electronics, Golden, CO 80401.

2   Loehman, R., "Interfacial Reactions in Ceramic-Metal Systems", Ceram Bull 68
        [4] (1989), pp. 891-896.

3   Loehman, R.E., and A.P. Tomsia, "Reactions of Ti and Zr with AlN and Al_2O_3",
        Acta Metall Mater 40 (suppl 1992), pp. 575-583.

4   Laurent, V., D. Chatain, and N. Eustathopoulos, "Wettability of SiO_2 and oxidized
        SiC by Aluminum", Mat Sci Eng A135 (1991), pp. 89-94.

5   Xue, X.M., J.T. Wang and M.X. Quan, "Wettability and Spreading Kinetics of
        Liquid Aluminum on Boron Nitride", J Mater Sci 26 (1991), pp. 6391-6395.

6   Shimbo, M., M. Naka, and I. Okamoto, "Wettability of Silicon Carbide by
        Aluminum, Copper and Silver", J Mat Sci Letters 8 (1987), pp. 663-666.

7   Kristalis, P., L.Courdier, and N. Eustathopoulos, "Contribution to the Study of
        Reactive Wetting in the $CuTi/Al_2O_3$ System", J Mat Sci 26 (1991), pp. 3400-3408.

8   Ambrose, J.C., M.G. Nicholas and A.M. Stoneham, "Dynamics of Braze
        Spreading", Acta Metall Mater 40 [10] (1992), pp. 2483-2488.

9   Kim, D.H., S.H. Hwang, and S.S. Chun, "The Wetting, Reaction and Bonding of
        Silicon Nitride by Cu-Ti Alloys", J Mat Sci 26 (1991), pp. 3223-3234.

10   Rhee, S.K., "Critical Surface Energies of Al_2O_3 and Graphite", J Amer Cer Soc 55
        (1972), pp. 300-303.

11   Nikolopoulos, P., "Surface, Grain Boundary and Interfacial Energies in Al_2O_3 and
        Al_2O_3-Sn, Al_2O_3-Co Systems", J Mat Sci 20 (1985), pp. 3993-4000.

12   Ambrose, J.C., M.G. Nicholas, N. Young, and S.L. Jenkins, "Wetting and
        Spreading of Ni-P Brazes: Effects of Workpiece and Braze Composition", Mat Sci
        Technol 6 (Oct 1990), pp. 1021-1031.

13   Nicholas, M.G., D.A. Mortimer, L.M. Jones, and R.M. Crispin, "Some
        Observations on Wetting and Bonding of Nitride Ceramics", J Mat Sci 25 (1990),
        pp. 2679-2689.

14   Xue, X.M., J.T. Wang, and Z.T. Sui, "Wettability and Interfacial Reaction of alumina and Zirconia by Reactive Silver-Indium Base Alloy at Mid-Temperatures", <u>J Mat Sci</u> <u>28</u> (1993), pp. 1317-1322.

15   Fujii, H., H. Nakae, and K. Okada, "Interfacial Reaction Wetting in the Boron Nitride/Molten Aluminum System", <u>Acta Met Mater</u> <u>41</u> [10] (1993), pp. 2963-2971.

16   Fujii, H., H. Nakae, and K. Okada, "Four Wetting Phases in AlN/Al and AlN Composites/Al Systems, Models of Nonreactive, Reactive and Composite Systems", <u>Met Trans</u> <u>24A</u> [10] (1993), pp. 1391-1397.

17   Naidich, Yu.V., "Wettability of Solids by Liquid Metals", <u>Prog in Surf Membrane Sci</u> <u>14</u> (1988), pp. 353, 388.

18   Ip, S.W., M. Kucharski, and J.M. Toguri, "Wetting Behavior of Aluminum and Aluminum Alloys on Al_2O_3 and CaO", <u>J Mat Sci Letters</u> <u>12</u> (1993), pp. 1699-1702.

19   Fisher, L.R., "Measurement of Small Contact Angles for Sessile Drops", <u>J Colloid Int Sci</u> <u>72</u> [2] (1979), pp. 200-205.

20   Meier, A., M.D. Baldwin, P.R. Chidambaram, and G.R. Edwards, "The Effect of Large Additions on the Wettability and Work of Ahhesion of Copper-Oxygen Alloys on Polycrystalline Alumina", accepted for publication in: <u>Mat Sci Eng A</u>.

21   Li, J.G., L. Courdurier and N. Eustathopoulos, "Work of Adhesion and Contact-Angle Isotherm of Binary Alloys on Ionocovalent Oxides", <u>J Mater Sci</u> <u>24</u> (1989) pp. 1109-1116.

22   MacDonald, J.E., and J.G. Eberhardt, "Adhesion in Aluminum Oxide-Metal Systems", <u>Trans AIME</u> <u>233</u> [3] (1965) pp. 512-517.

23   Iida, T. and R.I.L. Guthrie, <u>The Physical Properties of Liquid Metals</u> Clarendon Press, Oxford (1988), pp. 70, 134 and 183.

24   Kristalis, P., B. Drevet, N. Valignat, N. Eusthathopoulos, "Wetting Transitions in Reactive Metal/Oxide Systems", <u>Scripta Met</u> <u>30</u> [9] (1994), pp. 1127-1132.

25   Deneker, S.P., "Cohesion in Refractory Moncarbides Mononitrides and Monoxides", <u>J Less Common Metals</u> <u>14</u> (1968), pp. 1-22.

26   Mehorta, S.P., and A.C.D. Chaklader, "Interfacial Phenomena between Molten Metals and Sapphire Substrate", <u>Met Trans</u> <u>16B</u> (Sept 1985), pp. 567-585.

27   Nogi, K., K. Oishi, and K. Ogino, "Wettability of Solid Oxides by Liquid Pure
     Metals", Mat Trans JIM 30 [2] (1989) pp. 137-145.

28   Masslaski, T.B. (Ed.), Binary Alloy Phase Diagrams Vol. 1 and 2, ASM
     International, Materials Park, OH (1990), pp. 106 and 1495.

II. Role of Oxygen in the Cu-O-Ti/Sapphire Interfacial-Region Formation

In the Cu-Ti/sapphire system, modeling of the interfacial-region formed by Active Metal Brazing has traditionally been based on the reduction reaction between titanium (the active element) and oxygen (from the stoichiometric decomposition of the sapphire) to form TiO as the intermediate oxide layer that bridges between the filler metal and the ceramic substrate. Several investigations have been devoted to identifying possible contributions for the total Gibb's free energy change of the reaction which would validate this proposed path for the interfacial region formation.

However, the oxygen concentrations of the precursor alloy and gas phase are normally ignored in these investigations. Additionally, titanium atoms in solution are not known to be surface active in liquid copper and the observed titanium segregation to the metal/sapphire interface can only be attributed to a coadsorption phenomenon involving oxygen participation. Therefore, instead of accepting sapphire decomposition as the only source of oxygen in the brazing system, this research task has been devoted to elucidate the mechanistic role of the readily available oxygen in the Cu-Ti alloy and in the gas phase, in the formation of the Cu-Ti/sapphire interfacial region.

Sapphire single crystals were contacted with three Cu-Ti-O alloys containing 4 at. pct. titanium and 100, 860 and 2,400 ppm of oxygen in controlled oxygen partial pressure atmospheres (between 10^{-7} and 10^{-30} atm).

The Ti-O system stability diagram was extended to include all possible solid and gaseous oxides species. $TiO_{(g)}$ was shown to be a predominant gaseous species for the processing conditions. The development of thermodynamic equations describing the segregation of titanium in the Cu-O-Ti system indicated that the surface enrichment of titanium occurs simultaneously with oxygen. This observation of Ti-O coadsorption and the identification of the gaseous $TiO_{(g)}$ species were then used to model the formation of a sessile drop with Cu-Ti alloy in static equilibrium with the sapphire substrate. In this model, the reaction layer responsible for bonding between sapphire and Cu-Ti alloy is postulated to be the result of a solution reaction between the Ti-O coadsorbed interphase and the sapphire, and not the direct reaction between titanium and sapphire. XPS analyses conducted on the metal/gas atmosphere and metal/sapphire interfaces corroborate the proposed model. Measurements obtained by SEM analyses showed that the thickness of the Ti-O coadsorption interphase increased dramatically with increase in the bulk oxygen concentration from 100 to 2,400 ppm. These results provide definitive validation that the titanium segregation to the interfacial region is controlled by the bulk oxygen concentration.

A. Experimental

1. Ceramic-to-Oxide Bonding: Background

Oxygen is an inherent species in the Cu-Ti/Sapphire system. It is present in the sapphire (Al_2O_3), in the reaction interphase region, in the Cu-Ti filler alloy, and in the brazing atmosphere. Consequently, the demand for improvements in the performance and reliability of metal alloy-sapphire joints, when operating under extreme conditions, can be best achieved by developing a basic understanding of the role of oxygen in the complex interactions that occur at the metal/sapphire interface, in the bulk metal, and at the metal/gas atmosphere interface.

Oxide-metal interfaces of commercial and scientific interest are typically obtained by the formation of a reaction interphase region at the interface. As such, thermodynamics becomes the tool of choice for the development of models that would predict the possible formation (stability) of ceramic/metal interfaces and provide insight into the possible mechanisms. The first investigations based on equilibrium (thermodynamic) approach were conducted by Humenik and Kingery [1], and, McDonald and Eberhart [2] who proposed that bonding of Al_2O_3 to a metal could be directly established between the oxygen anions at the ceramic surface and the metal cations. The stability of this bond could be evaluated by a direct relation between the work of adhesion (W_{AD}) and the standard Gibb's free energy of formation of the metal oxides ($\Delta G°$). The direct bonding model between metals and sapphire is supported by the theoretical quantum-mechanical calculations conducted by Johnson and Pepper [3] and Ohuchi et al. [4](1991).

Blackburn [5] showed that wetting of Al_2O_3 by metals and subsequent adherence could be promoted through metal oxidation in slightly oxidizing atmospheres, followed by the dissolution of this oxide in Al_2O_3, even for metals that do not possess the necessary thermodynamic "potential" to reduce sapphire (e.g. copper and nickel). The strength of the interface is related to the reaction between this metal-oxide interlayer and Al_2O_3 to form a complex oxide.

The understanding of the role of oxygen in the interface region formation has advanced substantially with the advent of modern surface analysis techniques. In the system Cu-Al_2O_3 system, Ohuchi et al. [6,7] showed that when oxygen is absent in the atmosphere (under UHV conditions, with estimated oxygen partial pressures lower than 10^{-40} atm) copper will bond directly, even though weakly, to sapphire. This result was the first experimental confirmation of the direct bonding between copper and the oxygen atoms at the surface of Al_2O_3 as proposed by Johnson and Pepper [3]. In contrast, the deposition of copper in the presence of oxygen showed a much larger number of atoms involved in the interfacial bonding phenomenon, which also increased the adhesion significantly [8]. The same trend was also found in the Ni/Al_2O_3 system, where the expected formation of nickel aluminate spinel (NiAl_2O_4) [9] only occurred when oxygen partial-pressure was high enough to stabilize the spinel [7]. Under UHV conditions, however, an Ni-Al intermetallic compound was observed to form at the Ni/Al_2O_3 interface.

The results obtained from the surface analytical techniques clearly demonstrated that bonds with entirely different characteristics can be obtained depending on the amount of oxygen available in the system. In contrast, the differences in the nature of the interfaces observed in active metal brazing promoted two different thermodynamic approaches in modeling the oxide ceramic/filler metal interfacial reactions. The first approach considered the solution reaction, where the presence of oxygen (in a form of a surface oxide film) is required for the formation of an interphase [10,11]. The second approach is more applicable to metals with great affinity for oxygen (e.g. titanium, vanadium, chromium, yttrium, and niobium). In this case, the development of an interfacial product is considered to be the result of a reduction reaction, where the active element reacts directly with the ceramic oxide, establishing a bond between the two components.

In the case of the solution reaction, the presence of an initial oxide interlayer, such as would be formed under an oxidizing atmosphere is not essential. Instead, segregation of oxygen may occur as a result of coadsorption of metal atoms with the residual oxygen atoms (commonly found in copper and other metals) to form metal oxygen complexes. This coadsorption phenomenon has been identified with the simultaneous surface segregation of two species through an attractive interaction mechanism [12]. Thus, coadsorption refers to the process by which a non-surface active solute segregates to an interface because of its strong interaction with oxygen, which is surface active, in the liquid metal. In the case of Cu-Ti-O system, it is well documented in the literature that oxygen is highly surface active in liquid copper [13,14,15]. On the other hand, quantum-mechanical theoretical calculations such as those carried out by Mukherjee [16] have shown that titanium is not surface active in molten copper. Also, the very large negative interaction parameter for Ti-O [17] reflects the strong interaction between the titanium and oxygen atoms, which is consistent with the expectation of strong coadsorption of titanium-oxygen in liquid copper.

The concept of reduction reaction is easily incorporated in the active metal brazing process and consists of the addition of an active element, at low concentrations, to a metal solvent to promote the reduction of the oxide ceramic at the interface. However, even for those elements known as "reactive metals", the standard Gibb's free energy changes for the reduction of Al_2O_3 are positive, which implies that the reduction reactions are not thermodynamically feasible. This dilemma has been the focus of several investigations in search of favorable contributions to the total free energy change for the reduction reaction, which would then provide explanations for the experimentally observed interface formation, thereby validating the reduction reaction as the actual path for the interfacial region formation [18].

In the Cu-Ti/Al$_2$O$_3$ system, the formation of TiO has been identified by many authors as the interfacial product associated with the bonding process [19,15,20,21]. In fact, soon after Naidich and Zuralev [22] identified TiO as the only titanium oxide to be wetted by copper, the thermodynamic requirement for the stability of TiO in the interfacial-region has been used interchangeably as the conditions for the promotion of wetting in systems using titanium as the active element. The proposed path for the interface formation is as follow:

$$Al_2O_3 + \underline{Ti} \Leftrightarrow \underline{Al} + TiO \qquad \Delta G_R^o$$

where $\underline{Al}$ and $\underline{Ti}$ represent element solution in the copper matrix. Several investigators have shown that the above reaction can be stable if the contributions of the dissolution of the active elements such as oxygen and aluminum in the solvent metal are considered [20,21].

However, direct reduction of the ceramic oxide may not necessarily be the only path for the interphase region formation. In fact, several limitations to the direct reduction approach exists. For example, if the stoichiometric dissociation of the ceramic oxide is considered as the only source of oxygen, the initial oxygen content of the alloy, which could have a dramatic influence on the thermodynamic activity of the active element in the melt would be ignored. This limitation becomes even more apparent when it is recognized that all active elements interact readily with oxygen, and that the presence of oxygen at residual levels is common in most metals.

Additionally, the titanium concentration at the metal/Al$_2$O$_3$ interface can be much higher than that within the bulk liquid phase [15], indicating segregation of the element. Since titanium is not a surface-active element in copper [16], the observed surface enrichment can only be attributed to coadsorption of (Ti-O) complexes.

Ultimately, the separation between solution reaction and reduction reaction can only be considered academic, since there are no clear criteria that define the boundaries between the two approaches. The most compelling reason for refuting this division is that even in the most controlled brazing environments, the presence of oxygen can not be completely excluded from the system. Thus, there is the need of better understanding of the mechanism(s) of interfacial reaction layer formation.

2. Thermodynamic Considerations

a. Equilibrium Al-O-Ti Phase Distribution

In the Cu-Ti/Sapphire system, the formation of TiO, with 1 to 2 at. pct. dissolved aluminum, has been identified by several investigators [15,20,21,23,24,], in sessile drop experiments or in actual active-metal brazing experiments carried out in protective atmospheres or vacuum. This consistency in the identification of TiO as the reaction product is, in fact, surprising if one considers that the formation of TiO alone is not supported by the most recent isothermal section of the Al-O-Ti ternary diagram shown in Figure 1 [25]. This diagram shows tie-lines radiating from Al_2O_3 to all other phases present in the diagram, indicating that Al_2O_3 can be in equilibrium with any phase but the β-Ti. Using basic rules such as the charge and mass balance, and chemical potential gradient considerations, [25] showed that the phase distributions for the Ti-Al_2O_3 reaction can only follow the two sequences below:

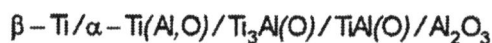

$$\beta - Ti / \alpha - Ti(Al, O) / Ti_3Al(O) / TiAl(O) / Al_2O_3$$

or

$$\beta - Ti / \alpha - Ti(Al, O) / Ti_3Al(O) / Al_2O_3$$

depending on the oxygen activity in the system. In addition to the α- and β-Ti solution, intermetallic compounds such as Ti_3Al and TiAl, with oxygen in solution, are the phases identified. These sequences of reaction products agree with the results of [26,27], and with the analyses conducted by Ohuchi [4], Chaug et al. [28], using advanced surface-sensitive techniques under extremely well controlled conditions.

The interfacial reaction products reported by the investigators engaged in phase diagram determination, Al_2O_3/$Ti_3Al(O)$ or TiAl(O)/TiO, are clearly different from that determined in active-metal brazing experiments, TiO. The distinction between the two processes is traced to fact that the phase diagram determination experiments were conducted at 1100°C with no melting of the alloys, while the brazing process involved the melting of the filler metal. In the solid-state reactions, oxygen may not be readily available for the reactions at the interface. The presence of coadsorbed Ti-O complexes in the interfacial regions in active-metal brazing is, most likely, the explanation for the different reaction products identified in the two experiments.

The controversy on the nature of the interfacial products extends also to the duplex layer (Ti₃Al(O) and TiAl(O)) identified in the solid-state diffusion couples and phase diagram experiments which is not always found in the reported interfacial analyses of active-metal brazed joints. Nevertheless, Kritsalis et al. and Bang [20,21] reported the presence of Cu_2Ti_2O and Ti_3Cu_3O between the reaction layer of TiO and the bulk filler metal.

b. Titanium-Oxygen Phase Stability

In addition to the Al-O-Ti ternary diagram, the Ti-O stability diagram is also useful in the investigation of interfacial reactions in $Cu-Ti/Al_2O_3$ bonding since it provides supplemental information on the stability of the different titanium oxides as a function of the partial pressures of gaseous species in the temperature range of interest. In the Ti-O system, vaporization experiments such as those conducted by Wahlbeck and Gilles [29] revealed that the non-stoichiometric $TiO_{(g)}$ is the main gaseous oxide species in the system, and should, therefore, be included in the calculations. Also, $TiO_{(s)}$ features a large range of nonstoichiometry and several phase transformations. Therefore, thermodynamic calculations of the partial pressures of oxygen, $TiO_{(g)}$ in equilibrium with $TiO_{(s)}$ and other titanium oxides (considered as stoichiometric line compounds) can provide semi-quantitative information on the interactions between the solid and gaseous species in the system. The final Ti-O equilibrium diagram, described in detail elsewhere [30], is shown in Figure 2. The calculations were made using the thermodynamic data provided by Pankratz [31] and considering the solid phases as in their standard state. The solid lines represent the calculated boundaries that define the stability regions of the different titanium oxides as a function of the temperature and oxygen partial pressure. The $TiO_{(g)}$ partial pressures in equilibrium with oxygen and the various titanium oxides are represented in Figure 2 in the form of isobar lines (dashed lines) ranging from 4.6 x 10^{-10} to 2.5 x 10^{-15} atm. These isobars must be interpreted by parts. In the region of the stability of $TiO_{(s)}$, the partial pressure of $TiO_{(g)}$ is obviously independent of the oxygen partial pressure and the isobars are represented by vertical lines. In the stability regions of the oxygen-rich oxides, however, the partial pressure of $TiO_{(g)}$ depends on the partial pressure of oxygen, with the isobars assuming a negative slope. This change in slope means that as the oxygen partial pressure increases in those regions, the system shifts to a new equilibrium position at higher temperatures to provide the same $TiO_{(g)}$ partial pressure. In a similar way, it can be said that for the same temperature, the $TiO_{(g)}$ partial pressure decreases significantly as the system moves to regions of stability of oxygen-rich titanium oxides. Also, it is important to notice that the $TiO_{(g)}$ partial pressures in equilibrium with $TiO_{(s)}$ (e.g. 10^{-14} to 10^{-10} atm), are many orders of magnitude higher than the corresponding equilibrium oxygen partial pressures (10^{-35} to 10^{-25} atm), in agreement with the titanium oxide vaporization experiments by Gilles et al. [32]. This analysis implies that the transport of oxygen between the solid and gaseous phase may be realized basically in the form of $TiO_{(g)}$ and not as oxygen itself.

c. Oxygen-Titanium Coadsorption in Molten Copper

The concept of coadsorption of copper-oxygen, most probably Cu_2O, as being responsible for the decrease in the liquid copper surface tension is well established in the literature [13,14,33]. More recently, the coadsorption of chromium-oxygen clusters has been invoked by Kritsalis et al. [34,35] as necessary to explain the wetting behavior of Cu-Cr and Ni-Cr alloys on sapphire under variable oxygen activities in the metallic alloy.

The modeling of interfacial segregation based on the regular solution type model for the thermodynamic equilibrium between the bulk and surface phases has been popular for quite some time [36]. Recently, Plessis and van Wyk [37] revisited the regular solution model for segregation in multicomponent alloys in terms of macroscopic thermodynamic properties and generalized the model derived by McMahon and Marchut [38] which based on the initial analysis of Guttmann [36]. The basic hypothesis of the model are as follow:

- Surface segregation is defined as the redistribution of solute atoms between the surface and bulk of a crystal which are both regarded as open systems.

- The surface region is finite and the bulk is infinite in size but the two together form a closed system for the crystal as a whole.

- Equilibrium is reached when the energy of the closed system is a minimum.

A complete and detailed derivation of the model is shown in the work of McMahon and Marchut [38]. Application of the model with detailed calculations has also been discussed recently by Camargo [30]. Here, only the final equations are presented for the determination of the equilibrium surface concentrations of the solutes in a ternary system. Species 1 and 2 are the solutes and species 3 is the solvent.

$$X_1^\Phi = \frac{X_1^B \exp\left(\frac{\Delta G_1}{RT}\right)}{1 - X_1^B + X_1^B \exp\left(\frac{\Delta G_1}{RT}\right) - X_2^B + X_2^B \exp\left(\frac{\Delta G_2}{RT}\right)} \tag{1}$$

and

$$X_2^\Phi = \frac{X_2^B \exp\left(\frac{\Delta G_2}{RT}\right)}{1 - X_2^B + X_2^B \exp\left(\frac{\Delta G_2}{RT}\right) - X_1^B + X_1^B \exp\left(\frac{\Delta G_1}{RT}\right)} \tag{2}$$

where

$$\Delta G_1 = \Delta G_1^\circ + 2\Omega_{13}\left(X_1^\Phi - X_1^B\right) + \Omega'\left(X_2^B - X_2^\Phi\right) \tag{3}$$

$$\Delta G_2 = \Delta G_2^\circ + 2\Omega_{23}\left(X_2^\Phi - X_2^B\right) + \Omega'\left(X_1^B - X_1^\Phi\right) \tag{4}$$

and

$$\Omega' = \Omega_{12} - \Omega_{13} - \Omega_{23} \tag{5}$$

In these equations, X_i^Φ and X_i^B correspond to the respective surface and bulk concentrations of the element i, the Ω_{ij} terms represent the binary interaction parameters in a quasi-chemical manner of the corresponding binary system. The term ΔG_i° represents the adsorption energy of the element i in the metal matrix (species 3) and it accounts for the change in non-configurational enthalpy (i.e. elastic, vibrational, etc.) when one solute atom replaces one solvent atom in the boundary [36].

The interaction parameters Ω_{12}, Ω_{13}, and Ω_{23} were obtained from the Gibb's free energy of the solution calculation considering each of the solutes in their respective solvent. The thermodynamic data for the Cu-O, Cu-Ti, and Ti-O binary systems were extracted from work of Schmid [39], Hoshino et al. [40], and Pajunen and Kivilathi [41], respectively. In addition to the binary interaction parameters, the adsorption energies of oxygen and titanium (as solutes) in the Cu-O and Cu-Ti binary systems (ΔG_1° and ΔG_2°) were determined. In particular, the adsorption energy in the Cu-Ti system was estimated using the Miedema [42] model, with the prediction of -47,500 J/mole for the adsorption of titanium in copper. For the adsorption of oxygen in copper, an adsorption energy of +420,000 J/mole was provided by Gerasimov [43] who based his calculation on experimental measurements and theoretical predictions. This value is an indication of the strong surface activity of oxygen in copper and correlates very well with the value of + 458,964 J/mol reported by Toyoshima and Somorjai [44]. Figure 3 shows the chemical composition profile for a system that contains 0.01 at. pct. oxygen and 1.0 at. pct. titanium. Note that approximately 50 to 100 ppm oxygen are typically found in OFHC copper and 1.0 at. pct. titanium represents the typical upper limit of the active metal in most commercial brazing alloys. Despite that titanium is not surface active in copper, it can cosegregate to the free surface due to the strong interaction with oxygen. Also, the equilibrium free surface composition is basically composed of oxygen with small concentrations of titanium and no copper. The complete absence of copper (the solvent) at the surface is because of the strong decrease in the total Gibb's free energy of the system promoted by the Ti-O coadsorption to the free surface. In addition, the equilibrium surface composition is temperature sensitive, with the general trend of oxygen enrichment and titanium depletion on the surface as the temperature increases. The time required to establish the equilibrium at the interface was estimated to be only fractions of a second for a temperature just below the melting point of the Cu-Ti alloy [30]. This calculation clearly indicates that equilibrium Ti-O surface concentration in the Cu-O-Ti ternary system can be established at the free surface much earlier than any interfacial reaction between the Cu-O-Ti alloy and the sapphire. According to these calculations, the modeling of the brazed Cu-O-Ti/sapphire interfacial region as a result of the interaction between the Ti-O coadsorption layer and sapphire is well based.

The equilibrium titanium-oxygen ratio, in addition to being temperature sensitive, was found to depend strongly on the bulk titanium and oxygen concentrations. As such, the physical properties of the Cu-O-Ti/sapphire interfacial region are expected to closely relate to the equilibrium Cu-O-Ti surface composition.

3. Model Conceptualization

Based on the Ti-O coadsorption model and the characterization of $TiO_{(g)}$ as the most important gaseous species in the Ti-O system, the equilibrium situation for a sessile drop in the Cu-Ti-O/sapphire system was proposed in Figure 4. The thermodynamic equilibrium between the bulk and the surface of the Cu-O-Ti alloy is achieved by the Ti-O coadsorption to both Cu-O-Ti/sapphire and Cu-O-Ti/gas atmosphere interfaces. As indicated earlier, the actual interfacial titanium-to-oxygen ratio is determined by the processing temperature and the titanium and oxygen bulk concentrations in the Cu-O-Ti alloy. At the metal/gas interface, the adsorbed layer is in equilibrium with oxygen and $TiO_{(g)}$. It is important to notice that the formation of $TiO_{(g)}$ would imply in a loss of oxygen and titanium from the alloy, and the formation of a $TiO_{(s)}$ film ahead of the gas/liquid/solid triple point by vapor deposition. Such a film would be an ideal substrate for the Ti-Cu alloy to wet and spread. It is important to recall from Figure 2 that $TiO_{(s)}$ features the highest equilibrium partial pressure of $TiO_{(g)}$ compared to other higher oxygen titanium oxides, suggesting that $TiO_{(s)}$ would de the ideal metal/gas interfacial product. Moreover, stronger interaction with the sapphire would be expected if the reacting titanium oxide film is an oxygen-deficient TiO as compared to a titanium-rich stoichiometric oxide such as TiO_2. The phase diagram in Figure 1 shows that TiO can be in equilibrium with Al_2O_3 and a limited solution would promote the formation of the interfacial region.

Considering that TiO represents the ideal Ti-O coadsorption product at both metal-gas and metal/sapphire interfaces, the optimal range of oxygen partial pressures for TiO stabilization is shown in Figure 2, between 10^{-27} and 10^{-35} atm at 1100°C.

In summary, the model proposes that the reaction layer, responsible for the interface formation between the Cu-Ti alloy and the sapphire, is not a result of direct reaction between titanium and sapphire. Instead, the interface formation is a result of the reaction between the adsorbed layer (Ti-O) and sapphire. At brazing temperatures, titanium aluminate ($TiAl_2O_4$) is not stable according to the Al-O-Ti ternary phase diagram presented in Figure 1. Nevertheless, a small amount of TiO may dissolve in sapphire to result in the Cu-Ti/sapphire bonding (formation of the "reaction layer" shown in Figure 4). In terms of bonding between the Cu-Ti alloy and the sapphire as a whole, TiO from Ti-O coadsorption represents the ideal surface reactant which interacts with sapphire to form the final interfacial region, and not the ideal reaction product itself as it is largely accepted in the current literature.

4. Experimental Verification

High purity copper and titanium metal were used to produce the three Cu-Ti filler metals with 4 at. pct. titanium and varying levels of oxygen (100, 860 and 2,400 ppm) for the sessile drop experiments. The substrates were high purity single crystal sapphire with (0001) orientation.

The sessile drop experiments were conducted in an environmental cell which was heated in a vertical tube furnace. The process temperatures and holding times were 1110, 1150 and 1190°C, and 5, 10 and 30 min., respectively. Details of the laboratory equipment has been described elsewhere [30]. Varying oxygen partial pressure was achieved by ppm-level hydrogen additions to the argon carrier gas and controlled by equilibrium between the hydrogen-oxygen-water vapor in the gas phase. Measurements of oxygen partial pressures in the range of 10^{-7} to 10^{-30} atm were made using a stabilized zirconia oxygen sensor. A titanium sponge reactor was also incorporated into the system to remove trace concentrations of nitrogen in the gas phase.

The characterization of the interfacial regions was performed using three techniques: scanning electron microscopy (SEM), energy dispersive spectroscopy (EDS), and X-ray photoelectron spectroscopy (XPS). The SEM equipment (JEOL JXA-840) provided general imaging, phase characterization of the interfacial region, and X-ray dot mapping of the interfacial region (JEOL 6400-JSM SEM). These analyses provided for quantification of the size of the Ti-O coadsorption interphase as a function of the processing conditions. X-ray photoelectron spectroscopy (XPS) analyses was the most important analysis technique in this research since it has the surface sensitivity necessary to unambiguously identify the composition of the interfacial region in terms of its chemical constituents and bonding states.

B   Results and Discussion

The analysis and discussion section will be presented in two parts, based on the results obtained at the filler metal/gas atmosphere interface and the filler metal/sapphire interface.

1.   The Filler Metal/Gas Interface

The presence of concentric rings around the sessile drops that were produced by vapor deposition from $TiO_{(g)}$ was consistently observed in the sessile drop experiments. Note that $TiO_{(g)}$ was identified from the Ti-O stability diagram in Figure 2 as the most important gaseous species in this system, featuring equilibrium partial pressures that are many orders of magnitude higher than the corresponding oxygen partial pressures (Figure 2). Figures 5a and 5b show the top and bottom (through the transparent sapphire crystal) views of the metal droplet, respectively. The top view shows two distinct rings with characteristic colors: the inner ring (closer to the metal droplet) has a yellowish color that can be associated with that of TiO; the

212

outer ring appears bluish which is characteristic of oxygen-rich titanium oxides. These colors are quite different from the grayish color of the Cu-Ti/sapphire interface when viewed from the bottom through the sapphire lens, which is characteristic of hypostoichiometric TiO. XPS analysis on both rings showed that titanium and oxygen are present confirming the presence of titanium oxides. However, nitrogen has also been detected, particularly in the inner ring, indicating the presence of TiN. In fact, the color of the inner ring matches the color of TiN (goldish-yellow), closer than it does with TiO. Since TiN is not volatile [45] and that the nitrogen content in the Cu-Ti alloy was extremely low, it must have formed between nitrogen in the gas phase and the titanium oxide film formed by vapor deposition on the sapphire substrate. The source of nitrogen in the gas phase is argon because nitrogen can be present in argon as a trace impurity (Grade 5 Argon - between 4 to 6 ppm). While this concentration may seem to be minute, it can result in nitrogen partial pressures in the gas phase at the order of 3 to 5×10^{-6} atm (at 0.84 atm ambient pressure), which is orders of magnitude higher than the nitrogen partial pressure in equilibrium with TiN at 1100°C (approximately 1×10^{-17} atm).

The range of oxygen partial pressures more significant for the vapor deposition of $TiO_{(g)}$ appears to be in the vicinity of 1×10^{-18} atm, which is much higher than the pressures that would stabilize $TiO_{(s)}$ at the metal-gas interface, and therefore promote the highest equilibrium pressures of $TiO_{(g)}$ as calculated in Figure 2. This difference may be attributed to the volatilization kinetics which could lead to conditions very different from those described by thermodynamic equilibrium considerations. Also, $TiO_{(s)}$ was considered as a line compound in this investigation and the effect of its large nonstoichiometry on the equilibrium with the gas phase was not included in the stability analysis. Nevertheless, the observation of the films formed by TiO vapor deposition agrees with the model proposed in Section 3. The experimental observation clearly corroborates the significant role that $TiO_{(g)}$ plays in the early stages of interface formation, by providing a precursory $TiO_{(s)}$ film that modifies the surface of the sapphire substrate. Although the range of oxygen partial pressures that maximize the $TiO_{(g)}$ partial pressure in the system also corresponded to the region where $TiO_{(s)}$ is stable, as

213

discussed in Section 2, $TiO_{(g)}$ will always be present in the system, to a lesser or greater extent, depending on which titanium oxide is stabilized by the prevailing oxygen pressure.

A series of experiments were also conducted at low oxygen partial pressures, approximately 1×10^{-27} atm measured in the outlet gas leaving the environmental chamber. Surprisingly, the sessile drops processed at this level of oxygen partial pressure practically preserved the original pancake shape of the solid metal pellet and exhibited a goldish-yellow color typical of TiN. In addition, these sessile drops were only weakly bonded to the sapphire substrate after processing and could be dislodged by the application of only a moderate force. Figure 6 shows two photographs of one of these sessile drops, as seen from the top (metal side of the metal/gas interface) and from the bottom of the droplet after debonding from the sapphire substrate (metal side of the metal/sapphire interface). Both sides of the sessile drop in fact presented approximately the same color. However, the metal side of the metal-sapphire interphase region is much smoother and reflective, and appears to be much darker in the photograph than it appears on visual inspection. These results indicate that at this level of oxygen partial pressure, TiN (already identified to be present and discussed before) may control the interfacial product formation and its interaction with the sapphire substrate would not leading to bonding. This fact is readily apparent since TiN has great thermodynamic stability and is a well known diffusion barrier for many elements.

One of these sessile drops was sectioned and analyzed using an SEM. The resulting image and X-ray dot maps are shown in Figure 7. The photomicrograph of the metal side of the metal/gas interface is shown at the top of the figure where a continuous and well defined interfacial product of thickness between 1.5 and 2 μm can be observed. This continuous phase encapsulated the molten sessile drop and prevented it from flowing, acquiring the characteristic semi-spherical cap shape. The bottom part of Figure 7 shows the corresponding X-ray dot maps for titanium and copper, where it can be observed that the interface region constitutes primarily of titanium and virtually no copper. The titanium content in the bulk Cu-Ti alloy was analyzed by EDS and showed approximately the same concentration as the starting material (4 at. pct.). Titanium appears to be absent in the bulk in the X-ray map because the bulk titanium concentration is relatively low compared to the much higher (segregated) concentration at the metal/gas interfacial region.

The capability of an SEM in conducting microanalyses is limited because the equipment cannot detect elements of low atomic numbers such as oxygen and nitrogen. Nonetheless, titanium is expected to coadsorb with oxygen to the free surface as discussed in the model proposed in Section 3. The observation of titanium enrichment at the metal/gas interfacial region is an important result since it is believed, in general, that the titanium interfacial enrichment would occur only at the metal/sapphire interface and the "driving force" for such enrichment would be the "desire" of titanium to react with the sapphire [46]. In addition, the Ti-O coadsorption phenomenon to the metal/gas interfacial region demonstrates that small titanium additions in copper completely changes the metal/gas interfacial region. This fact has been largely neglected in the Cu-Ti/sapphire sessile drop experiments reported in the literature, [15,20]where the surface tension of copper is normally assumed not to be affected by small titanium additions.

The metal side of the metal/gas interfacial region of the sessile drop shown in Figure 6 was also analyzed using XPS. The overall photoemission spectra are shown in Figure 8 for the as-received and sputtered conditions. The sputtering rate was 120 Å/min during all analyses. The very "clean" spectra obtained are strikingly evident in Figure 8. Small traces of calcium and silicon, found in the as-received condition, rapidly disappeared as the first layers of the surface was removed by short-time sputtering. The presence of nitrogen in the metal/gas interfacial-region is confirmed in the photoemission spectra, with the N1s peak increasing significantly from the as-received to the 3 minutes sputtering condition. The same behavior is followed by the $Ti2p_{3/2}$ peak under the same conditions. The O1s, in contrast, showed accentuated decrease as the first layers of the surface were sputtered away.

The expanded $Ti2p_{3/2}$ photoemission spectra are shown in Figure 9. In this figure, the range of the electron binding energies, characteristic of the different compounds, is given in the form of a label with the name and the corresponding energy range (solid black bar) as provided by Chastian [47]. For the as-received condition, Figure 9a shows that the surface titanium is in a bonding state that corresponds predominantly to TiO_2 and TiN. The formation of TiO_2 actually occurred when the sessile drop was in contact with the ambient air within the period (several days) after the sessile drop experiment and before the XPS analyses. This conclusion is also based on the fact that the gas phase oxygen partial pressure (1×10^{-27} atm) was not expected to stabilize TiO_2 in accordance with the stability diagram shown in Figure 2. The conclusion is further substantiated by the spectra illustrated in Figure 9b, where after the first surface layers are sputtered away, the $Ti2p_{3/2}$ peak shifts to lower binding energies indicating the presence of TiO and TiN, predominantly. The actual quantitative composition analyses are shown in Table 1. These data confirm that titanium is the major metallic element at the interface, with a virtual absence of copper as predicted in Figure 7.

Table 1 - Quantitative analysis corresponding to spectra shown in Figure 9(b).

Element	Area (cts-ev/s)	Sensitivity Factor	Concentration (at. pct.)
Ti2p	549000	168.731	60.04
N1s	58970	44.949	24.21
O1s	51710	66.598	14.33
Cu2p	28340	367.737	1.42

The identification of the titanium compounds at the interface can be further characterized by the expanded photoemission-spectra for nitrogen as shown in Figure 10. The nitrogen peak presents a very limited shift in the N1s electron binding energy for all possible nitrides. However, titanium forms only the line compound TiN with nitrogen and therefore, the nitrogen spectra displayed in Figure 10 can be interpreted as being all associated with the titanium. The spectra corresponding to $Ti2p_{3/2}$ and N1s also support the conclusion that the interfacial region is constituted primarily of TiN and TiO. Since TiN can be considered as a line compound, nitrogen can be related to titanium using the atomic ratio of 1:1. This estimation shows TiN as accounting for 40% of the interfacial atomic composition. The remaining titanium would then be associated with oxygen to form oxygen-deficient $TiO_{0.4}$ which would account for approximately 60% of the interfacial atomic composition.

The simultaneous surface enrichment of titanium and oxygen identified here clearly confirms the predictions of the Ti-O coadsorption in the Cu-O-Ti ternary system as proposed by the model presented in Section 3. The premise that the Ti-O coadsorption would be restricted to a monolayer is obviously not realistic in this system, since the Ti-O adsorbed interfacial region observed in Figure 7 was between 1 to 2 μm thick. The bulk oxygen concentration of 0.086 at. pct. in this experiment is close to that used to predict the Ti-O coadsorption in the Cu-O-Ti ternary system presented earlier in Figure 3 (0.1 at. pct. Oxygen). The difference between the experimental results ($TiO_{0.4}$) and the prediction from Figure 3 at 1150°C (approximately 10%) is due to the lower titanium concentration used in the calculations (titanium 1 at. pct.). It is important to remember that the titanium concentration in the Ti-O coadsorption predictions in Section 2 were limited to 1 at. pct. because of the limitation in the validity of the thermodynamic data.

Obviously, the presence of $TiO_{(s)}$ at the metal/gas interface can also be interpreted as being due to the oxidation of the bulk titanium by oxygen in the gas phase, since the prevailing oxygen partial pressure (1×10^{-27} atm) was in the TiO stability-range as indicated by Figure 2. However, titanium is a solute in copper and the virtual absence of copper at the surface would be inconsistent with the direct reaction of titanium with oxygen in the gas phase. Moreover, if reaction occurs between titanium and the gas phase, only TiN would be expected to be found at the surface since the nitrogen partial pressure in the gas phase (4 to 6 $\times 10^{-6}$ atm) is about ten orders of magnitude higher than the nitrogen partial pressure in equilibrium with TiN at the processing temperature used in these experiments. Therefore, the presence of TiO at the metal/gas interface provides supporting confirmation of the Ti-O coadsorption prediction in the Cu-O-Ti ternary system, as proposed in the model for the Cu-O-Ti/sapphire sessile-drop equilibrium presented in Section 3. It is relevant to note that, due to the high intrinsic rate of the adsorption phenomenon, the Ti-O coadsorption at the metal/gas interface, followed by the reaction of this adsorbed interphase with nitrogen, occurs in a time interval short enough such that no significant change in the original shape of the metal alloy occurs during the sessile drop experiment.

The analytical results for the metal side of the Cu-Ti alloy/gas interface indicate that when the oxygen partial pressure is decreased below a certain level, the nitrogen present in the Grade 5 argon would adversely affect the metal-sapphire interaction, including spreading of the molten filler metal over the sapphire substrate. While it may be tempting to interpret this finding as indicating that the oxygen partial pressure should be maintained at a level that is above the minimum where nitrogen would control the interface formation. However, within the context of exploring the mechanisms of the interfacial-region formation, it indicates instead that the nitrogen can have a confounding effect on the process and should be maintained at much lower levels than those found in standard commercial Grade 5 argon.

It is clear from the results presented above that the control of the nitrogen partial pressure in the system is mandatory. Because of this, a titanium sponge bed was incorporated into the gas train to getter the nitrogen. However, XPS analyses conducted at the interface showed that the titanium sponge nitrogen getter only lowered the nitrogen partial pressure in the gas phase to a limited extent since the amount of nitrogen identified at the interface was approximately 30% of the initial value corresponding to the condition of no nitrogen control.

It is also apparent that the Ti-O coadsorption at the metal/gas phase interface demonstrates that the presence of small concentrations of titanium and oxygen in copper can completely modify the nominal copper/gas phase interfacial region. Thus, the segregation at the metal/gas interface due to the Ti-O coadsorption, which has typically been neglected in the literature, should now be incorporated in the determination of the Work of Adhesion (W_{ad}) from Cu-Ti/sapphire sessile drop experiments.

The ppm-level hydrogen addition to the argon gas-phase which was intended to shift the water vapor/hydrogen/oxygen equilibria proved to be a very effective strategy in the control of extremely low oxygen partial pressure ranges ($< 10^{-20}$ atm).

2.   The Metal/Sapphire Interfacial Region

The Cu-O-Ti/sapphire interfacial region was subjected to extensive SEM and XPS analysis. Ultimately, it is this interface that is responsible for bonding between the Cu-Ti alloy and the sapphire.

a.   XPS Analyses of the Metal/Sapphire Interfacial Region

XPS analysis of the metal/sapphire interfacial region was conducted on the metal side interfacial region of the weakly bonded Cu-Ti/sapphire interface displayed in Figure 6. In other samples that featured much stronger metal-sapphire interaction, fracture was observed to occur within the near-surface region of the sapphire but not at the metal/sapphire interface itself. This sample was processed without the nitrogen gettering system. The formation of TiN on the metal side of the metal/gas interface, as discussed in the previous section, was responsible for encapsulating the "sessile drop" and maintaining it in the original pancake shape. The observed weak metal-sapphire interaction was attributed to the presence of nitrogen in the gas phase combined with an oxygen partial pressure below 10^{-27} atm. The overall photoemission spectra for this interface is shown in Figure 11 for the as-received and sputtered condition. The presence of nitrogen at the metal/sapphire interface is confirmed in the spectra shown in Figure 11a corresponding to the as-received condition. However, no nitrogen were detected in the sputtered interface (after 1 min.) as shown in Figure 11b. As expected, the nitrogen concentration at the metal/sapphire interface is much less than at the metal/gas interface which is in direct contact with the gas phase. Nevertheless, it is important to note that even a

219

relatively small concentration of nitrogen at the interfacial region could hinder the interaction between the metal and sapphire.

The overall photoemission-spectra in Figure 11 also indicated that titanium and oxygen represent the predominant elements on the metal side of the Cu-O-Ti/sapphire interfacial-region, with no copper present. The expanded titanium photoemission spectra displayed in Figure 12 show the same features as presented by the previously analyzed metal/gas interface, with TiO_2 being identified at the surface in the as-received condition. Again, the electron binding energies shifted to lower levels toward TiO as the first layers of the surface were removed. The intensities of oxygen, on the other hand, did not present any significant changes for both surface conditions as can be observed in Figure 3.

The weak metal-sapphire interaction can be further characterized by the very small aluminum peak detected in the interfacial-region (only for the sputtered condition) observed in Figure 11b. The important implication of this result is that the oxygen present at the interface is not due to the sapphire decomposition, since the small aluminum concentration identified at the surface is not commensurate with the large oxygen concentration at this location on the basis of the sapphire stoichiometry. The oxygen at the metal/sapphire interfacial-region is not transferred from the gas phase either, since this interface is not directly exposed to the gas phase. Thus, the titanium and oxygen concentrations on the metal side of the metal/sapphire interface are due to Ti-O coadsorption from the bulk of the Cu-O-Ti alloy in accordance with the model developed in Section 3. In fact, this conclusion is definitive in identifying that the titanium enrichment in the metal at the sapphire interface is due to the simultaneous oxygen enrichment from the bulk as proposed in Section 3 and 3. The reaction that is responsible for bonding between the sessile drop and sapphire is in fact the interaction of this Ti-O adsorption layer with the sapphire, and not of the solute titanium with sapphire.

With regards to the specific results of this experiment, the presence of the nitrogen hindered more extensive metal-sapphire interaction. In the absence of nitrogen, the unimpeded interaction at the interface will depend on the titanium-to-oxygen surface ratio as determined by the Ti-O coadsorption in the Cu-O-Ti system. To maximize the reactivity between the Ti-O adsorbed layer and the sapphire, a high titanium-to-oxygen ratio is necessary.

b.   SEM Analysis of the Metal/Sapphire Interfacial Region

The predictions for Ti-O coadsorption in the Cu-O-Ti ternary system presented above show that the interphase titanium and oxygen concentrations are highly dependent on the bulk compositions as well as the processing temperature. The model, however, does not provide details about the growth of the adsorption interphase as a function of these same variables.

In addition to the XPS analyses described previously, SEM analyses of cross-sections of the interfacial region were conducted so that the behavior of the interphase thickness could be quantified as a function of the bulk oxygen concentration, temperature and time. The X-ray dot maps for titanium, copper and aluminum for three selected at 1110°C and 15 min. holding time are shown in Figures 13, 14 and 15, corresponding to the bulk oxygen concentrations of 100, 860 and 2,400 ppm, respectively. The X-ray dot maps were generated separately for each element within the entire interfacial region. These maps indicate that the width of the Ti-O adsorption region (shown on the left of the top photograph and repeated on the right of the bottom photograph) increased significantly - from approximately 1 to 5 μm - with an increase in the bulk oxygen content in the Cu-Ti alloy, from 100 ppm (Figure 13) to 2,400 ppm (Figure 15). This result provide additional verification that the observed titanium enrichment (and simultaneous oxygen enrichment) in the coadsorption interphase is determined by the bulk oxygen concentration present in the system.

At lower bulk oxygen concentrations (e.g., 100 and 860 ppm), the boundaries between the Ti-O coadsoprtion layer and the bulk filler metal are extremely sharp and well defined, as indicated in the X-ray maps for copper, titanium and aluminum shown in Figures 13 and 14. Similarly, the boundaries between the Ti-O coadsorption layer and the sapphire are also sharp and well defined. The higher bulk oxygen concentration specimen (2,400 ppm), however, exhibited a different behavior. Only the Ti-O/sapphire interface boundary remained sharp, composed predominantly of titanium and oxygen. The Cu-Ti/Ti-O interface boundary became wide and diffuse, with a higher concentration of copper. The duplex character of this interfacial region is revealed more clearly in the X-ray dot maps shown in Figure 16.

The occurrence of this duplex Ti-O coadsorption layer has also been reported in the investigations of Bang and Kritsalis et al.[20,21]. For the alloy compositions they reported, it was demonstrated in Section 2 that a mixture of copper, TiO and the ternary compound Ti_3Cu_3O should be the equilibrium phases to solidify. Although the oxygen bulk concentration in the Cu-Ti alloy was not reported in either investigations, it is believed the copper used by Bang [21] to produce Cu-Ti filler metals contained between 4,000 to 5,000 ppm of oxygen, which would confirm that the formation of the duplex phase Ti-O coadsorption layer is associated with a high bulk oxygen concentration in the Cu-Ti alloy.

The results of the measurements of the adsorption layer thickness, corresponding to the processing temperature of $1110°C$, are summarized by the plots in Figure 17. The significant increase in the Ti-O coadsorption interphase with increased bulk oxygen concentration in the Cu-Ti alloy is clearly evident. The asymptotic behavior of the adsorption interphase thickness beyond a holding time of five minutes should also be noted, whereas the portion of the curve below five minutes is indeterminate. It was discussed in Section 2 that the Ti-O coadsorption at the interface occurs in a time that is of much smaller-scale, of the order of fractions of a second, before any intimate contact between the sessile drop and the sapphire can occur. In addition, development of what can now be identified as the Ti-O coadsorption layer has been claimed by some investigators [11,21] to be a diffusion-controlled reaction involving the titanium in the alloy and oxygen from the stoichiometric decomposition of the sapphire.

C CONCLUSIONS

The following conclusions are listed to summarize the accomplishments achieved in this research program:

- The currently accepted modeling of the bonding in the Cu-Ti/sapphire brazing system based on direct reaction between titanium and sapphire to produce $TiO_{(s)}$ as the interfacial product should be expanded to also consider titanium and oxygen cosegregation.

- The Cu-O-Ti ternary phase diagram indicate that $TiO_{(s)}$ is not an equilibrium phase that would occur naturally in the ternary system.

- Thermodynamic calculations using regular solution model determined that titanium is not surface active in molten copper.

- Titanium segregation to the Cu-O-Ti alloy/sapphire interface was shown to occur due to a coadsorption mechanism with oxygen.

- The predictions from the model for the Ti-O coadsorption in the Cu-O-Ti indicated that the titanium-to-oxygen equilibrium interfacial ratio is determined by both the bulk oxygen and titanium concentrations as well as the system temperature.

- A model for the static equilibrium for a Cu-O-Ti sessile drop in contact with sapphire was proposed. It incorporates the Ti-O coadsorption and the pertinent O_2 and $TiO_{(g)}$ species. The reaction responsible for the bonding between the sessile drop and the sapphire was demonstrated to be a solution reaction between the Ti-O coadsorbed layer and the Al_2O_3.

- The interfacial Ti-O coadsorption layer has been shown to increase monotonically with increases in the bulk oxygen concentration. In addition, the duplex phase Ti-O coadsorption layer has been shown to be linked to high oxygen bulk concentrations in the Cu-O-Ti alloy, in excess of 2,400 ppm.

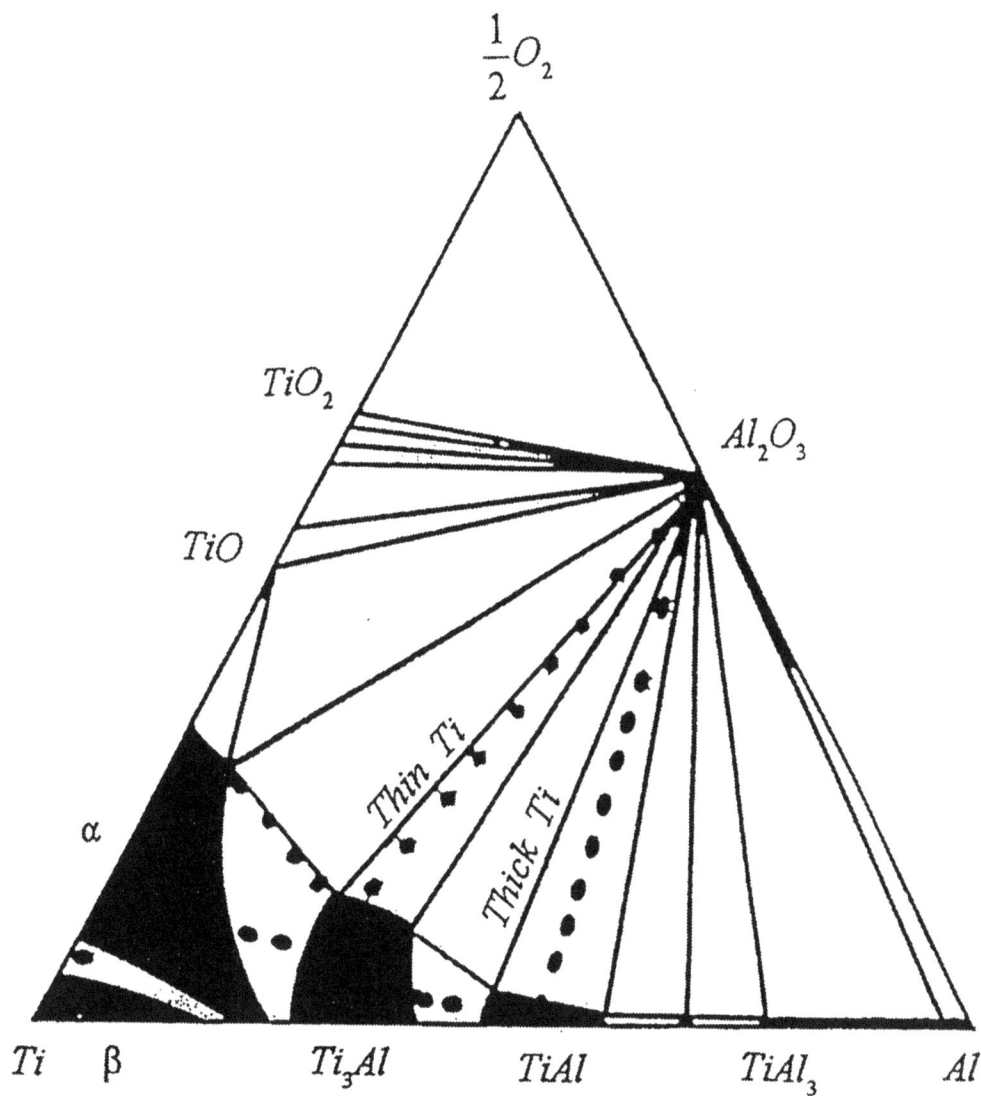

Figure 1: Isothermal section of the Al-O-Ti ternary phase idagram at 1100°C [26].

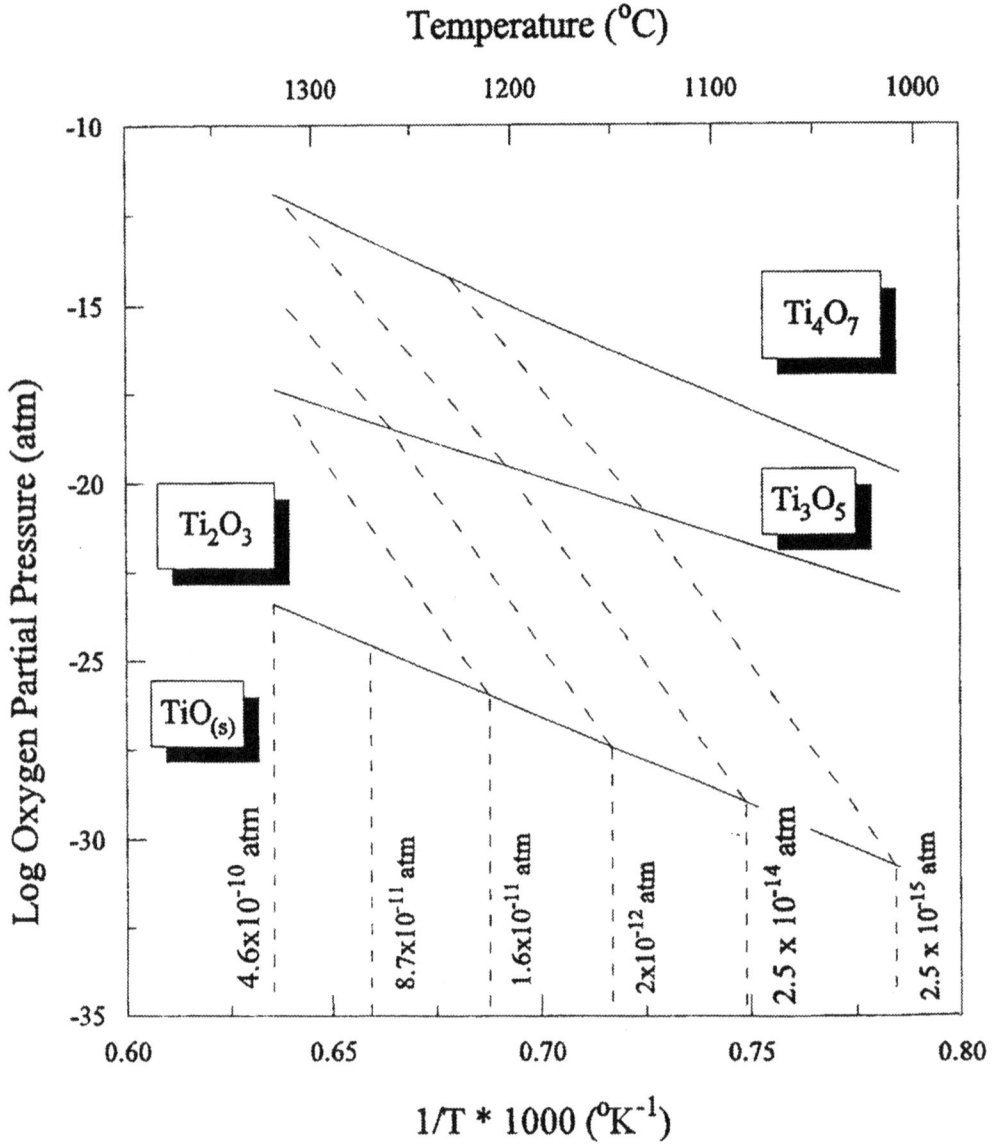

Figure 2: Ti-O stability diagram including the $TiO_{(g)}$ isobars.

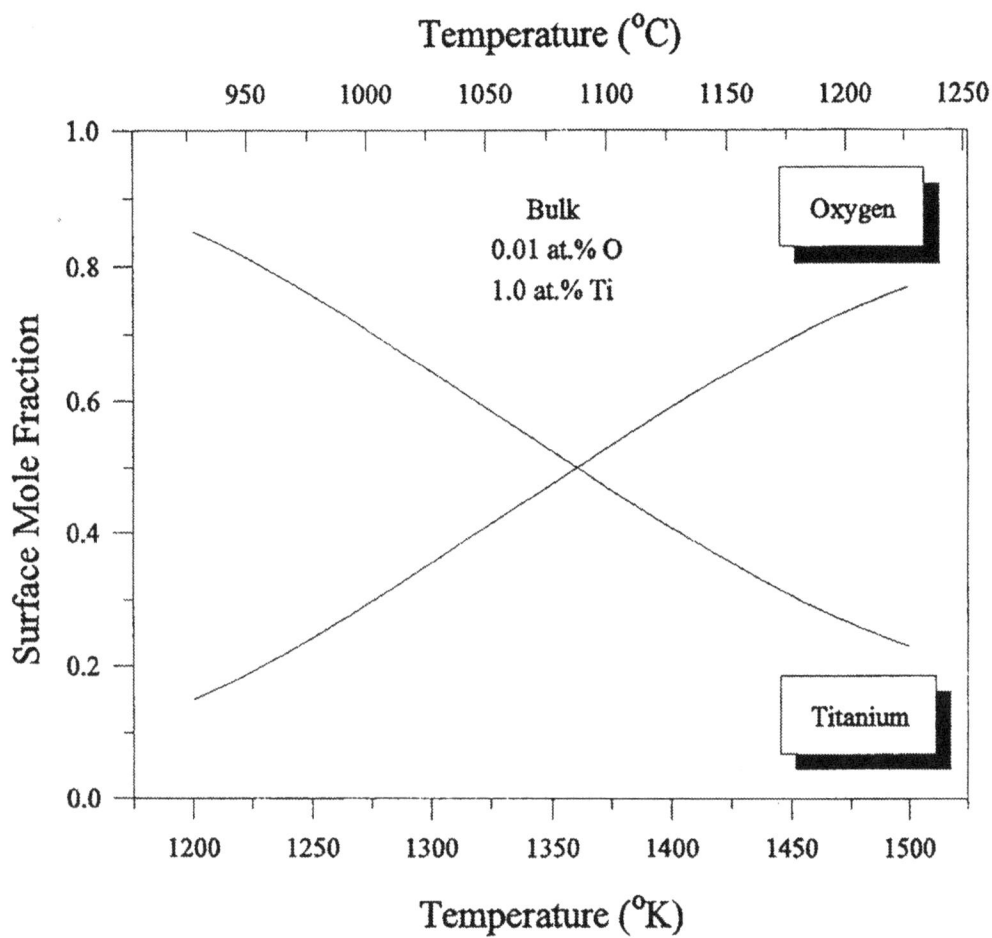

Figure 3: Equilibrium oxygen and titanium surface mole fraction as a function of the temperature (Bulk concentrations: 0.01 at. % O; 1.0 at. % Ti).

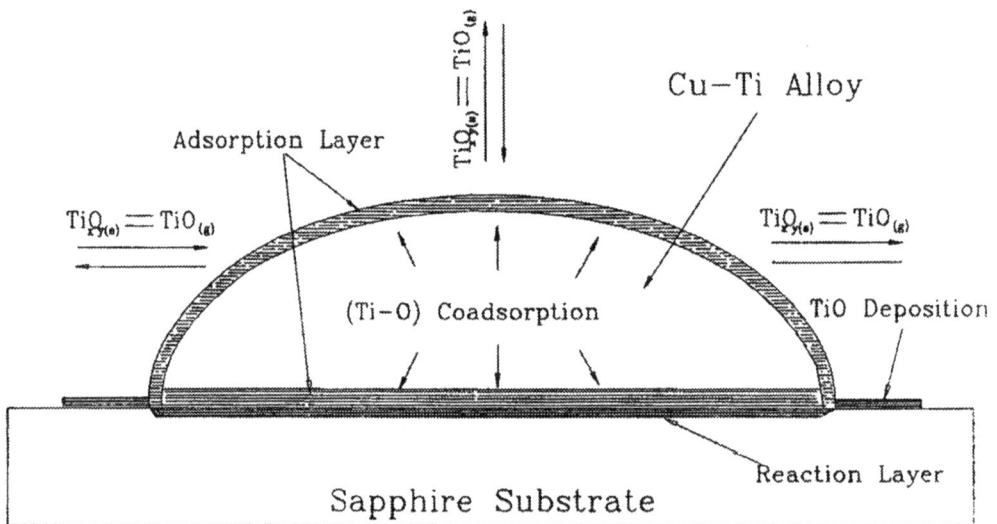

Figure 4: Modelling of the equilibrium situation for a Cu-O-Ti sessile drop experiment on sapphire incorporating the concepts of Ti-O coadsorption and the $TiO_{(g)}$ species.

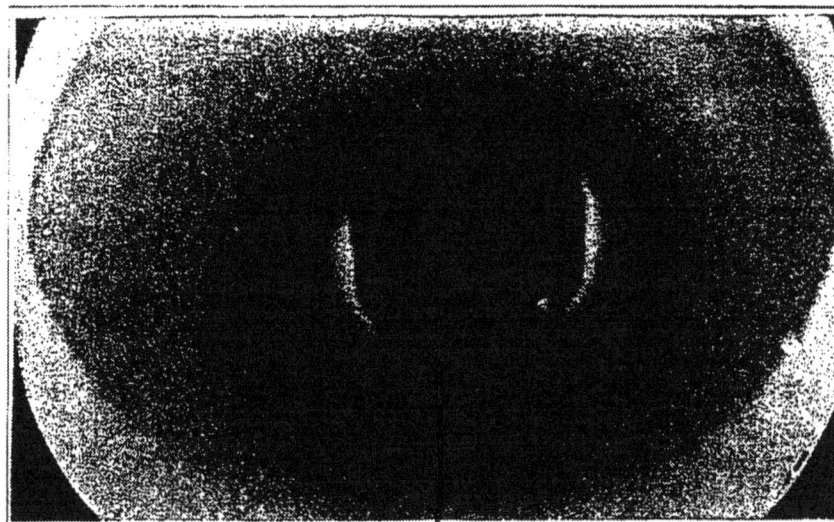

(a)

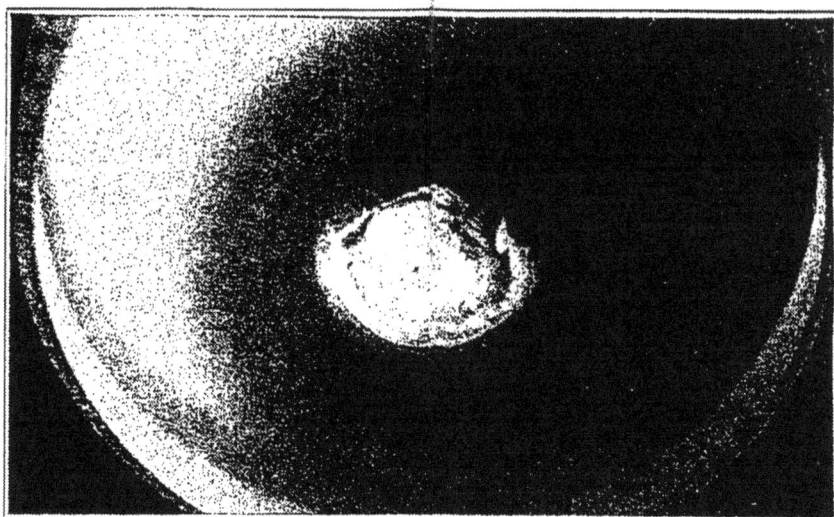

(b)

2 mm

Figure 5: Sessile drops on sapphire illustrating the features of vapor deposition of $TiO_{(g)}$ on the sapphire substrate. Processing conditions: Cu-Ti 1at. %, 1110°C, 30 min., oxygen bulk concentration = 500 ppm; (a) view from the top of the sessile drop; (b) view from the bottom through the sapphire.

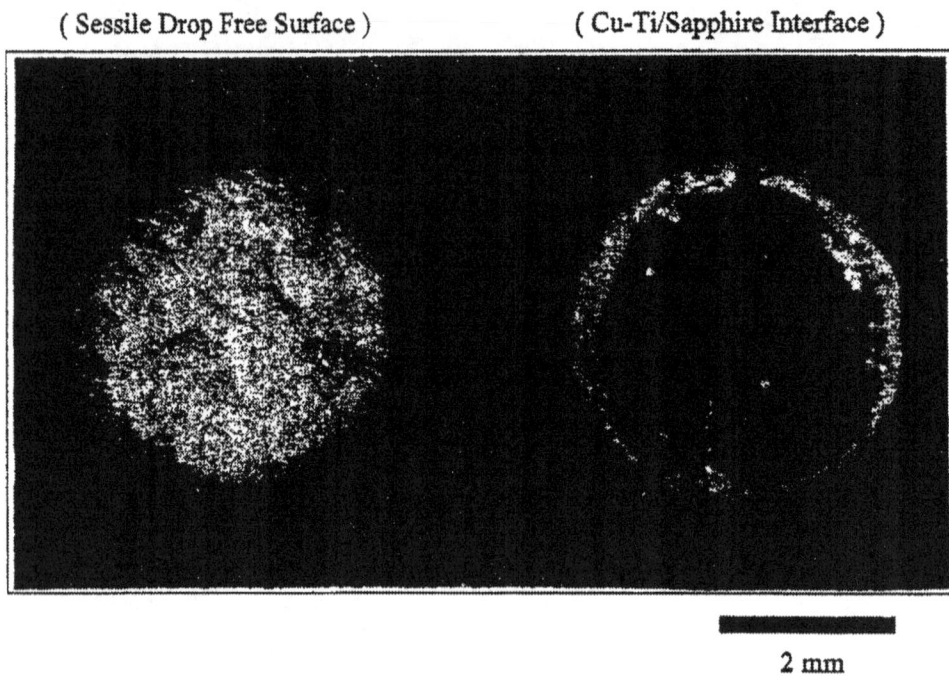

(Sessile Drop Free Surface)   (Cu-Ti/Sapphire Interface)

2 mm

Figure 6: TiN formation on Cu-Ti alloy sessile drop. TiN is present at both the gas-metal and "weakly bonded" Cu-Ti/sapphire interfaces. Processing conditions: Cu-Ti 4 at %, no nitrogen control, 1150°C, 15 min., oxygen bulk concentrations = 860 ppm, oxygen partial = 10^{-27} atm.

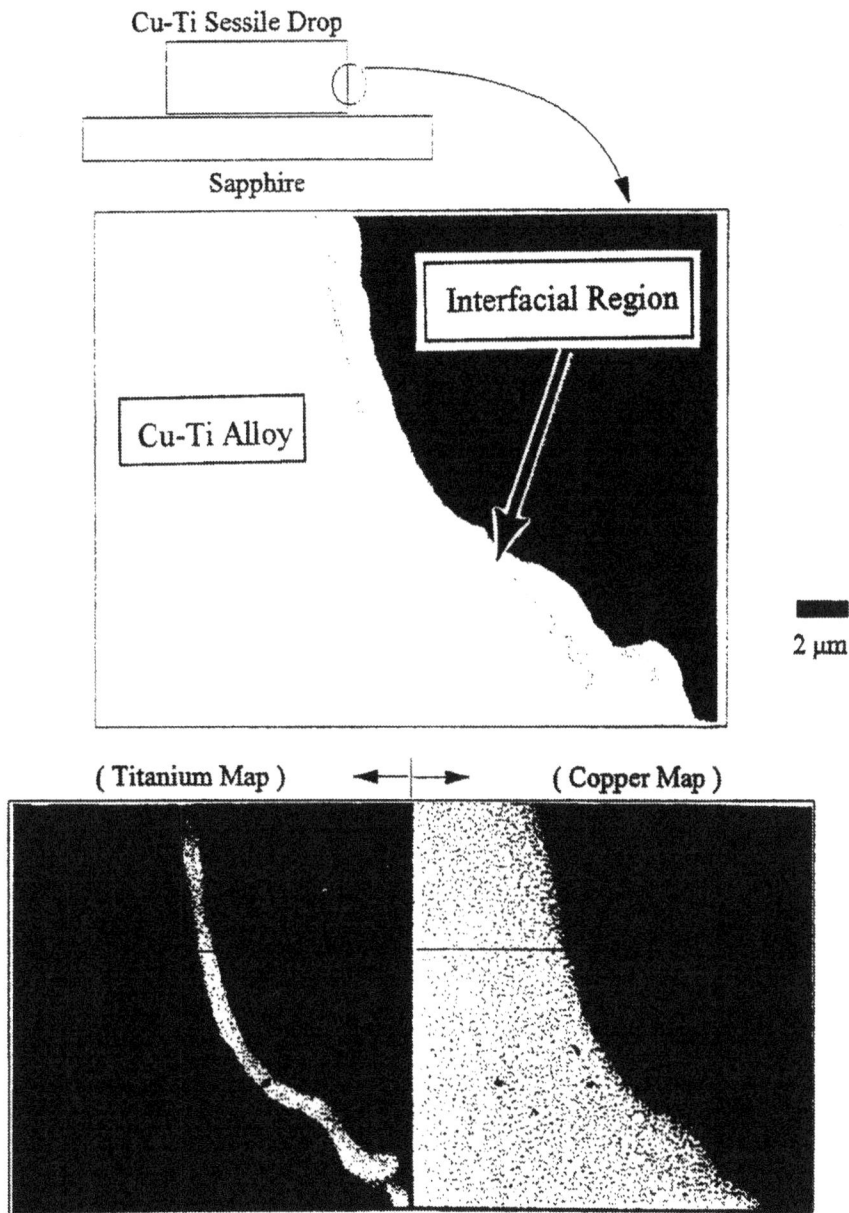

Figure 7: X-ray dot maps of Ti and Cu for the metal side interfacial region of the Cu-Ti alloy/gas interface. Processing conditions: Cu-Ti 4 at. %, no nitrogen control, oxygen partial pressure = 10^{-27} atm., 1150°C; 15 min., oxygen bulk concentration = 860 ppm.

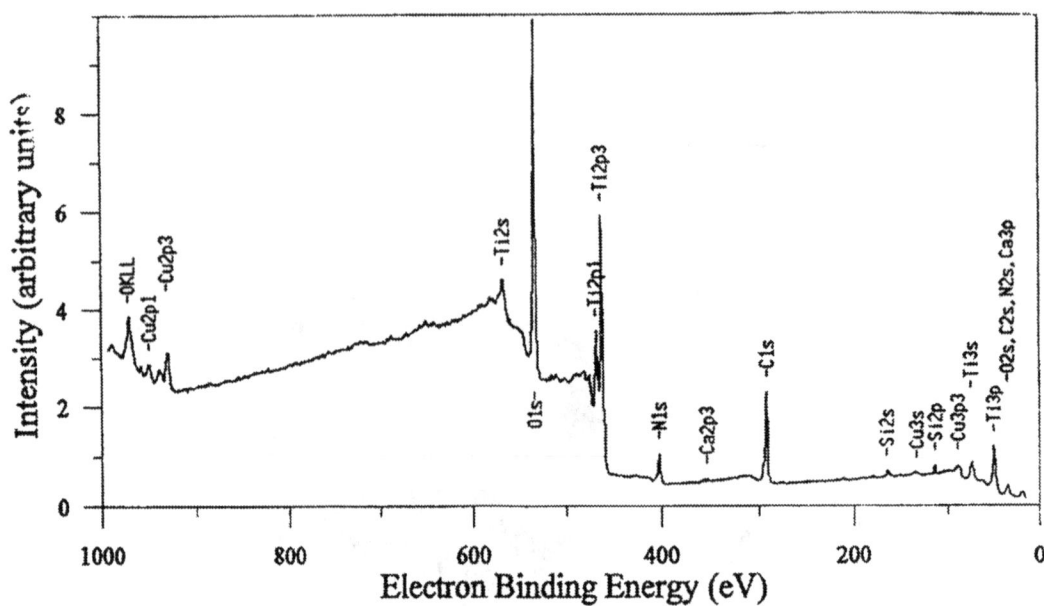

(a)

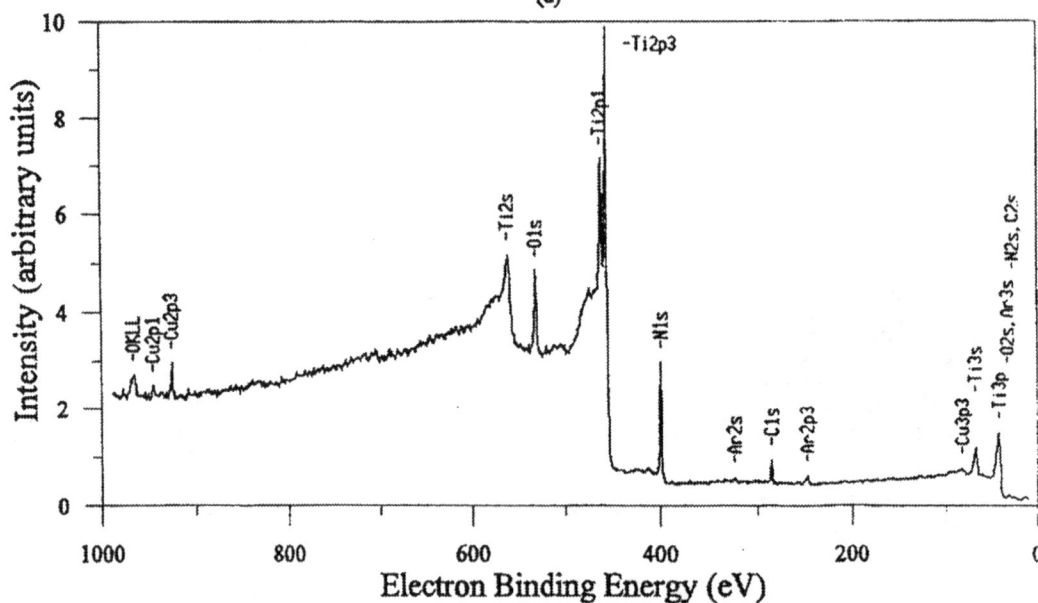

(b)

Figure 8: Overall photoemission spectra, corresponding to the metal side interfacial region of the Cu-Ti alloy/gas interface displayed in Figure 7. Processing conditions: Cu-Ti 4 at. %, no nitrogen control, oxygen partial pressure = 10^{-27} atm., 1150°C; 15 min., oxygen bulk concentration = 860 ppm; (a) as received, (b) after 3 min. sputtering.

231

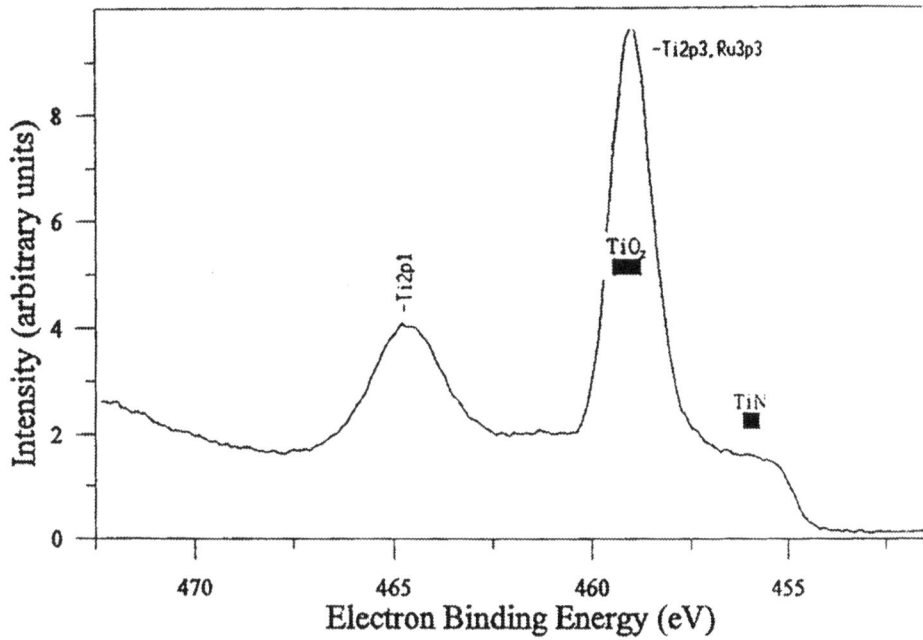

(a)

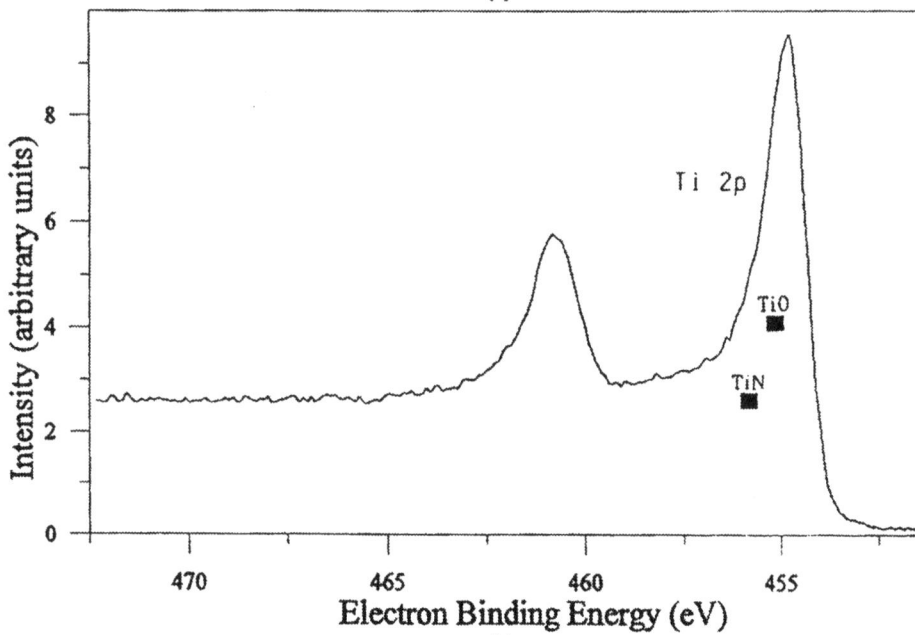

(b)

Figure 9: Titanium 2p3 photoemission spectra, corresponding to the overall spectra
displayed in figure 8. Processing conditions: Cu-Ti 4 at. %, no nitrogen
control, oxygen partial pressure = 10^{-27} atm., 1150°C; 15 min., oxygen bulk
concentration = 860 ppm; (a) as received, (b) after 3 min. sputtering.

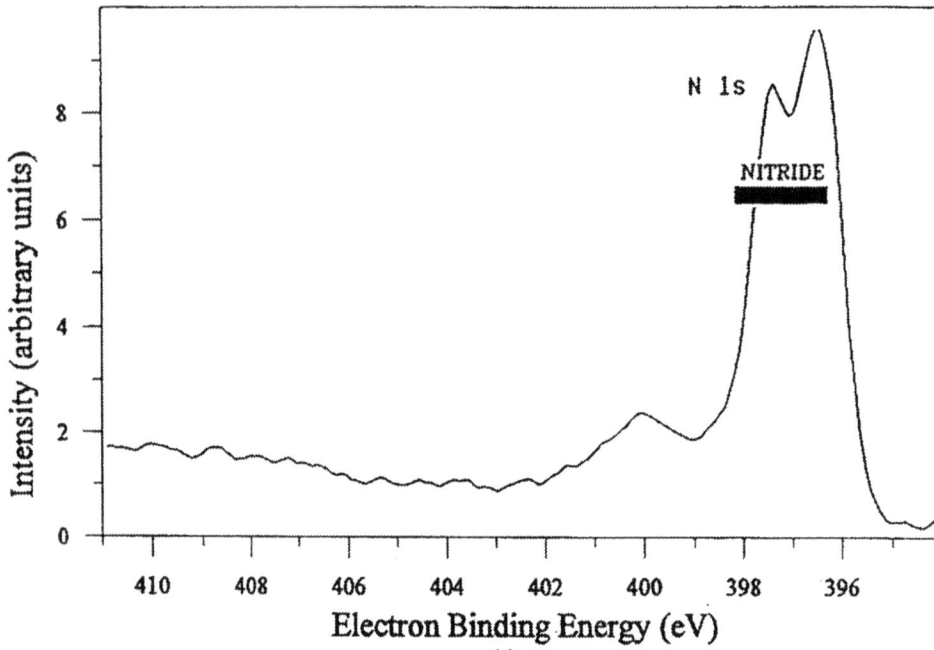

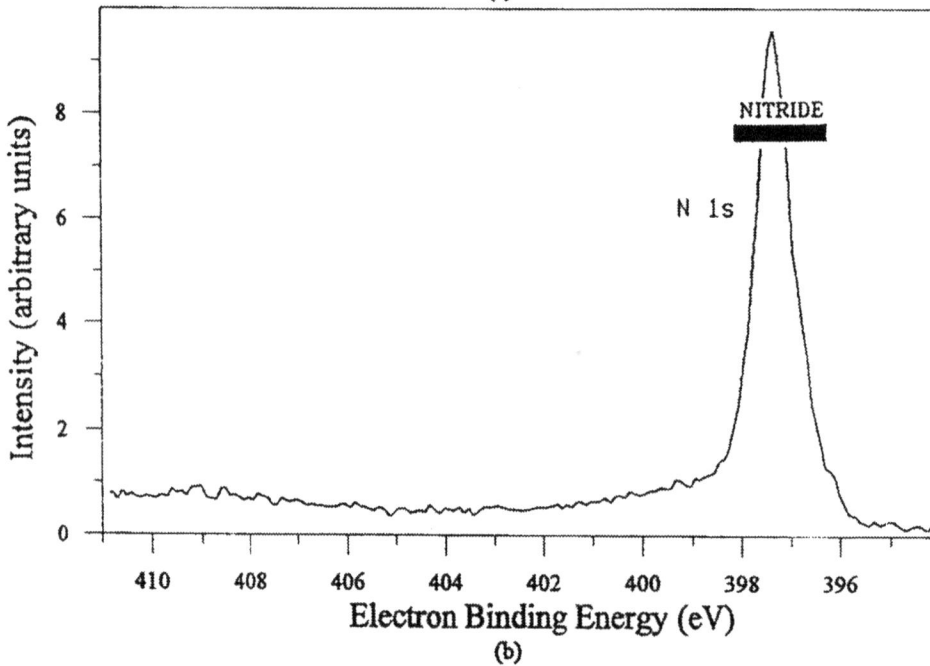

Figure 10: Nitrogen 1s photoemission spectra, corresponding to the overall spectra displayed in figure 8. Processing conditions: Cu-Ti 4 at. %, no nitrogen control, oxygen partial pressure = 10^{-27} atm., 1150°C; 15 min., oxygen bulk concentration = 860 ppm; (a) as received, (b) after 3 min. sputtering.

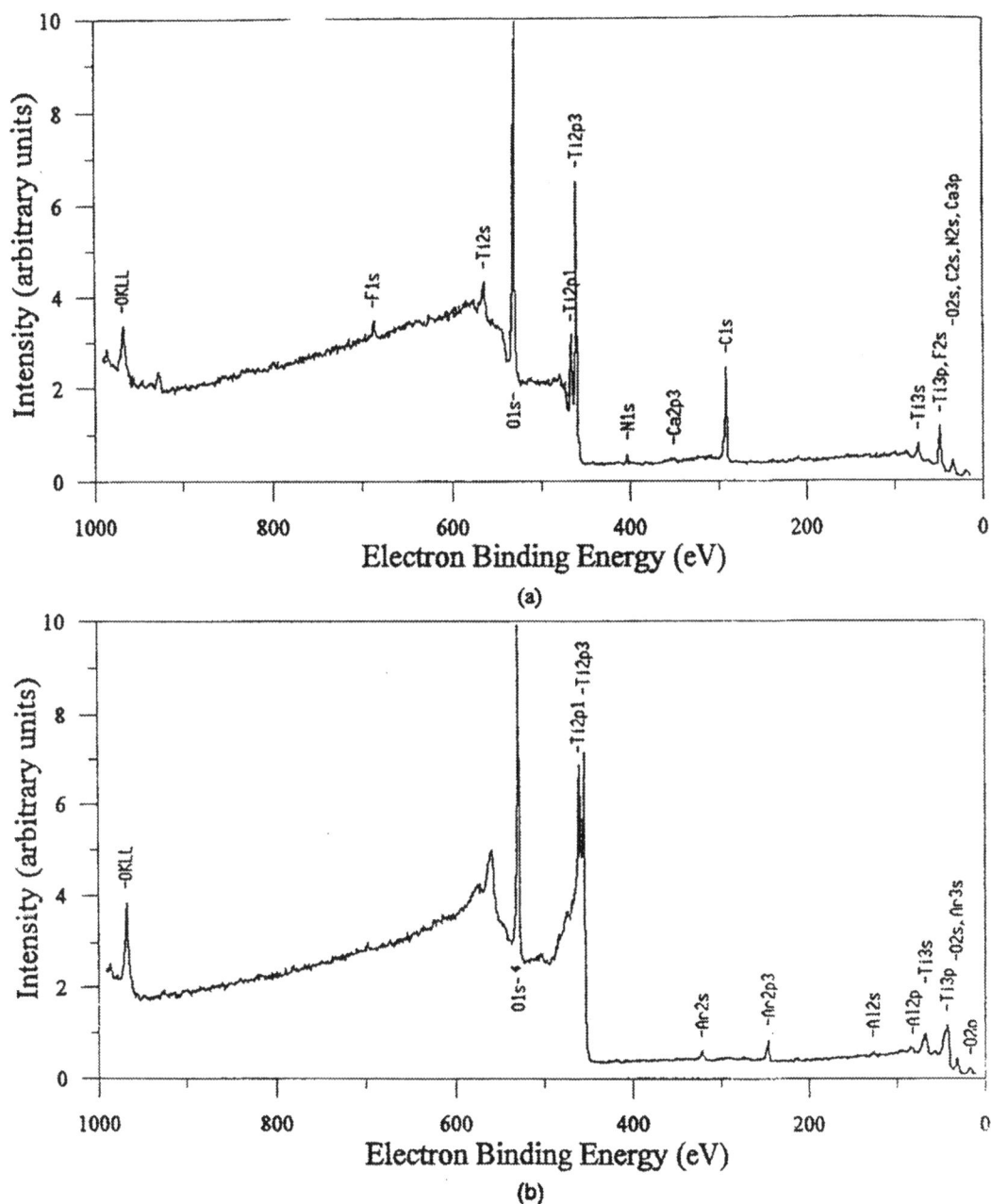

Figure 11: Overall photoemission spectra, corresponding to the metal side interfacial region of the "weakly bonded" Cu-Ti alloy/sapphire interface displayed in Figure 6. Processing conditions: Cu-Ti 4 at. %, no nitrogen control, oxygen partial pressure = 10^{-27} atm., 1150°C; 15 min., oxygen bulk concentration = 860 ppm; (a) as received, (b) after 1 min. sputtering.

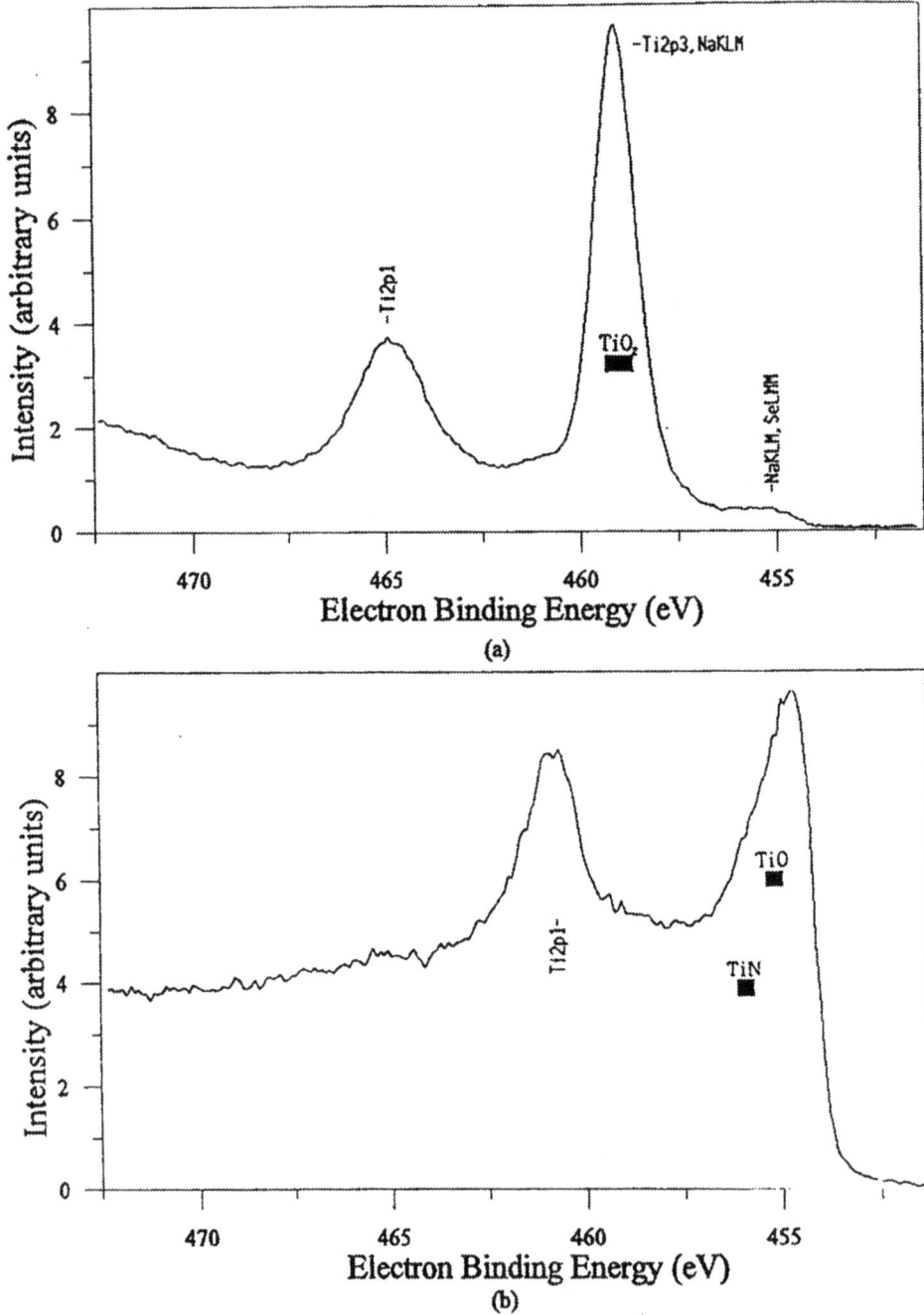

Figure 12: Titanium 2p3 photoemission spectra corresponding to the overall spectra displayed in Figure 11. Processing conditions: Cu-Ti 4 at. %, no nitrogen control, oxygen partial pressure = 10^{-27} atm., 1150°C; 15 min., oxygen bulk concentration = 860 ppm; (a) as received, (b) after 1 min. sputtering.

235

SEM Image
of
Interfacial Region

(Titanium Map)   ←|→   (Aluminum Map)

(Copper Map)   ←|→   (Titanium Map)

2 μm

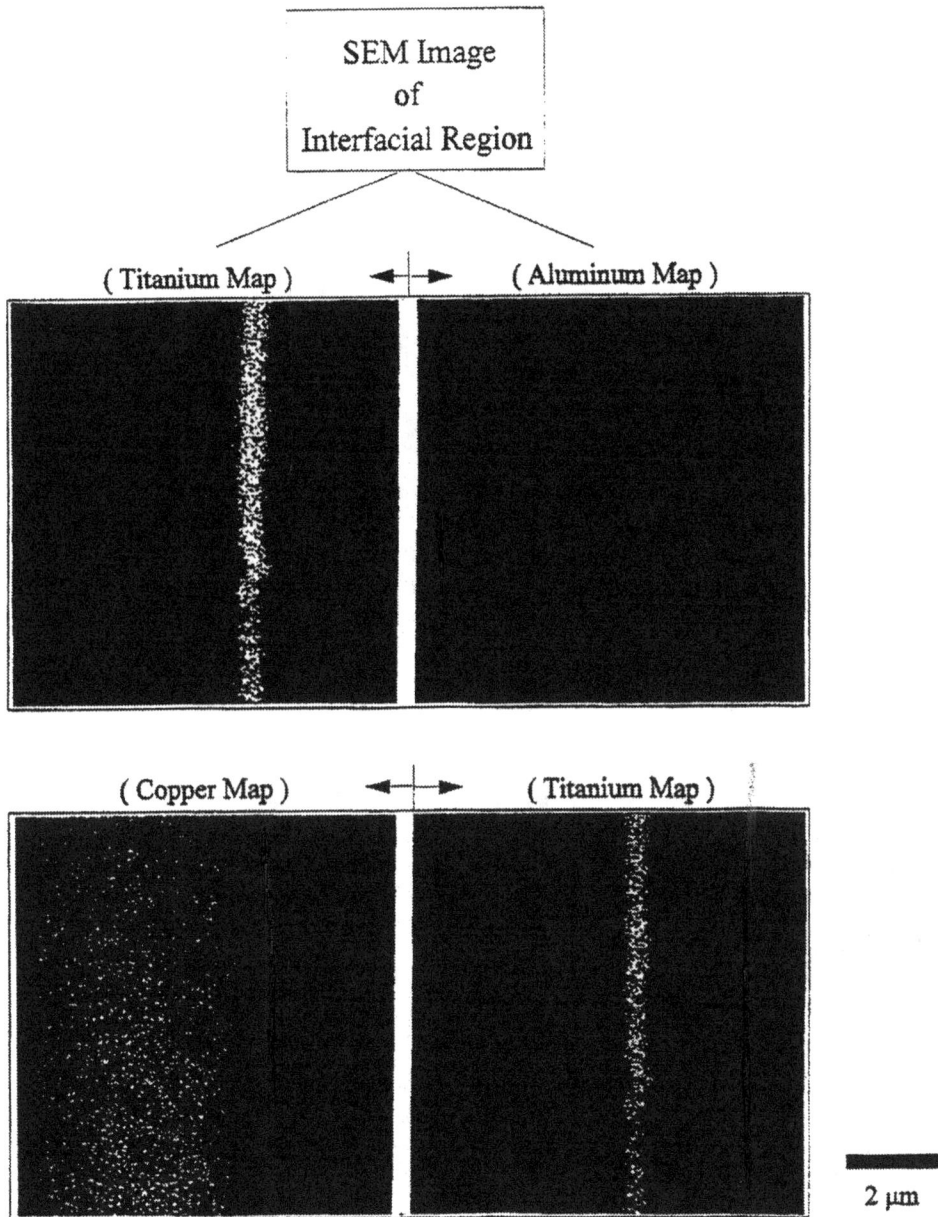

Figure 13: X-ray dot maps of Ti, Al, and Cu at the Cu-Ti/sapphire interface. Processing conditions: Cu-Ti 4 at. %, 1110°C, 15 min., titanium sponge nitrogen getter, oxygen partial pressure $= 10^{-21}$ atm., oxygen bulk concentration $= 100$ ppm.

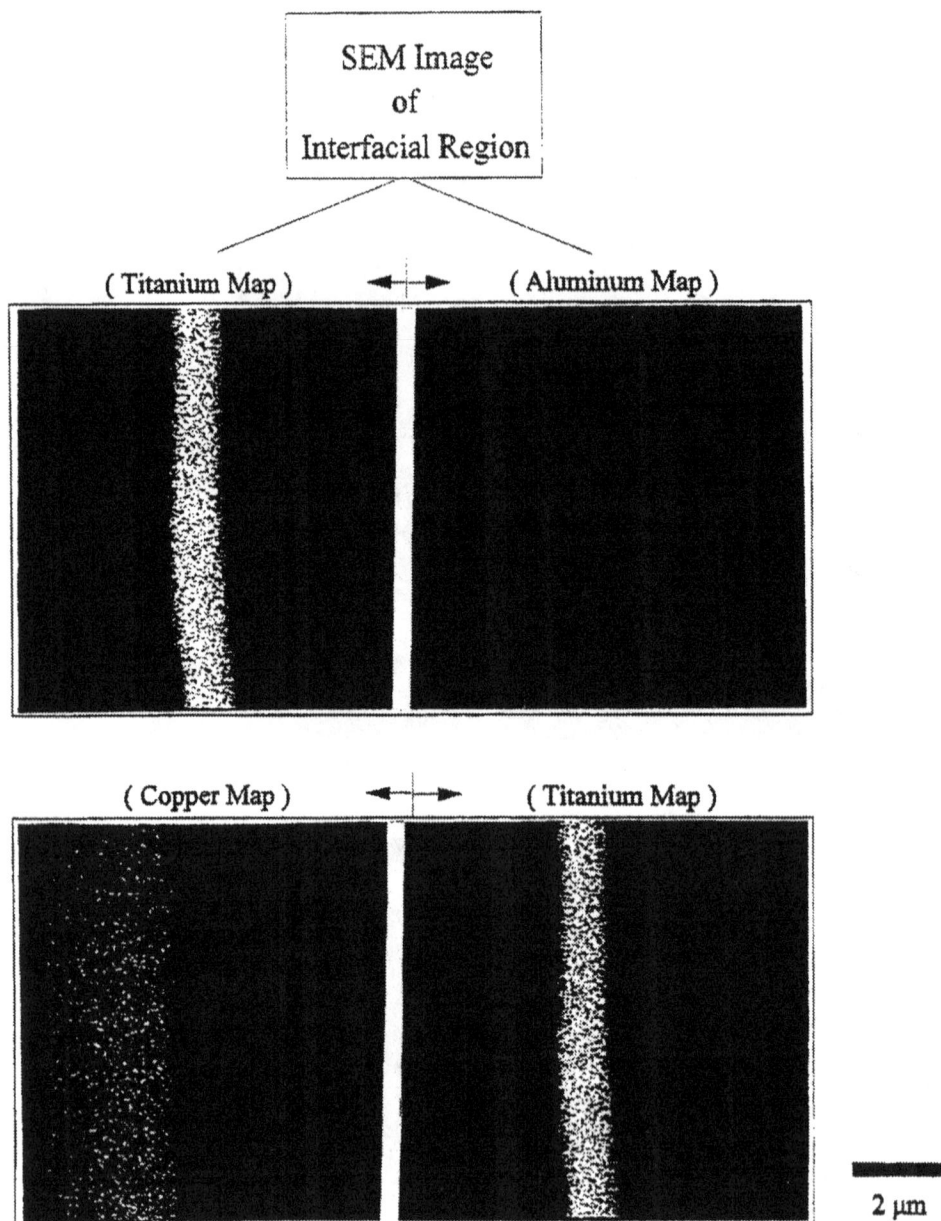

Figure 14: X-ray dot maps of Ti, Al, and Cu at the Cu-Ti/sapphire interface. Processing conditions: Cu-Ti 4 at. %, 1110°C, 15 min., titanium sponge nitrogen getter, oxygen partial pressure $= 10^{-21}$ atm., oxygen bulk concentration = 860 ppm.

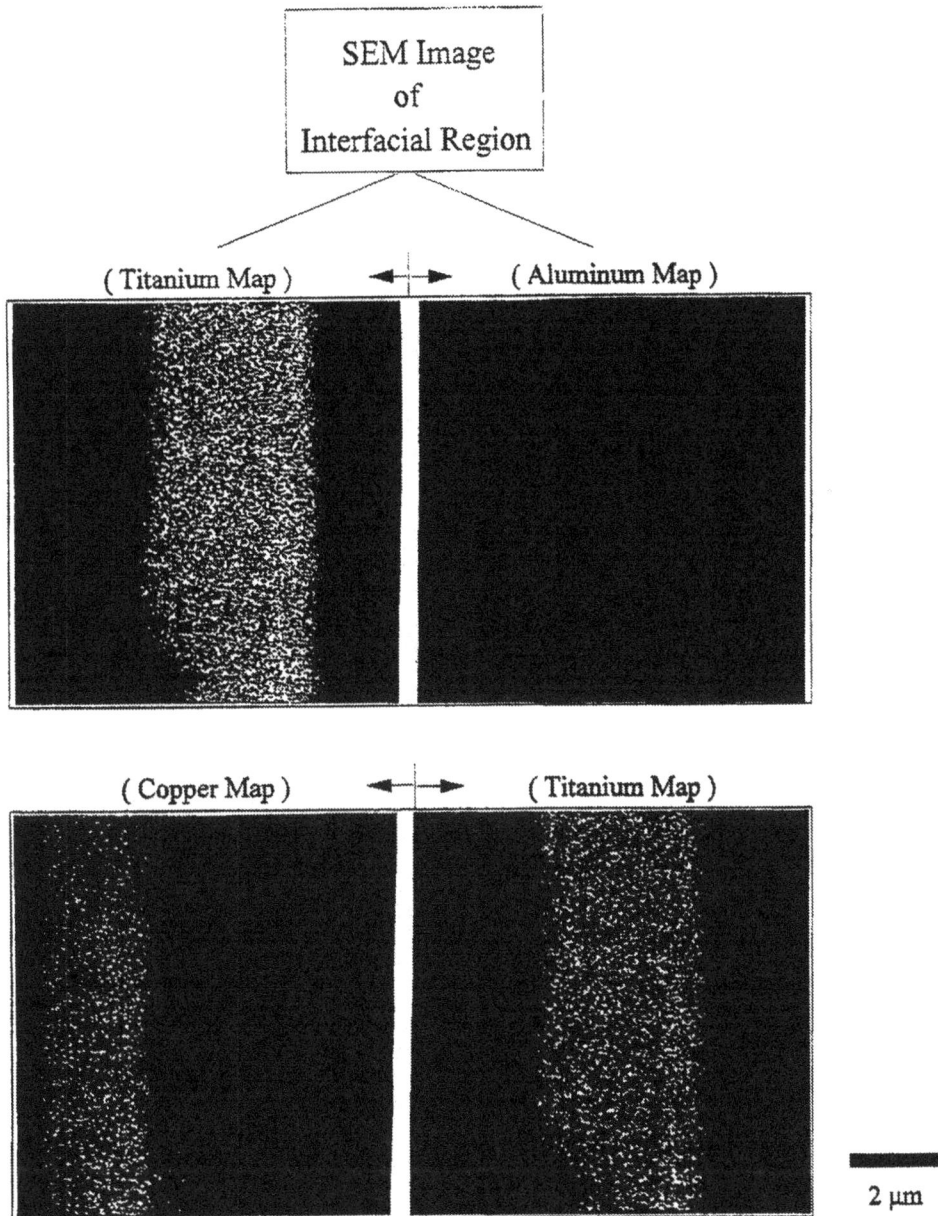

Figure 15: X-ray dot maps of Ti, Al, and Cu at the Cu-Ti/sapphire interface. Processing conditions: Cu-Ti 4 at. %, 1110°C, 15 min., titanium sponge nitrogen getter, oxygen partial pressure $= 10^{-21}$ atm., oxygen bulk concentration = 2,400 ppm.

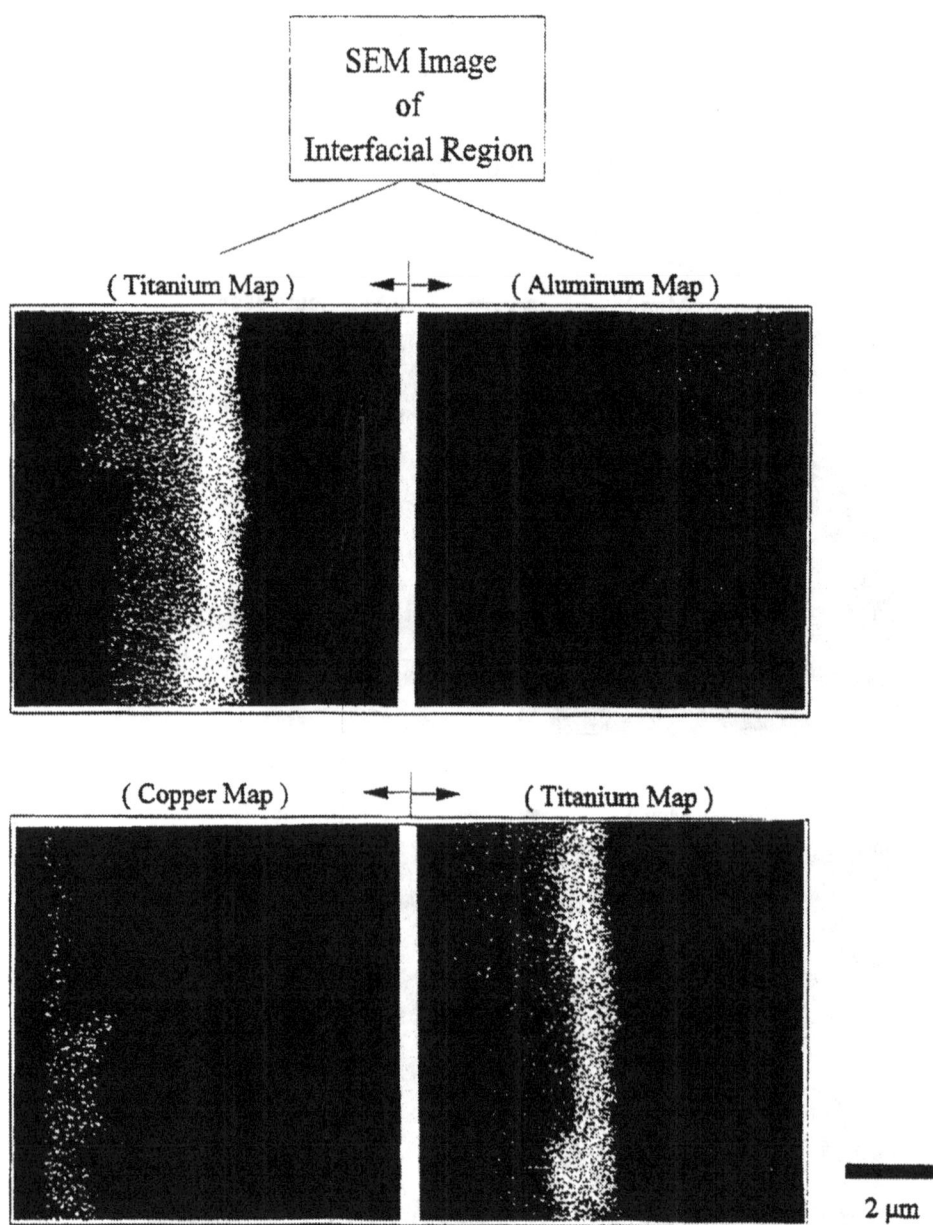

Figure 16: X-ray dot maps of Ti, Al, and Cu at the Cu-Ti/sapphire interface illustrating the formation of a duplex phase Ti-O coadsorption layer. Processing conditions: Cu-Ti 4 at. %, 1190°C, 5 min., titanium sponge nitrogen getter, oxygen partial pressure $= 10^{-21}$ atm., oxygen bulk concentration $= 860$ ppm.

D. REFERENCES

1   Humenik, M.; Kingery, W.D. (1954), "Metal-Ceramic Interactions: III, Surface Tension and Wettability of Metal-Ceramic Systems", J. Am. Ceram. Soc. V.37 pp. 18-23.

2   McDonald, J.E.; Eberhart, J.G. (1965), "Adhesion in Al Oxide-Metal Systems", Transactions of Metallurgical Society of AIME V.233 pp. 512-517.

3   Johnson, K.H.; Pepper, S.V. (1982), "Molecular-Orbital Model for Metal-Sapphire interfacial Strength", J. Appl. Phys. V.53(10) pp. 6634-6637.

4   Ohuchi, F.S. (1991), "Electronic Structure and Chemical Reactions at Metal-Al_2O and Metal-Al Nitride Interfaces", J. Am. Ceram. Soc. V.74(6) pp. 1163-1187.

5   BlackBurn, A.R.; Shevlin, T.S.; Lowers, H.R. (1949), "Fundamental Study; and Equipment for Sintering and Testing of Cermet Bodies, I-III", J. Am. Cer. Soc., V.32(3) pp. 81-98.

6   Ohuchi, F.S.; French, R.H.; Kasowski, R.V. (1987), "Cu Deposition on Al_2O_3 and AlN Surfaces: Electronic Structure and Bonding", J. Appl. Phys. V.62(6) pp. 2286-2289.

7   Ohuchi, F.S. (1989), "A Surface Science Investigation of Metal-Ceramic Interfacial Reactions", Proc. of Japan International SAMPE Symposium, pp. 1404-1411 Nov.28-Dec.1.

8   Ohuchi, F.S.; Zhong, Q. (1990), "Electronic Structure of Metal-Ceramic Interfaces", ISIJ International V.30(12) pp. 1059-1065.

9   BlackBurn, A.R.; Shevlin, T.S.; Lowers, H.R. (1949), "Fundamental Study; and Equipment for Sintering and Testing of Cermet Bodies, I-III", J. Am. Cer. Soc. V.32(3) pp. 81-98.

10 Suganuma, K. (1990), "Recent Advances in Joining Tecnology of Ceramic to Metals", ISIJ International, V. 30(12) pp. 1046-1058.

11 Chidambaram, P.R (1993), "Thermodynamic and Kinetic Aspects of Reactive Metal Liquids in Contact with Oxide Ceramics", Ph.D. Thesis, Colorado School of Mines.

12   Plessis, J. du; van Wyk, G.N. (1989a), "A Model for Surface Segregation in Multicomponent Alloys - Part III: The Kinetics of Surface Segregation in a Binary Alloy", J. Phys. Chem. Solids, V.50(3) pp. 237-245.

13   Gallois, B.; Lupis, C.H.P. (1981), "Effect of O on the Surface Tension of Liquid Cu", Metallurgical Transactions B V.12B pp. 549-557.

14   O'Brien, T.E.; Chaklader, A.C.D. (1974), "Effect of O on the Reaction Between Cu and Sapphire", J. of Am. Ceram. Soc. V.57(8) pp. 329-332.

15   Naidich, J.V. (1981), "The Wettability of Solids by Liquid Metals", in Progress in Surface and Membrane Science, V.14 pp. 353-485.

16   Mukherjee, S. (1987), "Surface Segregation in Transition-Metal Alloys and in Bimetallic Alloy Clusters", Surface Science 189/190 pp. 1135-42.

17   Pajunen, M.; Kivilahti, J. (1992), "Thermodynamic Analysis of the Titanium - Oxygen System", Z. Metallkd, V. 83(1) pp. 17-20.

18   Klomp, J.T. (1987a), "Interface Chemistry and Structure of Metal-Ceramic Interfaces", in Fundamentals of Diffusion Bonding, Tokyo, Elsevier Science Publishers, pp. 3-24 .

19   Nicholas, M.G. (1988), "Interactions at Oxide-Metal Interfaces", Materials Science Forum V.29 pp. 127-150.

20   Kritsalis, P.; Coudurier, L.; Eustathopoulos, N. (1991), "Contribution to the Study of Reactive Wetting in the $Cu-Ti/Al_2O_3$ System", J. of Mat. Sci. V.26 pp. 34003408.

21   Bang, K. (1991), "Kinetics of Interfacial Reactions in Reactive Metal Brazing", Ph.D. Thesis, Colorado School of Mines, Golden Colorado.

22   Naidich, Y.V.; Zhuravlev, V.S. (1973), "Adhesion, Wetting, and Formation of Intermediate Phases in Systems Composed of a Ti-Containing Melt and an Oxide", Poroshkovaya Metallurgiya V.11(131) pp. 40-46.

23   Nicholas, M.G.; Crispin, R.M. (1989), "Brazing Ceramics with Alloys
     ContainingTi", Ceram. Eng. Sci. Proc. V.10(11-12) pp. 1602-1612.

24   Nicholas, M.G. (1991), "Reactive Metal Brazing of Ceramics", Scandinavian Journal
     of Metallurgy V.20 pp. 157-164.

25   Li, X.L.; Hillel, R.; Teyssandier, F.; Choi, S.K.; Van Loo, F.J.J. (1992), "Reactions
     and Phase Relations in the Ti-Al-O System", Acta Metallurgica and Materialia,
     V.40(11) pp. 3149-3157.

26   Choi, S.K; Froyen, L.; Brabers, M.J. (1987), in High Tech Ceramics, Elsevier
     Science Pub, Amsterdam, p. 407.

27   Zhang, M.X; Hsieh, K.C., DeKock, J.; Chang, Y.A. (1992), "Phase Diagram of Ti-
     Al-O at 1100 °C", Scripta Metallurgica et Materialia, V.27 pp. 1361-1366.

28   Chaug, Y.S.; Chou, N.J.; Kim, Y.H. (1987), "Interaction of Ti with Fused Silica and
     Sapphire During Metallization", J. Vac. Sci. Technol. A, V.5(4) pp. 1288-1291.

29   Wahlbeck, P.G.; Gilles, P.W. (1966), "Dissociation Energy of $TiO_{(g)}$ and the High
     Temperature Vaporization and Thermodynamics of the Titanium Oxides. II.
     Trititanium Pentoxide", The Journal of Chemical Physics, V.46(7) pp. 2465-2473.

30   de Camargo, Paulo R., Role of Oxygen in the Cu-O-Ti/Sapphire Interfacial-Region
     Formation, Thesis (Ph. D.), Colorado School of Mines 1995

31   Pankratz, L.B. (1984a), Thermodynamic Properties of Elements and Oxides, United
     States Department of Interior, Bureau of. Mines, V. 672 pp. 427-437.

32   Gilles, P.W.; Carlson, K.D., Franzen, H.F.; Wahlbeck (1967), "High Temperature
     Vaporization Characteristics of the Crystalline Phases", The Journal of Chemical
     Physics, V.46(7) pp. 2461-2465.

33   Ownby, P.D.; Liu, J. (1988), "Surface Energy of Liquid Cu and Single-Crystal
     Sapphire and the Wetting Behavior of Cu on Sapphire", J. Adhesion Sci. Technol.
     V.2(4) pp. 255-269.

34   Kritsalis, P.; Li, J.G.; Eustathopoulos, N. (1990), "Role of Clusters on the Wettability and Work of Adhesion of the Cu-Cr/Al$_2$O$_3$ System", J. of Mat. Sci Letters V.9 pp. 1332-1335.

35   Kritsalis, P.; Merlin, V.; Eustathopoulos, N. (1992), "Effect of Cr on Interfacial Interaction and Wetting Mechanism in Ni Alloy/Al$_2$O$_3$ systems", Acta Metall. Mater. V.40(6) pp. 1167-1175.

36   Guttmann, M. (1975), "Equilibrium Segregation in a Ternary Solution: A Model for Temper Embrittlement", Surface Science, V.53 pp. 213-227.

37   Plessis, J. du; van Wyk, G.N. (1988a), "A Model for Surface Segregation in Multicomponent Alloys - Part I: Equilibrium Segregation", J. Phys. Chem. Solids, V.49(12) pp. 1441-1450.

38   McMahon, Jr. C.J.; Marchut, L. (1978), "Solute Segregation in Iron-Based Alloys", J. Vac. Sci. Technol., V.15(2) pp. 450-466.

39   Schmid, R. (1983), "A Thermodynamic Analysis of the Cu-O System with an Associated Solution Model", Metallurgical Transactions B V.14B pp. 473-481.

40   Hoshino, H.; Shimada. T.; Yamamoto, M.; Iwase, M. (1992), "Activities of Titanium in Molten Copper at Dilute Concentrations Measured by Solid-State Electrochemical Cells at 1373 K", Metallurgical Transactions B, V. 23B pp. 169-173.

41   Pajunen, M.; Kivilahti, J. (1992), "Thermodynamic Analysis of the Titanium - Oxygen System", Z. Metallkd, V. 83(1) pp. 17-20.

42   Miedema, A.R. (1978), "Surface Segregation in Alloys of Transition Metals", Z. Metallkunde, V.69(7) pp. 455-461.

43   Gerasimov, V.V. (1990), "Estimation of the Energies of Adsorption of Oxygen on Metals", Russian Journal of Physical Chemistry, V. 64(12) pp. 1822-1823.

44   Toyoshima, I.; Somorjai, G.A. (1979), "Heats of Chemisorption of O$_2$, H$_2$, CO, CO$_2$, and N$_2$ on Polycrystalline and Single Crystal Transition Metal Surfaces", Catalysis Review - Science Engeneering, V.19(1) pp. 105-159.

45   Pankratz, L.B.; Stuve, J.M.; Gokcem, N.A. (1984c), Thermodynamic Data for Mineral Technology, United States Department of Interior, Bureau of . Mines, V. 677 pp. 258.

46   Loehman, R.E., (1989), "Interfacial Reactions in Metal-Ceramic Systems", Ceramic Bulletin, V.68(4) pp. 891-896.

47   Chastian, J. (1992), Handbook of X-ray Photoelectron Spectroscopy, Perkin-Elmer Corporation.

III. DIFFUSION BONDING CERAMIC TO METAL USING DUCTILE, MULTILAYER REACTIVE METAL COATINGS:

The main objectives of this study are to investigate and model the behavior of ductile, multilayer reactive metal coatings as filler metal for bonding ceramics to metals. Activities for the past year were directed towards characterization of the bond layer, understanding the interface formation, variation of the initial ductile and reactive metal deposition conditions and their relation to process parameters. Detailed chemical analysis of the bond layer cross section was also necessary for the development of a multilayer interdiffusion model. Multilayer Ni/Ti coatings of 1 and 25μ total thickness, with variations in the total number of layers, were produced for joining specimens. They undergo heat treatment where time at temperature, temperature and bonding pressure are the process variables. The composition and structure of the multilayer coatings and bonded cross section were analyzed using x-ray diffraction, SEM/EDS. TEM work is also planned for the bond region characterization. Using literature data for comparison, the measured composition profiles are being modeled to understand the interdiffusion in the Ni-Ti multilayers.

A. EXPERIMENTAL PROCEDURE

As previously reported, characterization of the sputtering deposition rates were done by depositing Ti and Ni onto glass slides at 0.25, 0.5 and 1.0 kw, for 1, 5 and 10 minutes in an argon atmosphere at 1 mtorr. The film thickness were measured with surface tracing profilometer. The thickness measurements were used to derive an empirical model for use in controlling individual metal layers. Once the empirical model was derived, layers having a total thickness of approximately 10,000Å were deposited on glass and Coors ADS 995 ceramic substrates. The number of layers were varied in an odd numbered sequence from 3 to 15 layers ($i = 1$ to 7). The Ti thickness was varied between 309×10^{-9}m and 77×10^{-9}m, the Ni thickness was varied between 382×10^{-9}m and 55×10^{-9}m. Sputtering power for the multilayer films was 0.5kw for the Ni and 0.3kw for the Ti. Sputtering pressure was constant at 1mtorr.

To determine the interdiffusion characteristics of Ti-Ni and Al_2O_3, thick (25μm total) multilayer films were deposited on alumina substrates. Two thick film layering arrangements, composed of Ti/Ni/Ti (3 layers) and Ti/Ni/Ti/Ni/Ti (5 layers) were used as diffusion couples. Titanium layers were 61.92% of the total thickness, nickel layers made up the balance. The layer thickness were designed to produce equiatomic reaction product, NiTi. These were then heat treated in vacuum (10^{-6}Torr, flushed with 99.999 Ar) at 500°C for 10 and 20 hours. Specimens were examined by SEM/EDS, x-ray diffraction and some limited TEM prior to and following heat treatment.

B.   Results and Discussion

Prior to heat treatment, x-ray diffraction measurements of both the 1μm and the 25μm films showed only the structures of Ni, Ti and alumina. As previously reported [1], when heating the 1μm films at 500°C for 5 hours, both NiTi and Ni_3Ti could be seen to form and was shown in the x-ray diffraction spectra. In the thicker, 25μm films, only NiTi was observed after heating for 10 or 20 hours at 500°C in both the 3 layer and 5 layer specimens. This was in contrast to the 1μm films where Ni_3Ti was observed to form at 500°C in five hours, but was consistent with the previous observation [1] that coarser (thicker) metal layering diminished the Ni_3Ti formation. As was the case with the 1μm films where heat treatment at low temperatures (200-300°C) showed incomplete reaction in the form of primary Ni and Ti diffraction peaks, the 25μm films heat treated for 10 hours also showed there was some unreacted Ni and Ti. this was due to the diffusion distance required at the given temperatures. Thinner specimens (1μm) heat treated at lower temperatures (200-300°C) for short times (2-10 hours) show some of the same reaction product formation characteristics as the thicker (25μm) specimens heat treated at higher temperature (500°C) for longer times (10-20 hours). Assuming a thin film solution to the diffusion distance, and applying some of the Ni-Ti interdiffusion data from the literature [2,3,4] the observations seemed reasonable.

The composition profiles of the 25µm, 3 layer specimens are shown in figure 1. Areas approximately 1µm apart were examined and the compositions plotted against the distance from the alumina/metal interface. With a diffusion distance relationship such as:

$$x = 2\sqrt{Dt}$$

where x is the diffusion distance (cm), D is the diffusion coefficient (cm²/s) and t is time (seconds). Using the appropriate D's the diffusion distance for the 500°C-10 hours specimen, diffusion distances of around 6µm would have been expected, similarly, at the same temperature for 20 hours, a distance of around 8µm coincides with the observed distance (fig. 1). Not enough data has been taken to determine the diffusion coefficient for Ni and Ti across the alumina interface and the effect of metal loss into the alumina would have implications for the thin film assumption. Still, as preliminary observations, the system seems fairly well behaved.

The limited amount of TEM observations have been consistent with the x-ray diffraction and EDS work so far. The initial Ni/Ti layering on alumina has been observed, and the NiTi intermetallic could also bee seen. There appears to be, and the analysis was not yet complete, a small amount of undefined material at the alumina/metal interface. This would be expected to be some form of nickel aluminide, titanium aluminide spinel or one of the other Ti-Ni intermetallics. The methodology for good cross section preparation has not been completed so careful characterization remains a priority.

C.   Conclusions

The behavior of the thin (1μm) multilayer films can be compared to the thick (25μm) multilayer films with regards to NiTi formation for given time and temperature heat treatments. With consideration of the Ni-Ti interdiffusion coefficients the system responds in a predictable manner. The effects of metal solute loss into the ceramic substrate during long time exposure to high temperatures (in excess of 500°C) remains to be determined. Additional work on the interfacial product formation and identification is planned to coincide with determination of the metal-ceramic interdiffusion behavior. Thus far, the number and thickness of metal layering and the formation of the desired NiTi intermetallic as well as the formation of other Ni-Ti intermetallics have been determined for a range of temperatures (200-800°C) and times (2-20 hours).

The primary significance of the work was in the determination that a desired intermetallic compound could be formed through the use of sputter deposited multilayer metal films. The relation to the diffusion bonding process shows that it is possible to increase or control the maximum service temperature of the bonded joint through the use of this layering to produce a particular intermetallic product. With additional chemical analysis of the bonded cross section, development of a multilayer inter diffusion model and further evaluation of bonding pressure as related to mechanical properties several things are possible. A fundamental understanding of the process kinetics would have application not only to this system but to other alloy systems as well. That fundamental model correlated with process variables and product properties would then allow for greatly

improved process development and materials selection.

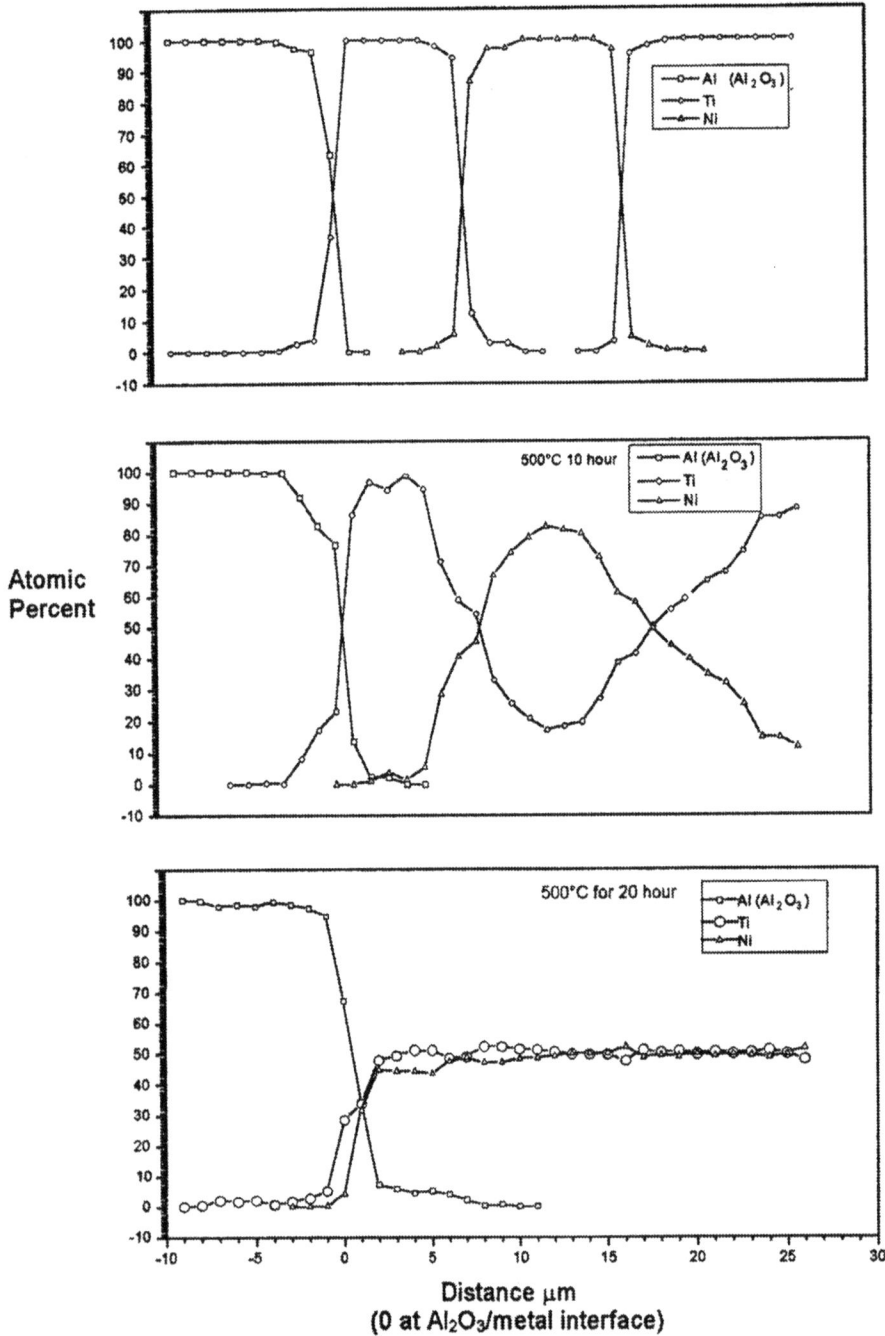

Figure 1: EDS composition profile of 25μm thick 3 layer (Ti/Ni/Ti) film deposited on alumina. Top profile is as-deposited; middle profile, heat treated for 10 hours at 500°C; bottom profile, heat treated for 20 hours at 500°C.

D.   REFERENCES

1   Edwards, G. R., Liu, S.: "Fundamental Aspects of Metal-to-Ceramic Brazing", MT-CWJR-94-037, (Oct. 1994), pp.94-100.

2.   Burminskaya, L. N.: "Diffusion Processes in a Ti-Ni Composite", Metalloved. Proch. Mater, n.3 (1971), pp.251-5.

3   Marinkovic, Z.: "Comparative Analysis of Interdiffusion in Some Thin Film Metal Couples at Room Temperature", Thin Solid Films, v. 127, n. 1/2, (Sept. 1992), p.26.

4   Maza, M., Sella, C., Ambroise, J. P., "Determination of Diffusion Coefficient, D, and Activation Energy, Q_a, of Nickel into Titanium I Ni-Ti- Multilayers by Grazing-Angle Neutron Reflectometry", Journal of Applied Crystallography, v. 25, n. 3, (Jun. 1993), p.334.

ULTRASONIC TECHNIQUES

FOR THE

EVALUATION OF

CERAMIC JOINTS

by

W.A. Simpson, Jr.
and
R.W. McClung

CONTENTS

ULTRASONIC TECHNIQUES FOR THE EVALUATION OF CERAMIC JOINTS*

W. A. Simpson, Jr., and R. W. McClung

ABSTRACT

The increasing use of structural ceramics in high-temperature applications has led to the need for nondestructive evaluation techniques to ensure the integrity of the ceramic materials and the quality of joints consisting of ceramics bonded to ceramics or to metals. We describe the development of ultrasonic techniques for the characterization of ceramic materials and for the detection of flaws in these materials and at ceramic joints. This work has led to the ability to determine which face of a 60-μm-thick layer of braze filler material is unbonded, thus providing information about the integrity of the ceramic—filler metal bond. We also describe the development of a rapid technique using Lamb waves to probe the bond between alumina coupons in flexure-strength specimens, whose geometry makes conventional ultrasonic evaluation of the bond difficult.

INTRODUCTION

The excellent high-temperature thermal, mechanical, and physical properties of structural ceramics make them likely choices for use in advanced engine designs to allow higher combustion temperatures and therefore higher thermodynamic efficiencies. In addition, the lower weight of such materials relative to that of high-temperature structural alloys should increase an engine's ratio of power to weight. Unfortunately, the low fracture toughness, and hence the small critical flaw size, of structural ceramics precludes the use of standard nondestructive evaluation (NDE) techniques that have been developed over the last 30 years for detection and characterization of critical flaws in metals. For example, flaws of critical size in most structural metals can be detected with ultrasonic waves whose frequencies lie in the range from 1 to 10 MHz. Consequently, most development in ultrasonics has encompassed this frequency range, with little activity occurring above 15 to 20 MHz. In conventional monolithic ceramics, however, the critical flaw size is about

20 µm, and for such small flaw sizes the use of frequencies of 50 MHz and
higher is required. In addition, detection of such small flaws requires
the use of focused radiation, and the propagation of such energy through
the ceramic surface introduces severe aberration into the beam, thus
limiting to a few millimeters the depth at which effective focus can be
maintained. Finally, several heat engine designs require that a ceramic
be joined to a metal to take advantage of the physical and/or mechanical
properties of each. These applications require the development of technology for joining ceramics to other ceramics and to metals and, no less
importantly, for inspecting such joints nondestructively to ensure bond
integrity. This report describes work performed by the Nondestructive
Testing Group of the Metals and Ceramics Division on the development of
equipment and techniques for detecting small flaws in ceramic-ceramic and
ceramic-metal joints.

The objective of this program is to investigate methods for NDE of
ceramic joints leading to recommendations and development of techniques
for evaluating important properties and characteristics that affect the
serviceability of joints.

EXPERIMENTAL PROCEDURES AND RESULTS

CERAMIC MATERIALS

In order to develop techniques for the inspection of ceramic joints
it is first necessary to characterize the ceramic materials themselves,
since they will often be the determining factor in the choice of test
parameters. This is in contrast to the ultrasonic inspection of metals,
where the properties other than the wave velocities of the host can
frequently be neglected. The engineering constants (Young's modulus,
Poisson's ratio, etc.) can easily be determined nondestructively for
ceramics by well-known techniques, but this information does not determine
the basic inspectability of the sample for a given flaw size. For
example, two specimens may have identical engineering constants, but one
may be inspectable with 100-MHz ultrasonic energy while the second may not
transmit energy above 20 MHz. This behavior results from the fact that

attenuation is highly sensitive to the microstructure of the host; for example, one can glean some information about the average grain size from an attenuation measurement. Therefore, to the determination of the standard engineering constants should be added the attenuation behavior of the particular specimen for ultrasonic energy in a frequency range commensurate with the flaw size of interest.

Transfer Curve

Although the measurement of attenuation at discrete frequencies is a standard procedure in ultrasonics, the use of such an approach at the high frequencies employed in ceramic inspection would be prohibitively expensive because of the high cost of transducers. A much better approach is to use a single broadband transducer to transmit a range of frequencies and, using a computer, to analyze each frequency component separately. This approach offers the additional advantage of automating the application of corrections to compensate for such nonspecimen losses as beam spread, acoustic impedance mismatch, etc. In addition, the effects of the transducer response and system electronics can be removed from the acquired data to yield a true attenuation versus frequency response for the specimen. This response is termed the transfer curve of the specimen, in analogy to the similar function in optics. In our work, all these corrections have been incorporated into a single computer program that computes the transfer curve from input data consisting of the signals reflected from the front and rear surfaces of a planar sample of the material under test.

When we first began computing transfer curves for ceramics, several anomalies arose that could not be attributed to the samples. In particular, the attenuation in all samples at frequencies above about 30 MHz was much higher than would be expected from optical measurements of the microstructure. This behavior was ultimately traced to the very thin (<1-µm) layer of water used to couple the ultrasonic energy into the ceramic part. For frequencies less than about 15 MHz, the thickness of this coupling layer is such a small fraction of the ultrasonic wavelength that it is negligible. As the frequency increases, however, the coupling layer becomes a larger fraction of a wavelength, and its presence is no longer negligible. When we analyzed this configuration, treating it as a three-layer rather than a two-layer problem, it became obvious that a correction would have to be added to account for the presence of the coupling layer. Figure 1 shows how the reflection coefficient varies as a function of frequency for a water coupling layer 1 μm thick. At lower frequencies the coefficient approaches the two-layer value, where the effect of the water is negligible. Our computer program was subsequently modified to actually measure the thickness of the coupling layer during a test and provide a correction in the data. The transfer curve so obtained is now highly repeatable and in agreement with destructive analysis.

ORNL-DWG. 87-6606

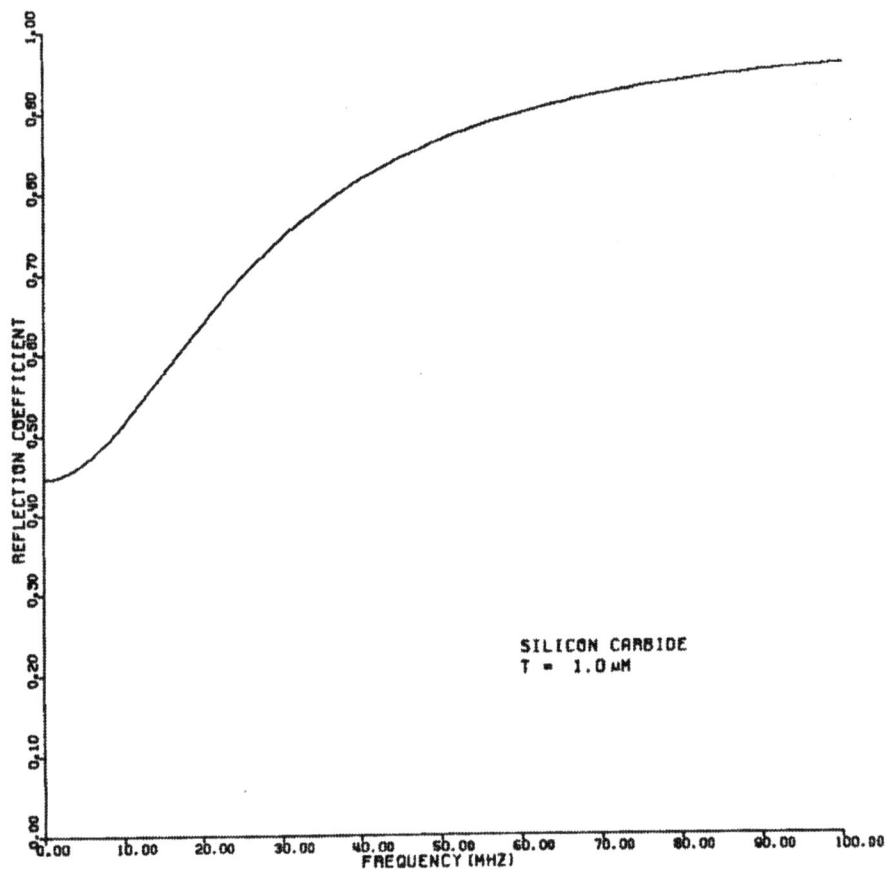

Fig. 1. Reflection coefficient at a water-silicon carbide interface for a 1-μm water layer.

Figures 2 and 3 illustrate the effect of the microstructure on the transfer curve. In Fig. 2, the sample is a piece of tetragonal zirconia polycrystalline (TZP) ceramic with a microstructure of primarily 10-μm tetragonal-phase grains. In Fig. 3, however, the sample is a partially stabilized zirconia (PSZ) with predominantly cubic grains about 100 μm in diameter. It is difficult to propagate through the PSZ an elastic wave whose frequency exceeds about 30 MHz because of severe scattering losses. For such material, the minimum detectable void diameter is probably also about 100 μm.

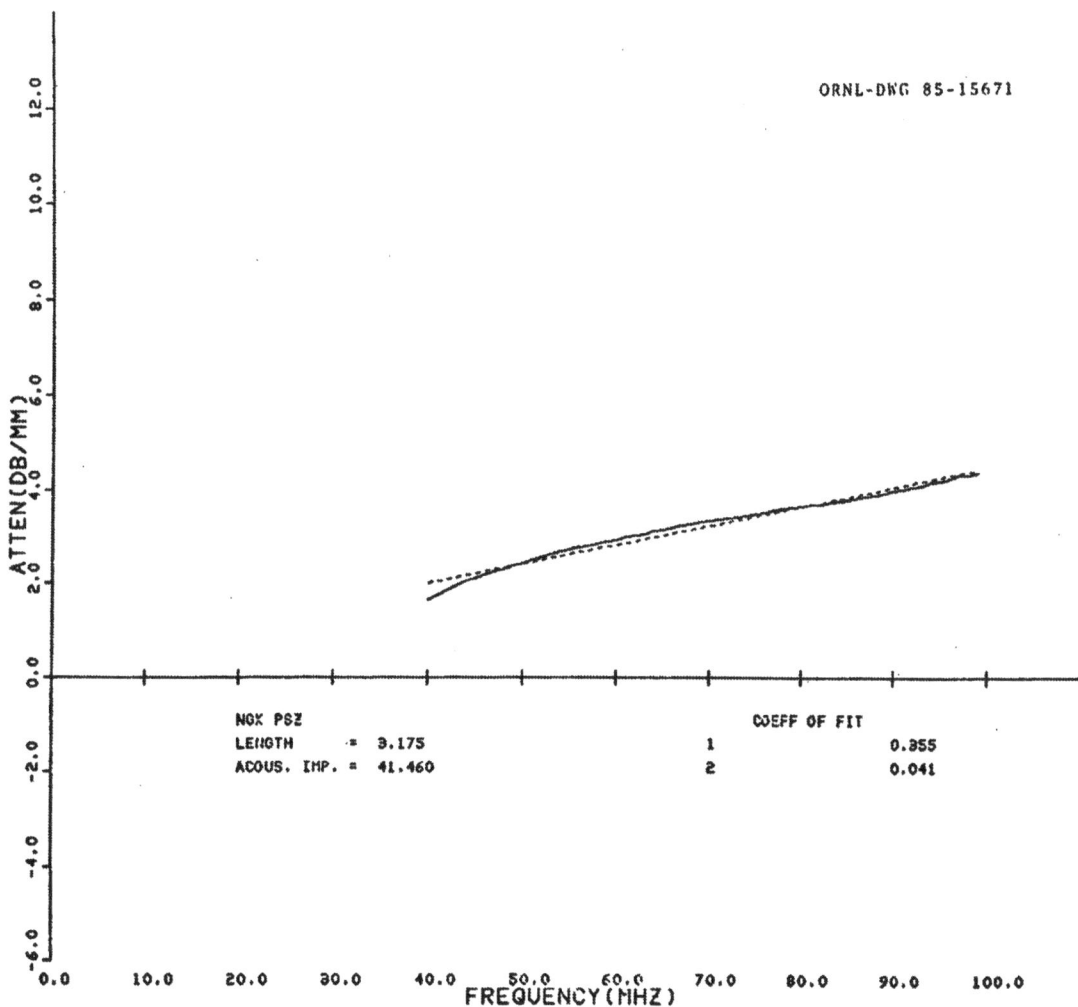

Fig. 2. Transfer curve of tetragonal zirconia polycrystalline ceramic.

ORNL-DWG 85-15661

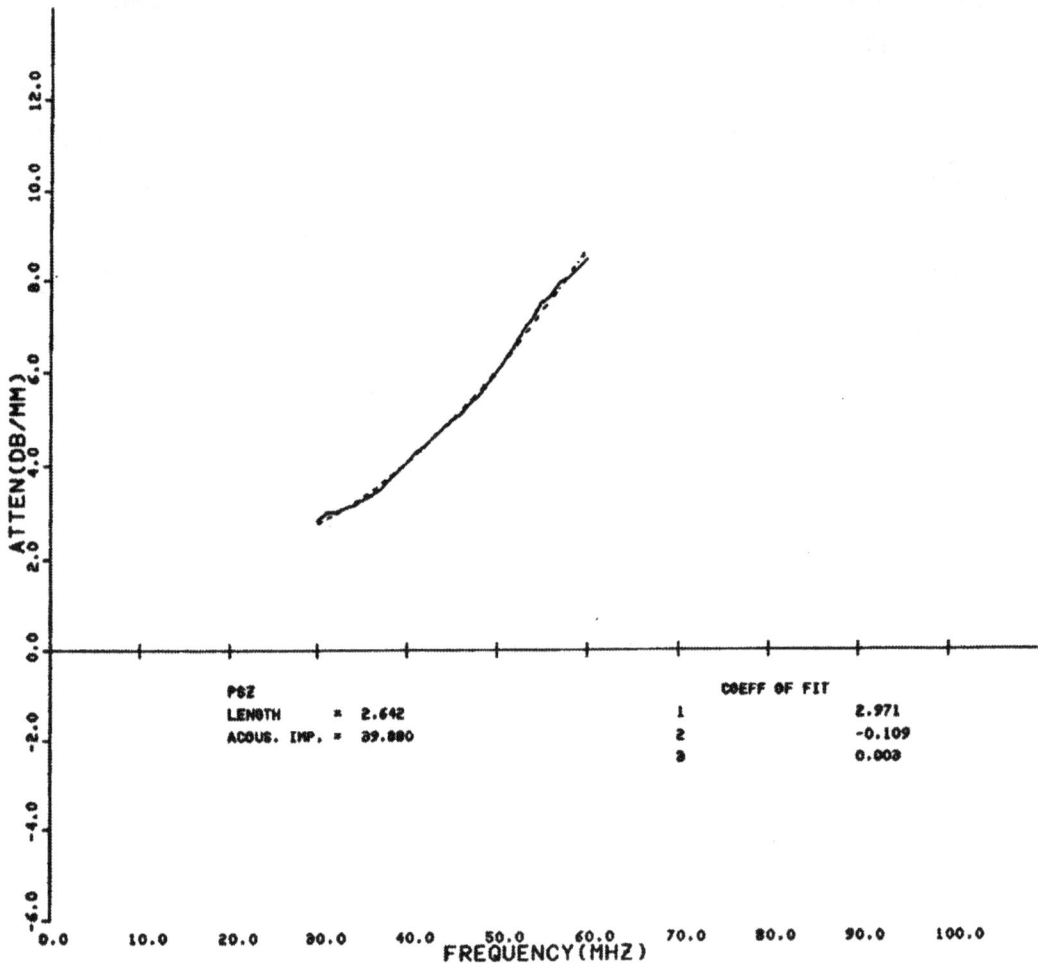

Fig. 3. Transfer curve of partially stabilized zirconia.

Detection of Flaws in Ceramics

Also of interest in characterizing the ultrasonic response of a
ceramic material is the detection of flaws comparable to or larger than
the critical flaw size, particularly if they occur in the vicinity of a
joint. From published destructive analyses of flaw initiators in ceramic
materials, a common flaw shape is the quasi-spherical void or inclusion.
This is fortuitous, since a model exists for the scattering of elastic
waves from spherical cavities and inclusions.[1] For a planar crack, we
have previously demonstrated a successful model for measuring crack
size.[2]

263

7

Figure 4 shows the calculated response to elastic waves of a
spherical void in silicon nitride. The abscissa is the dimensionless
product of the wave number, k (i.e., $2\pi/\lambda$, where λ is the ultrasonic
wavelength in centimeters), of the incident radiation and the void radius,
a. For a given flaw size, the abscissa is thus proportional to frequency.
The ordinate is the differential scattering cross section. The response
may be divided into three regions: (1) the low-frequency region
(Rayleigh scattering) in which the scattering increases as the fourth
power of the frequency, (2) the transition region (located at $ka \approx 1$), and
(3) the high-frequency region in which the scattering cross section is
oscillatory with an approximately constant average value. The latter
behavior can be traced to interference between two waves scattered by the
sphere. The first is the direct wave reflected from the near point of the
sphere, and the second is the so-called "creeping" wave, which, after
tangential incidence, propagates around the sphere and is reemitted after

ORNL-DWG 85-15662R

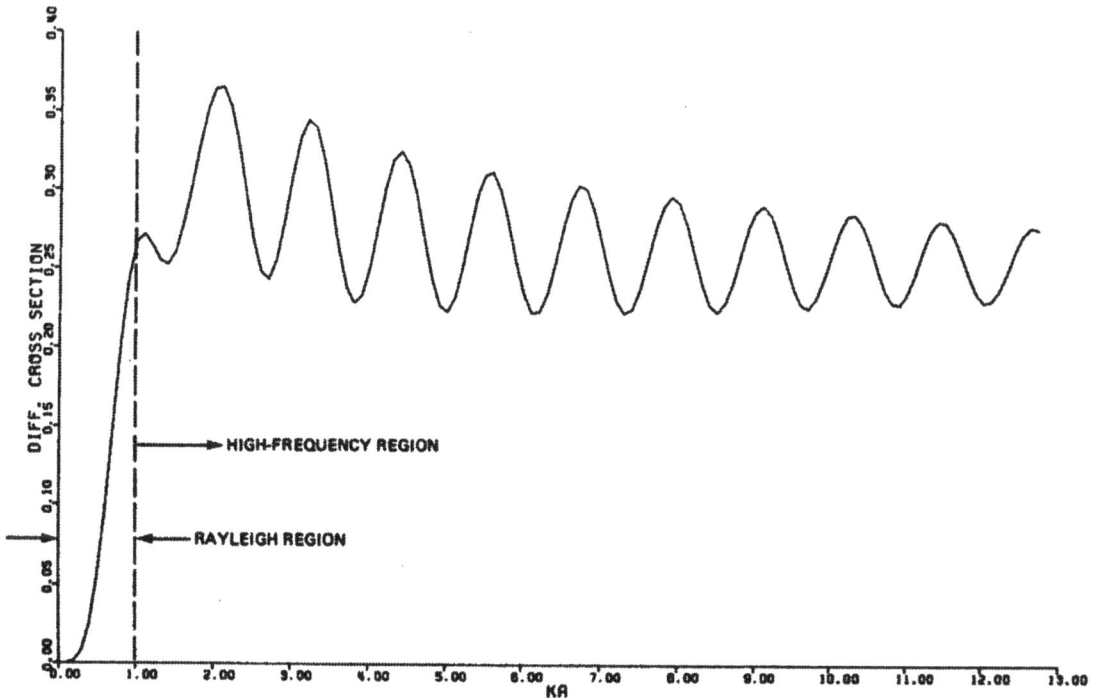

Fig. 4. Differential cross section for scattering from a spherical
void.

264

traversing half the circumference. The periodicity of the oscillation is
thus related to the size of the void. For natural flaws, however, even
those whose shape is very nearly spherical, the surface roughness may
rapidly attenuate the latter wave, and the high-frequency scattering
depicted in Fig. 4 may be replaced by a near-constant value. In this
case, the size of the scattering center may be approximated by noting that
the transition region occurs at $ka \approx 1$; thus if the frequency of the
turning point, that is, the transition from Rayleigh to high-frequency
scattering, is known, the radius of the sphere may be estimated.

Figure 5 shows experimental data obtained on a natural flaw in TZP
ceramic. Here the turning point occurs at about 89 MHz. Based on the
measured velocity, the flaw diameter was estimated to be approximately
25 µm. That the flaw size is of this order of magnitude is also supported
by the fact that it could not be detected reliably with a scan increment
of 50 µm but could with an increment of 25 µm.

ORNL-DWG 85-15660

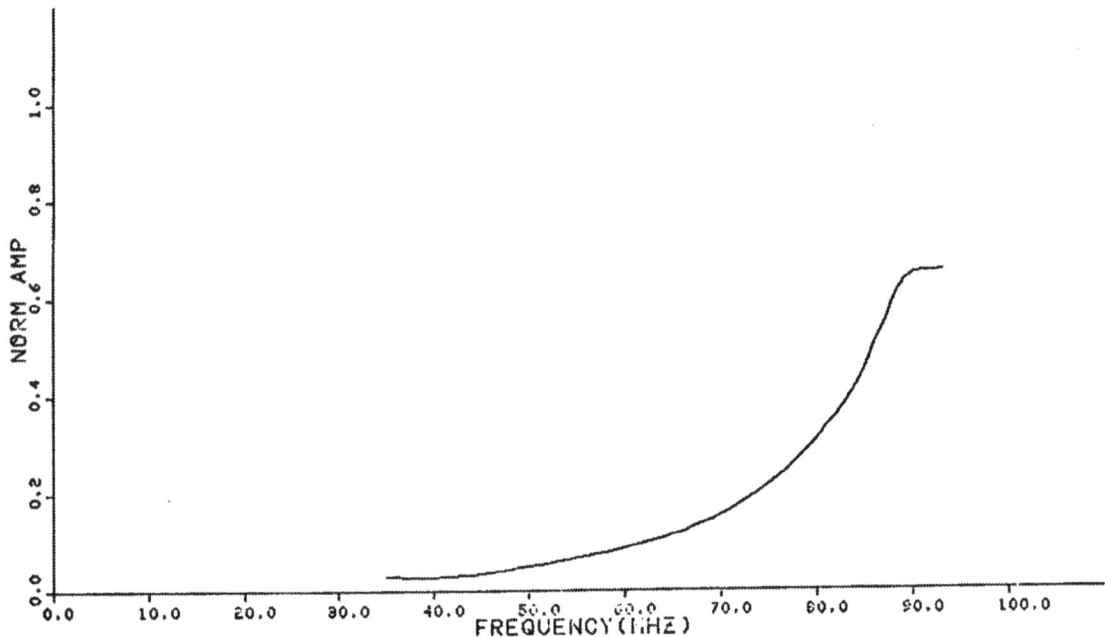

Fig. 5. Experimental data for scattering from a natural flaw in
tetragonal zirconia polycrystalline ceramic.

PLANAR JOINTS

Ceramic-Cap Piston Specimens

Two large ceramic-to-metal brazements that mock up the attachment
of a ceramic cap to a diesel engine piston were made available for
nondestructive testing studies. These specimens are 111 mm (4 3/8 in.)
in diameter and consist of a 6.4-mm-thick (0.25-in.) PSZ cap brazed to
a nodular cast iron (NCI) disk via a titanium transition piece and a
commercial Ag-Cu-Sn brazing filler metal.[3] The surface of the ceramic had
been vapor coated with titanium to promote wetting by the filler metal.
Figure 6 shows the geometry of the joint and Fig. 7 one of the specimens.
The thickness of each braze layer was about 60 μm and that of the titanium
transition piece about 0.6 mm.

ORNL-DWG 86-1862

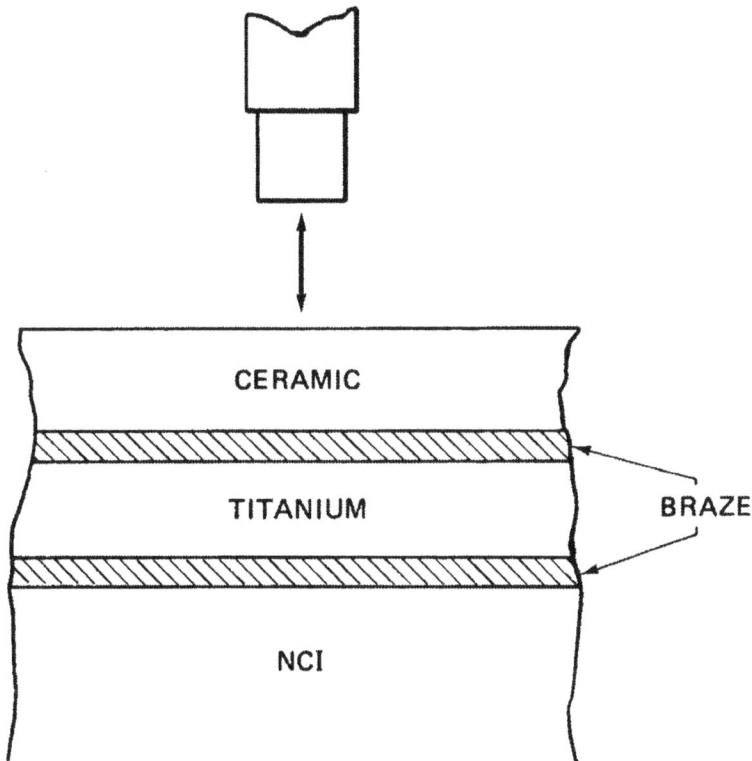

Fig. 6. Diagram of a transition joint for bonding a ceramic to nodular cast iron (NCI) to simulate a diesel-engine piston with a ceramic cap.

YP419

CENTIMETERS

0　1　2

Fig. 7.　Ceramic-cap diesel-engine piston specimen (one of two).

Initial ultrasonic examination of each specimen indicated a relatively high attenuation for elastic waves in the ceramic cap, restricting the maximum usable frequency to about 25 MHz. From our previous studies of the attenuation characteristics of PSZ material, this behavior was consistent with the scattering losses produced by the relatively large grain size (≈100 μm) of this ceramic.

Both specimens were scanned in our high-frequency (100-MHz) ultrasonic system, which permits variations in the transmission of the ceramic-metal bond to be displayed as a gray-scale image. A flat (unfocused), broadband transducer was used, and both samples were found to contain nonbonded regions near the center. Figure 8 shows the result for one of the brazements. Here lighter areas indicate relatively better transmission through (less reflection from) the bond, and darker areas indicate relatively poorer transition (greater reflection). The dark area near the center is completely unbonded. (In the original data this area was uniformly dark; it did not reproduce well photographically, however.) The dark ring near the periphery of the sample is an edge effect caused by the unfocused transducer.

ORNL-DWG 85-15664

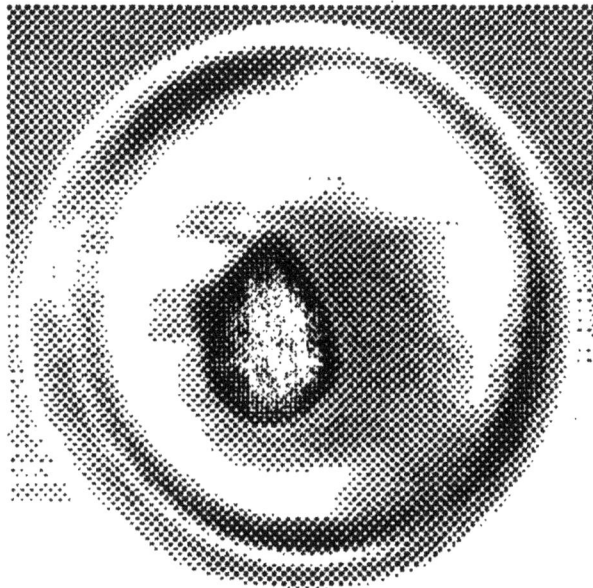

Fig. 8. Gray-scale presentation of ultrasonic transmission through the bond in piston specimen 1.

The samples were scanned a second time using a focused transducer.
Focusing minimizes edge effects and more sharply delineates the region of
unbond. Figure 9 shows the raw gray-scale data for the second sample,
which is seen to be unbonded in the center and at several locations around
the periphery.

ORNL-DWG 85-15658

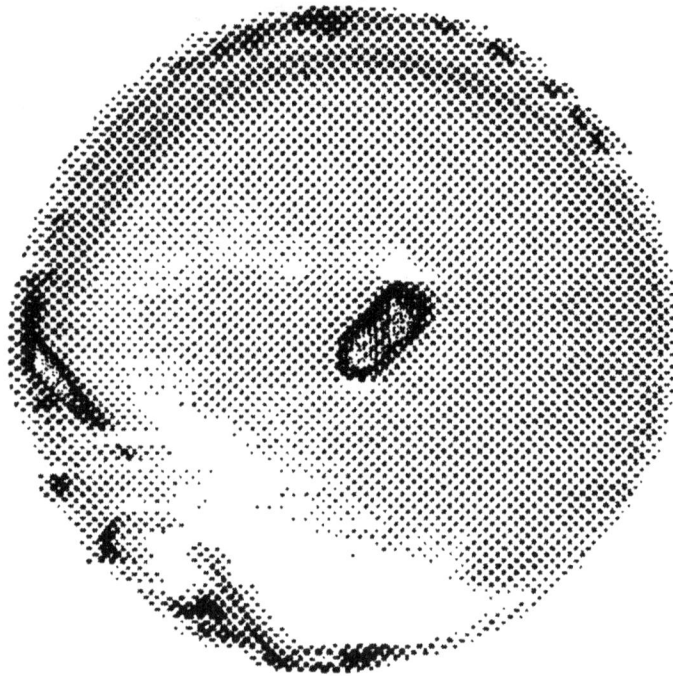

Fig. 9. Gray-scale presentation of ultrasonic transmission through
the bond in piston specimen 2.

The raw pixel data from these gray-scale images were next computer
processed to permit expansion of a selected range of gray-scale values.
Such processing allows minor (<3-dB) bond variations to be ignored and
fluctuations within the unbonded areas to be enhanced. Figures 10 and 11
show the processed images for the two specimens. As expected, the areas
delineated show virtually no variations, which is indicative of complete
unbonding. The shape, size, and location of the affected areas are also
precisely determined by this processing.

ORNL-DWG 85-15657

Fig. 10. Enhanced gray-scale presentation of ultrasonic transmission through the bond in piston specimen 1.

ORNL-DWG 85-15659

Fig. 11. Enhanced gray-scale presentation of ultrasonic transmission through the bond in piston specimen 2.

Because there are several interfaces in the PSZ-NCI joint (Fig. 6), one would like to know at which of these interfaces the unbond occurs, since this information could provide insight into the origin of the problem. In addition, while the precise location of the unbond may be immaterial from an accept-reject point of view, such information is important in the development of ceramic brazing technology. For example, if the unbond occurs at the ceramic-braze interface, it could indicate a failure of the braze filler metal to wet the ceramic surface, and wettability is a major consideration of the brazing development program. Fortunately, the location of the unbond can be determined nondestructively from the ultrasonic scattering data. Figure 12 shows the rf waveform of signals scattered by the transition joint in three areas of the specimen whose data were shown in Figs. 9 and 11. In Fig. 12, the upper trace (trace A) is that obtained from a well-bonded region. The two signals are from the faces of the 600-μm-thick titanium transition piece, with the PSZ to the left and NCI to the right. The ultrasonic wave is incident from the PSZ. In trace B, the second signal is much larger, and the third signal, resulting from reverberation in the titanium, indicates that the unbond occurs on the NCI side of the titanium. In trace C, the first signal is much larger, and the absence of the signal from the opposite face of the titanium indicates that the sample is unbonded on the PSZ side of the titanium.

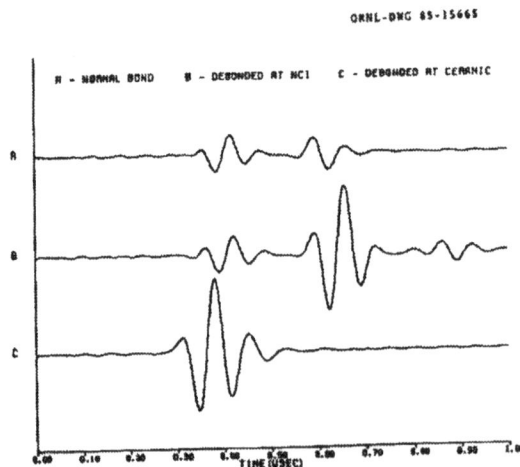

Fig. 12. Radio-frequency waveform of the ultrasonic signal scattered by the interface of piston specimen 2 for various bond conditions.

271

If the ceramic material will support elastic waves of a sufficiently high frequency, the signals from opposite faces of the nominal 60-μm-thick braze layer can be resolved (a time resolution of only 30 ns), and it is then possible to tell directly whether the unbond occurs at the ceramic-braze or braze-titanium interface. For example, Fig. 13 shows the waveform obtained from the joint region of a sample containing TZP (wide bandwidth) ceramic. The first two signals originate at the faces of the nominal 60-μm-thick braze layer between the TZP and the titanium transition piece. The second two are from the similar braze layer between the titanium and the NCI substrate. Note that the signals from the braze layers are resolved, which permits the condition of any of the four interfaces to be monitored. The restricted bandwidth of the PSZ is not sufficient to resolve these signals (frequencies above about 30 MHz are not transmitted). However, we have recently developed a technique to enhance the resolution of closely spaced ultrasonic signals.[4] Use of this technique will generally produce delta functions in the output data at the location of each individual wave center in the input data, even when the waves are too closely spaced to resolve in the time domain. When this technique was used to process the signals from the data of Fig. 12, the signals from either face of the braze layers could be resolved, as shown in Fig. 14. Since all four signals are present, the region is one in which the sample is well bonded. For the central unbond region of the simulator shown in Fig. 11, however, the processed data show only a single

YP1588

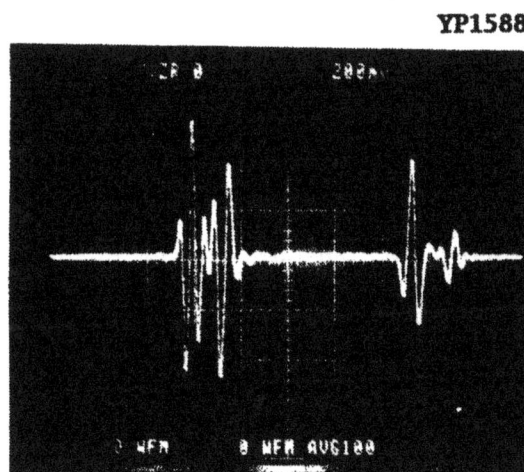

Fig. 13. Radio-frequency waveform of the ultrasonic signal scattered by the interface in tetragonal zirconia polycrystalline ceramic.

16

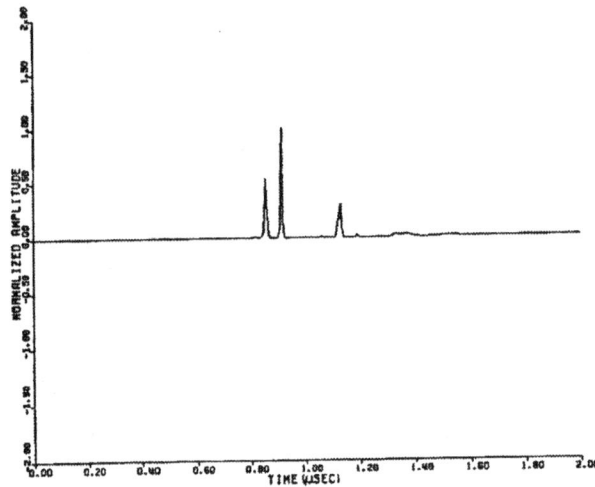

Fig 14. Processed data showing recovery of four interface signals in the bonded region of tetragonal zirconia polycrystalline ceramic.

delta function at the location of the PSZ-braze interface, indicating that the problem is failure of the braze to adhere to the ceramic.

These results indicate that for both samples, the central unbond occurs at the PSZ interface. For the specimen of Figs. 9 and 11, some of the edge indications are unbonded at the PSZ interface and some at the NCI interface.

Following NDE, the Materials Joining Group sectioned the sample of Figs. 9 and 11 through the unbonded region delineated by the central indication. When the region was cut out, the ceramic cap actually fell off. Subsequent metallographic examination of the ceramic-braze interface revealed the presence of extreme porosity, possibly due to trapped gas. The nondestructive results were thus validated.

Shear Specimens

Following examination of the piston-cap specimens, some smaller ceramic-metal brazements were made available. These samples were approximately 9-mm-square pieces of 3.5-mm-thick zirconia ceramic brazed to an NCI substrate with a 0.6-mm-thick titanium transition piece. In this case, however, the ceramic was a fine-grained TZP material that could support elastic waves at frequencies in excess of 100 MHz. In such

273

samples, effective focus could be maintained at depths up to 5 mm for these frequencies: accordingly, a 100-MHz focused transducer with a focal length of 25.4 mm in water was obtained to exploit this condition. At this frequency and with the transducer focused at the PSZ-braze interface, the signals from either side of the 60-μm braze layer could be resolved, even though these signals are separated by only about 30 ns. Therefore, we were able to gate selectively the signal from the TZP-braze interface for analysis. This signal is an indicator of the quality of the bond that exists between the braze filler metal and the TZP. Figure 15 shows two of the samples and Fig. 16 the results of scanning the bond region. In these gray-scale presentations, lighter areas indicate relatively better bonding while darker regions depict relatively poorer bonding. The sample on the right has numerous small regions where there is lack of bonding. These regions average perhaps 100 μm in diameter and are probably caused by bubbles of trapped gas in the braze material. The presence of these bubbles has been previously demonstrated by destructive analysis and inferred from examination of the fracture surface, but, until now, we have not been able to detect their presence nondestructively.

YP2041

Fig. 15. Tetragonal zirconia polycrystalline ceramic joint specimens.

Fig. 16. Gray-scale presentation showing the detection of pores in the braze layers of the ceramic joint specimens shown in Fig. 15.

A diagonal band, running from lower left to upper right of the sample on the right, can be seen in the data of Fig. 16. The cause of this darkening is not known, but possibilities include the presence of microporosity or residual stress in the braze layer. The dark area in the upper left of the sample on the right is caused by severe thinning of the braze material.

The generally darker nature of the sample on the left in Fig. 16 is possibly attributable to use of a different grade of ceramic. Small variations in the acoustic properties alter the reflection coefficient at the TZP-braze interface and vary the average brightness of the reproduced data.

The sample on the left also exhibits a periodic variation in the interface that may be caused by machining marks on the TZP. No such marks are detectable on the visible surface, but they sometimes occur on one or more surfaces of the blanks.

The above results indicate that a great deal of information about the nature of the PSZ-braze bond can be gleaned from the ultrasonic scattering if the signals from the braze layer can be resolved. However, if they cannot, as is the case at typical ultrasonic frequencies (1-10 MHz), the signal generated by subtle variations at the filler metal—PSZ interface will be swamped by the signals from the filler metal—titanium interface.

Manufactured Flaws

We have considerable interest in determining the minimum detectable area of unbond in a ceramic joint. From the backscattering spectrum obtained from discrete flaws in ceramics, we earlier showed (see Fig. 5) the ability to infer a minimum measurable flaw size of about 25 μm (the minimum detectable flaw size will be still smaller) with our present system. It is difficult to fabricate discontinuities in this size range. However, we obtained a standard consisting of a bonded couple between a P-leg and an N-leg of a silicon-germanium thermoelectric sample. The bond contains three manufactured flaws having diameters of 250, 650, and 125 μm, and the acoustic properties (i.e., velocity and density) of the silicon-germanium are similar to those of typical ceramics. Figure 17

ORNL-DWG. 86-8668

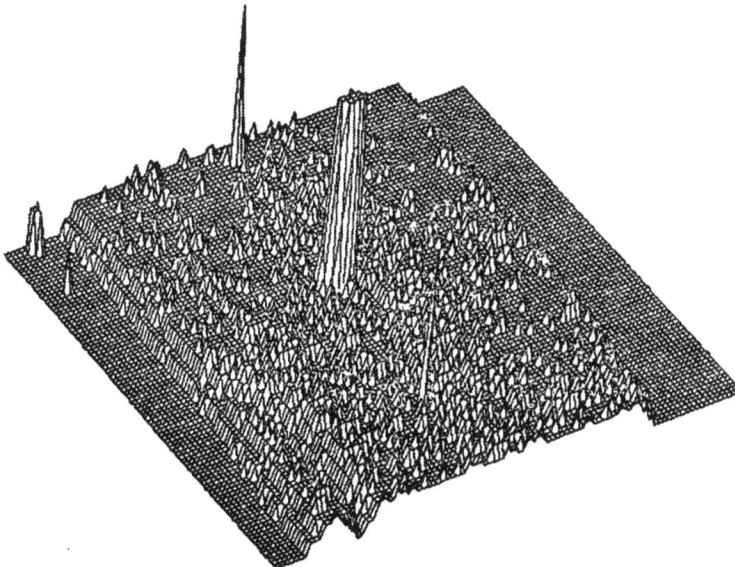

Fig. 17. Three-dimensional presentation of the ultrasonic detection of, left to right, 250-, 635-, and 125-μm flaws at the interface between two silicon-germanium thermoelectric samples.

276

is a pseudo three-dimensional view of the ultrasonic scattering from the bond line showing the detection of all three flaws. In the lower left of the figure is a natural crack that extended through the substrate and terminated at the interface. Figure 18 is a view of the data from a lower angle, making the smallest flaw more visible. Figure 19 is an expansion of the region around the 125-μm flaw. Note that the flaw signal is much larger than the background from the interface; thus, we should be able to see flaws considerably smaller than 125 μm reliably.

ORNL-DWG. 86-8669

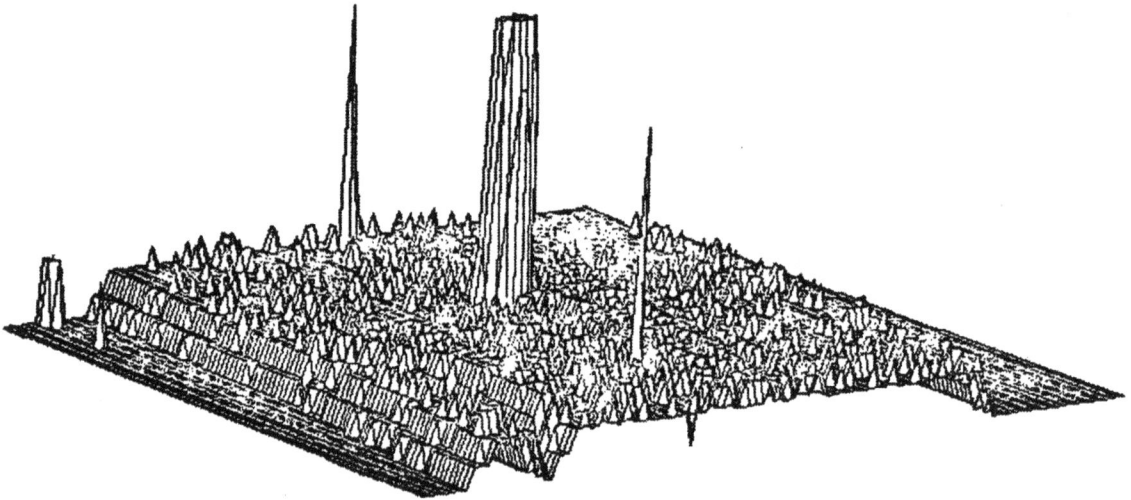

Fig. 18. Three-dimensional presentation (from a lower angle then in Fig. 17) of the ultrasonic detection of, left to right, 250-, 635-, and 125-μm flaws at the interface between the two silicon-germanium thermoelectric samples.

ORNL-DWG. 86-8671

Fig. 19. Three-dimensional presentation of the region around the 125-μm flaw at the interface between the two silicon-germanium thermoelectric samples.

BUTT JOINTS

Lamb Wave Studies

For the samples described earlier in this report, the specimen geometry was such that bulk-wave probing of the joint region (albeit at frequencies far above those used in conventional ultrasonic testing) could be performed at normal incidence. We demonstrated earlier the ability to resolve signals from either side of the 60-μm-thick layer of brazing filler metal when the ultrasonic waves were incident normally on the joint. This capability then allowed us to determine whether a lack of bond occurred at the ceramic or metal interface of the joint and to measure the variation in filler metal thickness across the specimen.

For some specimen configurations, normal-incidence testing of the braze region using bulk waves is not possible. For example, a butt-braze joint between thin plates requires angle beams if bulk waves are used to interrogate the interface. While this configuration may or may not be important in the final application of ceramic joining to heat engines, it is of considerable importance to the development of joining technology in that many test specimens use such joint geometries. For example, specimens to measure flexural strength of ceramic-ceramic or ceramic-metal brazements are of this type (see Fig. 20), and inspection techniques must be developed so that correlations between NDE indications and mechanical properties can be determined. Only in this way can one differentiate between significant and ignorable indications.

ORNL—DWG 85-7800

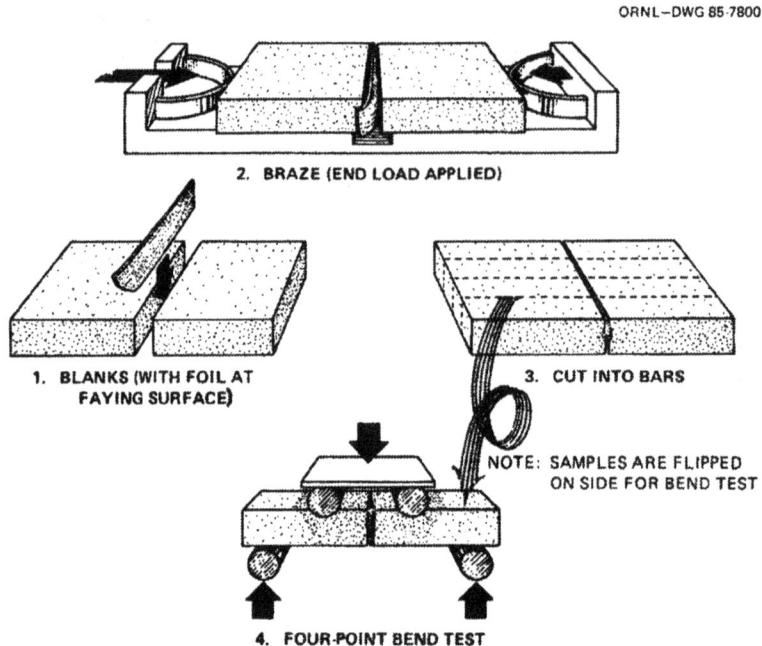

Fig. 20. Fabrication of ceramic flexure-strength specimens with butt-braze joints.

We have examined several butt-braze joints in 25.4 × 14.2 × 2.9 mm alumina coupons using high-frequency angle-beam techniques. While this approach would appear to have considerable potential for evaluating bond quality, the results thus far have been more difficult to interpret than the results from normal-incidence tests, and they suffer from the fact that the focal point of the transducer cannot be maintained exactly on the bond across the full height of the joint. This test is also relatively slow, since the bond must be scanned using a small linear increment.

A second approach for evaluating bond quality in butt-joint specimens relies on measurement of the transmission coefficient of so-called plate or Lamb waves propagated through the bond. In this technique, a Lamb wave, which is the elastomechanical analog of guided electromagnetic waves, is excited on one side of the joint and the amplitude of the wave on the second side measured after propagation through the bond. This approach has the advantage of requiring scanning only along a single axis, making it much faster than conventional testing. The frequencies involved are also quite low, typically 1 to 5 MHz. The quantity determined by this measurement is the relative area of the bonded region, which has been related to the prediction of shear strength for spot welds in metals.[5]

As is the case for electromagnetic guided waves, Lamb waves are highly dispersive. Since the propagation of such a wave along a plate produces (microscopic) flexure of the plate, Lamb waves can produce either symmetric or antisymmetric modes according to the symmetry of the flexure with respect to the center line of the plate. Figures 21 and 22 show the calculated dispersion curves for the first four symmetric and antisymmetric modes in 2.9-mm-thick alumina. Since the ordinate gives the Lamb wave phase velocity (normalized by the shear wave velocity), it is also related to the angle of incidence necessary to generate the given Lamb mode. Thus a horizontal line will intersect the various possible modes, whose abscissas give the frequencies necessary to establish the modes at the indicated angle of incidence of the exciting energy.

ORNL-DWG. 86-8672

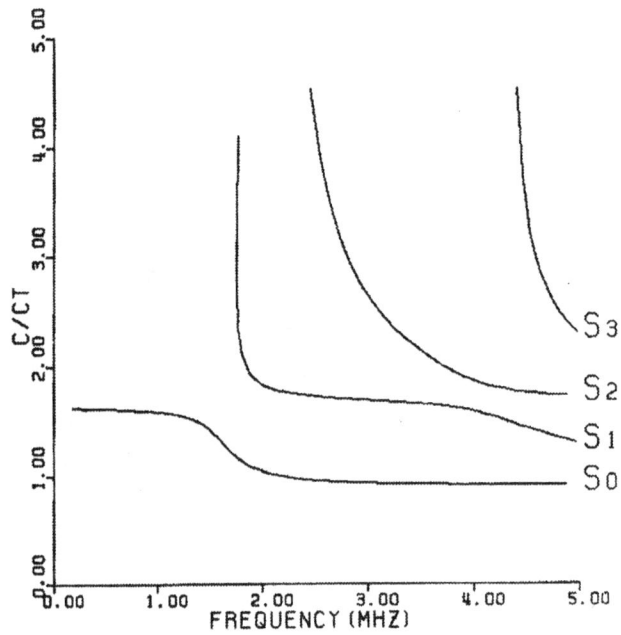

Fig. 21. Dispersion curves for symmetric Lamb wave modes in alumina.

ORNL-DWG. 86-8673

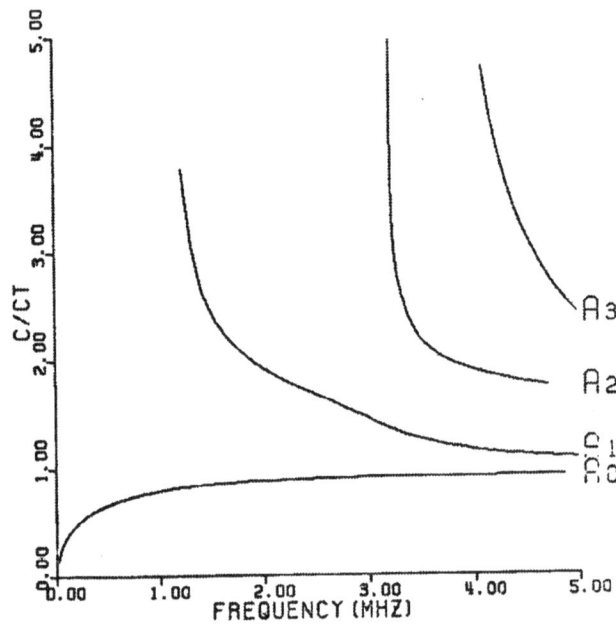

Fig. 22. Dispersion curves for antisymmetric Lamb wave modes in alumina.

Figure 23 is an experimental result showing the excitation of five Lamb wave modes in a 2.9-mm-thick alumina coupon. The normalized velocity of each wave is 1.25, and the mode frequencies are in good agreement with the values shown in Figs. 21 and 22. Note that this normalized velocity does not intersect the dispersion curve of the lowest-order antisymmetric mode; thus, this wave is not present in the spectrum. The relative amplitudes of the various waves are determined primarily by the response of the exciting transducer, which, in these studies, is a broadband 5-MHz unit.

ORNL-DWG. 86-8674

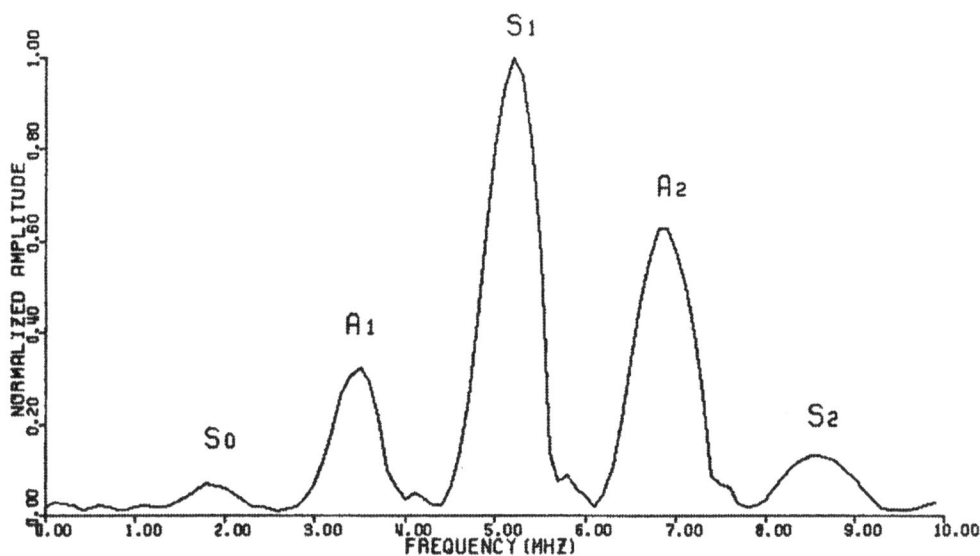

Fig. 23. Spectrum of Lamb wave modes in an alumina coupon.

A pure-mode response can be obtained by using single-frequency excitation of an appropriate transducer. Alternatively, one can use a relatively narrow-band transducer and select a normalized velocity (incident angle) such that only one mode exists within the passband of the transducer. Figure 24 shows the excitation of a single mode, the lowest-order symmetric mode, in an alumina coupon by a 2.25-MHz transducer using pulse excitation.

ORNL-DWG. 86-8675

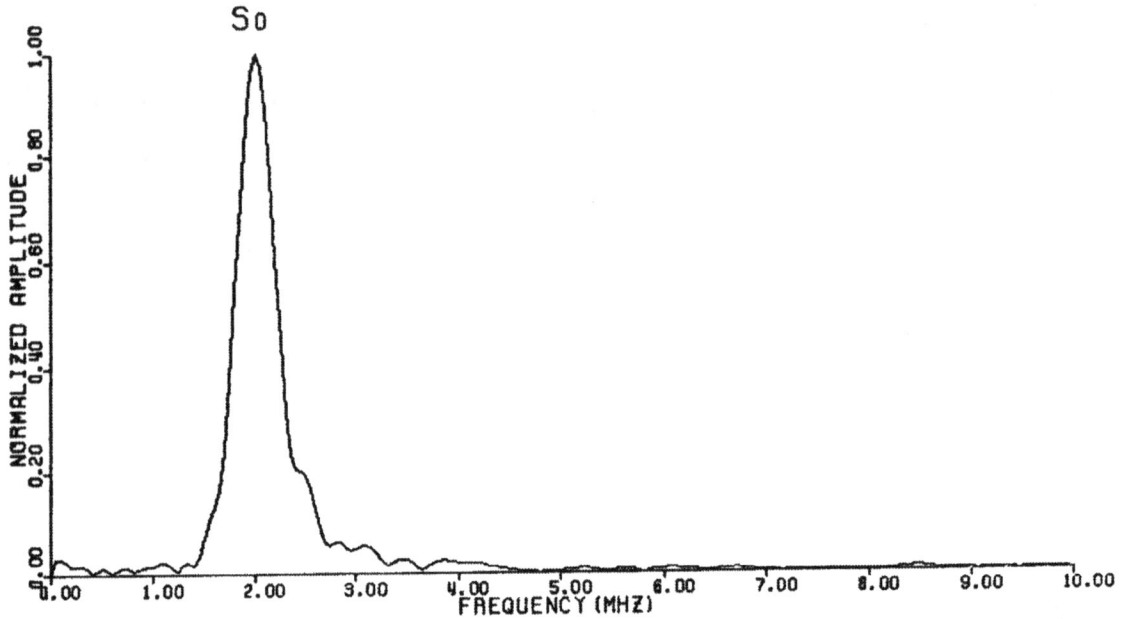

Fig. 24. Excitation of a single Lamb mode in an alumina coupon.

In order to test the effectiveness of using Lamb waves to interrogate the butt-braze joint in alumina flexure-strength specimens, the specially fabricated sample shown in Fig. 25 was made. A piece of tantalum foil at each end of the joint provides the proper separation of the ceramic halves during melting of the braze. A third piece of foil was placed in the center of this particular specimen to prevent bonding in that region and to simulate a nonbond of known dimensions. The brazing filler metal was one of several under development at Oak Ridge National Laboratory that will wet oxide ceramics directly with no pretreatment of the ceramic surface.

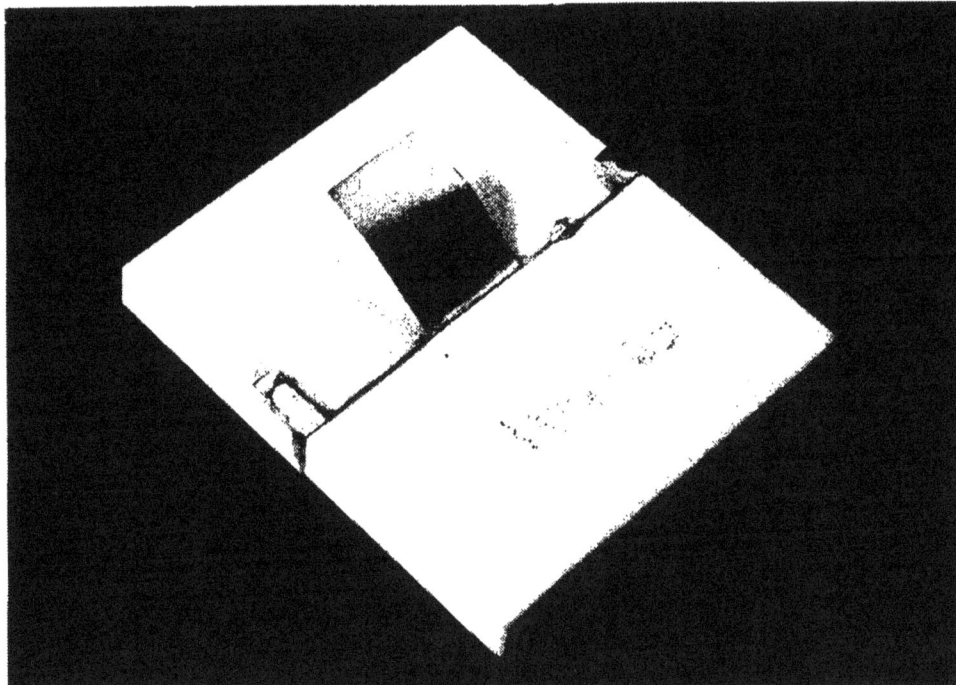

Fig. 25. Flexure-strength specimen containing a simulated nonbond.

The standard was first examined by conventional angle-beam
through-transmission techniques using 25-MHz transducers. The joint was
scanned in an *x-y* pattern using an index of 100 μm. As expected, the
central nonbond was easily detected using both flat and focused
transducers. For the latter, the sensitivity varied from top to bottom of
the interface because of variations in the beam profile with depth. In
both cases, however, numerous indications were generated at the top and
bottom of the interface by a slight vertical misalignment of the ceramic
coupons. This particular approach is sensitive to misalignment and is a
disadvantage of angle-beam testing of butt joints.

We next examined the standard using Lamb waves. Two transducers were
used with the transmitter and receiver located on opposite sides of the
interface. Figure 26 shows the transducer configuration. The transmitter
was driven with a tone burst to ensure excitation of a single Lamb mode.

ORNL-DWG 86-1861

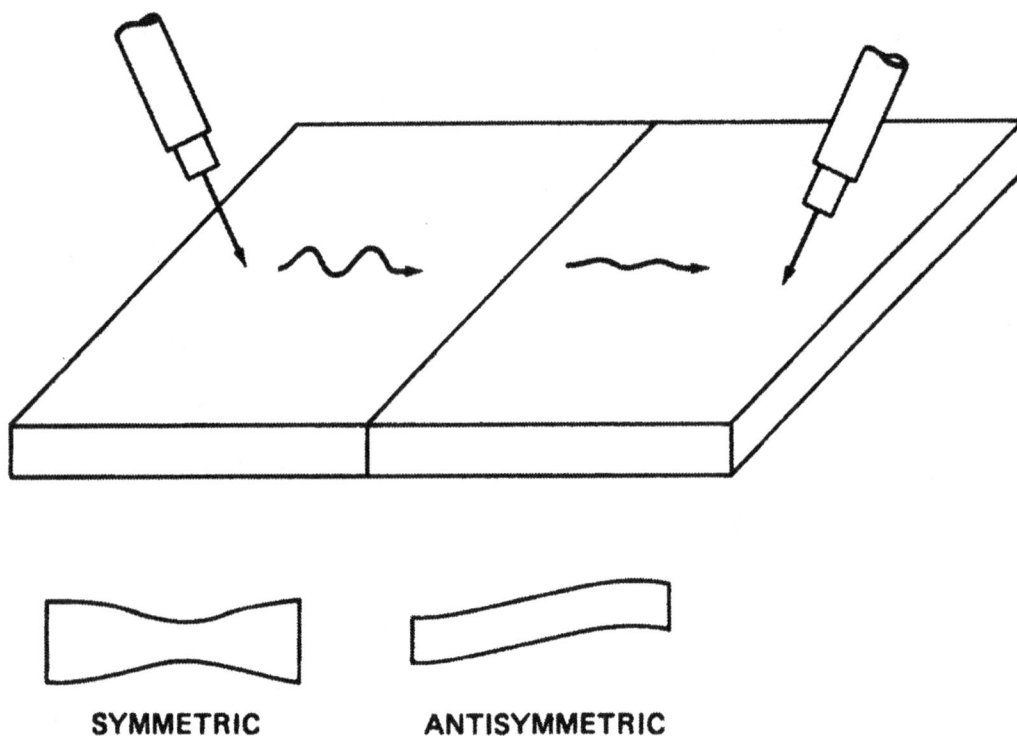

SYMMETRIC   **ANTISYMMETRIC**

Fig. 26. Transducer configuration for excitation and detection of
Lamb waves in the butt-brazed alumina coupon illustrated in Fig. 25.

As expected, virtually no change in transmission amplitude was detected
when the transducers were translated perpendicularly to the joint.
Nevertheless, an x-y scan (parallel and perpendicular to the interface)
was made. The results are shown in Fig. 27, where the height of the
surface represents the Lamb wave transmission amplitude. The central
nonbond corresponds to the large dip in the surface. Obviously, a single
scan parallel to the interface would have sufficed, so this test can be
performed very rapidly.

A second butt-braze specimen was also available; it was a standard
flexure bar consisting of titanium-coated PSZ coupons brazed with a commercial Ag-Cu-Sn filler metal. Only the two end pieces of tantalum foil
were present in this sample. It was first tested by conventional angle-
beam techniques with no indications detected other than the usual ones
attributable to vertical misalignment of the ceramic plates. In particular, no regions of unbonding were found.

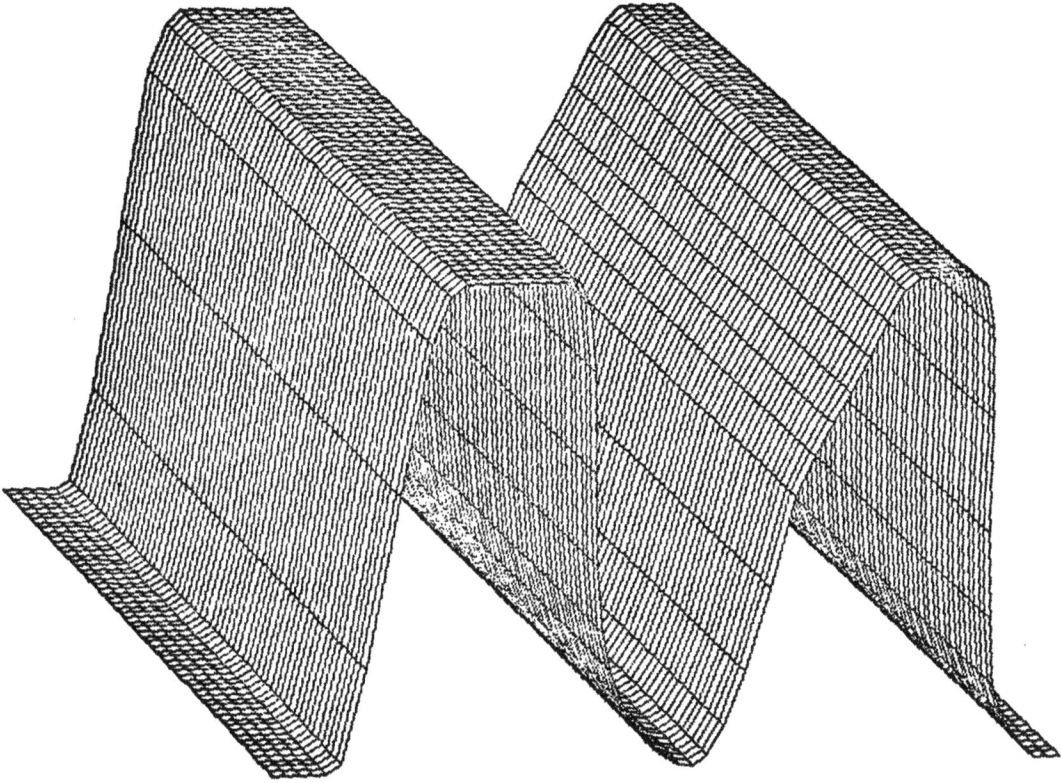

Fig. 27. Lamb wave transmission through the bond of the alumina flexure-strength specimen illustrated in Fig. 25, showing detection of the simulated nonbond.

Figure 28 shows the results obtained with Lamb waves on that second specimen. There is a small region near the center of the braze joint where the Lamb wave transmission dips noticeably. We then switched to a pulse-echo mode; that is, we monitored the amplitude of the Lamb waves reflected by the interface using first one and then the other transducer as a transmitter. In each case, an enhanced reflection was found at the location of the dip in Fig. 28. This eliminates problems in the ceramic (e.g., cracking) away from the interface as the source of the anomaly and establishes that some variation in the region of the interface engendered the signal.

Radiographic examination of the specimen revealed no detectable flaws in the interface.

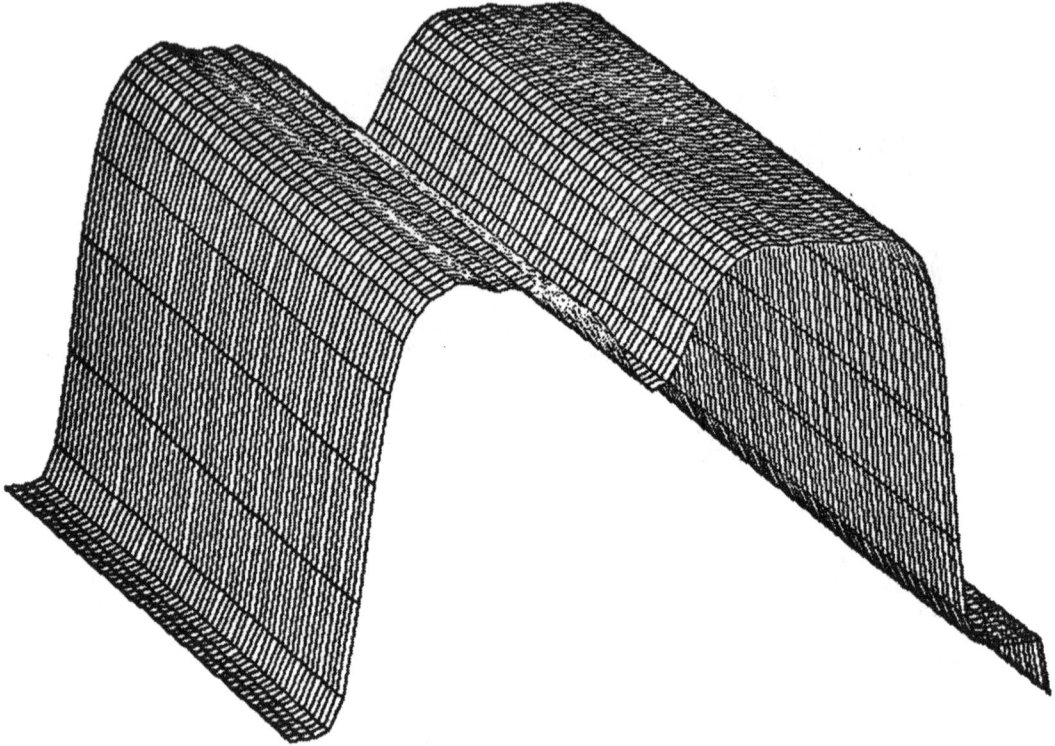

Fig. 28. Lamb wave transmission through the bond of another PSZ flexure-strength specimen having a simulated nonbond, showing a small indication near the center of the bond.

These results indicate that Lamb waves are sensitive to some characteristics of the braze joint that conventional nondestructive test techniques do not detect. The specimen will be cut into flexure bars and subjected to bend tests to try to develop a correlation between the inspection results and the strength of the bond.

SUMMARY AND CONCLUSIONS

The small critical flaw size in conventional monolithic structural ceramics has placed severe burdens on the development of ultrasonic NDE techniques for ensuring the integrity of parts fabricated of these materials. Since the detectability of a given flaw increases as the wavelength of the interrogating radiation decreases (at least until the wavelength becomes comparable to the flaw size), frequencies of at least 50 MHz are required to detect critical flaws in many ceramics. In addition, unlike structural metals, the interrogating ultrasonic radiation must be focused in order to ensure that a sufficiently large fraction of the incident radiation is intercepted by the flaw. The necessity of introducing focused radiation through the surface of the ceramic part (mandated by the desire to have rapid scanning capability) leads to severe spherical aberration of the ultrasonic beam within the part and limits the depth to which critical flaws can be detected to a few millimeters. Within these constraints, however, we have had considerable success in detecting flaws in the critical size range.

Since a common flaw type in ceramics is a quasi-spherical void or inclusion, a model was discussed for the scattering of ultrasonic waves from spheres. This model was applied to scattering from natural flaws in partially stabilized zirconia, and a flaw diameter of about 25 µm was inferred.

Since the attenuation behavior of structural ceramics for ultrasonic waves is critical to determining the minimum detectable flaw size, a technique was developed that permits a material transfer curve, or attenuation versus frequency response, to be determined for any ceramic. This process corrects for all known material-independent losses, such as diffraction (beam spread), acoustic impedance mismatches at the surfaces of the part and at internal interfaces, and frequency-dependent coupling losses between the transducer and the surface of the part. The transfer curve is highly sensitive to changes in the microstructure of the ceramic, and examples of the differences in TZP and PSZ ceramics were given.

Since current heat engine designs contain components consisting of a ceramic material clad on a metallic substrate to achieve both high-temperature tolerance and toughness, additional evaluation techniques were required for assurance of the quality of a ceramic-metal bond. The braze layer in such a bond is typically of the order of 60 µm thick, and, if nonbonding occurs, the tester needs to determine which face of this layer is nonbonded. For high-frequency ceramics, this condition could be determined by interrogating the bond with frequencies sufficiently high ($\approx$100 MHz) to resolve the layer thickness. Examples were given of the detection of 100-µm-diam pores in the braze layer using this approach. When the ceramic host would not support ultrasonic waves of such a high frequency, however, advanced signal processing techniques had to be developed to recover the desired information. An example was given of the identification of the correct interface in a braze joint in PSZ.

Although no specimens were available for study that contained manufactured flaws of known size in the bond layer of a ceramic-to-metal joint, a sample containing such flaws at the interface between a P-leg and an N-leg of a silicon-germanium specimen was obtained. The detection of a 125-μm flaw was demonstrated, and the signal-to-noise ratio for this flaw indicates that considerably smaller flaws could be detected reliably.

In testing joints between ceramic coupons, it is often not possible to introduce the ultrasound at normal incidence to the bond because of specimen geometry. In such cases, angle-beam testing techniques can be used, but it is difficult to maintain transducer focus across the full width of the interface. In addition, angle-beam testing was found to be unduly sensitive to slight vertical misalignment of the two plates being joined. For these configurations, a second approach, one using plate or Lamb waves, was studied. Two transducers were used: one launches a Lamb wave in either plate of the specimen, and the second receives the wave in the other plate after propagation through the joint. The amplitude of the received wave is an indicator of the condition of the bond in the region between the transducers.

Dispersion curves for Lamb waves in alumina coupons were calculated, and a system was assembled for evaluation of the bond by these waves. The system was shown to detect a simulated nonbond easily, and inspection of a second sample by this technique revealed an indication in the bond area that was not detected by angle-beam testing.

While bonds between ceramic components and between ceramic and metallic parts present a number of difficult problems to the researcher, considerable progress has been made in developing techniques that will allow the integrity of these bonds to be assured nondestructively. Many problems remain, however. Perhaps the most common criticism directed at ceramic inspection is that it is slow in comparison with inspection of metals. Rarely is it noted, however, that most ceramic inspection systems do not approach the state of the art in data acquisition rates. This is understandable, since technique development is of more importance at the present time in ceramic evaluation than it is for metals, where the techniques have a long history of development. Inspection rates will undoubtedly increase for ceramic evaluation as the approaches become established.

A more fundamental limitation in ceramic or ceramic-joint evaluation is related to the requirement for focused radiation. The severe aberrations introduced into the beam by propagation through the sample surface limit the depth to which effective focus can be maintained. This depth can be increased somewhat by increasing transducer frequency and focal length, but the relationship does not appear to favor this approach. Some form of this limitation will likely persist in the foreseeable future.

The techniques presented here do not exhaust the potential tools for ceramic-joint evaluation by a wide margin. For example, some form of direct interface wave (i.e., a wave that propagates in the bond material but not in the ceramic) may possibly yield information about the strength of the bond itself. This result would indeed be valuable, since there is currently no known correlation between bond strength and measurable acoustic properties.

ACKNOWLEDGMENTS

The authors thank A. J. Moorhead and M. L. Santella for the design and fabrication of the ceramic-cap piston specimens and the flexure-strength samples used in this study. The shear specimens were kindly provided by J. P. Hammond. The authors gratefully acknowledge helpful discussions of ceramic flaw types and sizes with T. N. Tiegs and P. F. Becher. Thanks are also due to J. L. Bishop for preparing the draft of this report, to O. A. Nelson for editing it, and to A. R. McDonald for final preparation of the report.

REFERENCES

1. C. F. Ying and R. Truell, "Scattering of a Plane Longitudinal Wave by a Spherical Obstacle in an Isotropically Elastic Solid," *J. Appl. Phys.* **27**, 1086 (1956).

2. W. A. Simpson, Jr., L. Adler, K. V. Cook, and R. W. McClung, *Ultrasonic Flaw Characterization Techniques for Stainless Steel Welds*, ORNL-6175, January 1986.

3. M. L. Santella, J. P. Hammond, S. A. David, and W. A. Simpson, "Zirconia to Cast Iron Brazing for Uncooled Diesel Engines," pp. 235–41 in *Proceedings of the Twenty-Third Automotive Technology Development Contractors' Coordination Meeting, P-165, Dearborn, Michigan, Oct. 21–24, 1985*, Society of Automotive Engineers, Warrendale, Pa., 1985.

4. W. A. Simpson, Jr., "Time-Domain Deconvolution: A New Technique to Improve Resolution for Ultrasonic Flaw Characterization in Stainless Steel Welds," *Mat. Eval.* 44(8), 998–1003 (1986).

5. S. I. Rohklin, "Interface Properties Characterization by Elastic Guided Waves," presented at the 1985 Review of Progress in Quantitative NDE, Williamsburg, VA., June 23–28, 1985.

www.ingramcontent.com/pod-product-compliance
Lightning Source LLC
Chambersburg PA
CBHW061344210326
41598CB00035B/5879